FRITJOF CAPRA received his Ph.D. from the University of Vienna and has done research in high-energy physics at several European and American universities. In addition to his many technical research papers, Dr. Capra has written and lectured extensively about the philosophical implications of modern science. He is the author of *The Tao of Physics*, an international bestseller that has sold a half million copies and has been translated around the world.

"A powerful book . . . Informative, provocative, and radical. With devastating clarity Capra shows how, in every field of science, and in the health field as well, our methods and our theories are leading to our own destruction . . . A book for every intelligent person."

—Carl Rogers, Ph.D. author of *A Way of Being*

"Every few years a book with the potential to radically change our lives is published . . . *The Turning Point* is the latest."

—*West Coast Review of Books*

"Capra's future has not yet quite happened. Drawing on a mix of straight science and 'alternative' research, he calls on scientists to *make* it happen; that is, to round the great turn from hard, mechanistic, reductionistic science to soft, organic, systems-view science."

—*Los Angeles Times*

"*The Turning Point* is a well written and compelling explanation of why so many things seem to be going wrong in the world."

—*The Washington Post*

Fritjof Capra

THE TURNING POINT

SCIENCE, SOCIETY, AND THE RISING CULTURE

BANTAM BOOKS
TORONTO • NEW YORK • LONDON • SYDNEY • AUCKLAND

THE TURNING POINT
A Bantam Book / published by arrangement with
Simon and Schuster

PRINTING HISTORY
Simon & Schuster edition published March 1982
Bantam edition / March 1983
6 printings through February 1988

New Age and the accompanying figure design as well as the statement "the search for meaning, growth and change" are trademarks of Bantam Books.

Book designed by Eve Metz.

For information address: Simon & Schuster, a Division of Gulf + Western Corporation, 1230 Avenue of the Americas, New York, N.Y. 10020.

Library of Congress Cataloging
in publication data
Capra, Fritjof.
The Turning Point.

Bibliography: p. 431
includes index.
1. Science—Philosophy. 2. Physics—Philosophy.
3. Science—Social Aspects. I. Title.
Q175.C246 501 81–16584
ISBN 978-0-553-34572-8 AACR2

Published simultaneously in the United States and Canada

Bantam Books are published by Bantam Books, a division of Bantam Doubleday Dell Publishing Group, Inc. Its trademark, consisting of the words "Bantam Books" and the portrayal of a rooster, is Registered in U.S. Patent and Trademark Office and in other countries. Marca Registrada. Bantam Books, New York, New York.

PRINTED IN THE UNITED STATES OF AMERICA

30 29 28 27

To the women in my life,
and especially
to my grandmother and my mother,
for their love, support, and wisdom

AUTHOR'S NOTE

It is my privilege and pleasure to acknowledge the help and advice of

Stanislav Grof,
Hazel Henderson,
Margaret Lock, and
Carl Simonton.

As special advisers in their fields of expertise, they wrote background papers for me that were incorporated into the text of the book, and spent time with me in discussions that were tape-recorded and transcribed for the same purpose. In particular, Stanislav Grof contributed in this way to chapters 6 and 11, Hazel Henderson to chapters 7 and 12, and Margaret Lock and Carl Simonton to chapters 5 and 10.

Before I began the actual writing, all five of us met for four days, together with Gregory Bateson, Antonio Dimalanta, and Leonard Shlain, to discuss the contents and structure of the book. These discussions, which did not lack their dramatic moments, were extremely stimulating and enlightening for me and will remain among the high moments in my life.

I am deeply indebted to all the people mentioned above for helping me with advice and information throughout the writing of the book, and for their critical reading of various parts of the manuscript. My special thanks go to Leonard Shlain for clarifying many questions related to medicine, and to Antonio Dimalanta for introducing me to recent developments in family therapy.

I am also especially grateful to Robert Livingston, whom I met during a later stage of my writing, and who gave me invaluable advice concerning the parts of the book that deal with biology.

Gregory Bateson exerted a major influence on my thinking throughout this work. Whenever I came up with a question that I could not associate with any discipline or school of thought, I would make a note in the margin of the manuscript, "Ask Bateson!" Unfortunately, some of these questions are still unanswered. Gregory Bateson died before I could show him any part of the manuscript. The first paragraphs of Chapter 9, which was strongly influenced by his work, were written the day after his funeral, at the cliffs on the Big Sur coast where his ashes had been scattered over the ocean. I will always be grateful for the privilege of having known him.

Acknowledgments

I would like to express my deep gratitude to the many people who have given me their help and support during the four years I worked on this book. It is impossible for me to mention all of them by name. However, I am especially grateful

—to Geoffrey Chew for an ongoing exchange of ideas, which has been my richest source of knowledge and inspiration, and to David Bohm and Henry Stapp for stimulating discussions of fundamental questions of physics;

—to Jonathan Ashmore, Robert Edgar, and Horace Judson for helpful discussions and correspondence about contemporary biology;

—to Erich Jantsch for inspiring conversations and for generously sharing his knowledge and resources with me;

— to Virginia Reed for opening my eyes to the expressive movements of the body, and for broadening my ideas about health and healing;

—to Martha Rogers and her students at New York University, with special thanks to Gretchen Randolph, for enlightening discussions of the role of nursing in the healing arts;

—to Rick Chilgren and David Sobel for their generous help with medical literature;

—to George Vithoulkas for introducing me to the theory of homeopathy and for his generous hospitality, and to Dana Ullman for helpful advice and resources;

—to Stephen Salinger for stimulating discussions of the relations between physics and psychoanalysis;

—to Virginia Senders, Verona Fonté, and Craig Brod for clarifying numerous questions regarding the history of psychology;

—to R. D. Laing for fascinating conversations about mental illness and the nature of consciousness, and for challenging my scientific thinking to its very core;

—to Marie-Louise von Franz and June Singer for illuminating discussions about Jungian psychology;

— to Frances Vaughn, Barbara Green, Frank Rubenfeld, Lynn Kahn, and Mari Krieger for enriching discussions about psychotherapy;

—to Carl Rogers for his inspiration, support, and generosity;
—to James Robertson and Lucia Dunn for helpful conversations and correspondence about economics;
—to E. F. Schumacher for a beautiful afternoon of discussions covering a broad range of topics, from economics and politics to philosophy, ethics, and spirituality;
—to my T'ai Chi teacher, Master Chiang Yun-Chung, who is also my doctor, for the experience of Chinese philosophy, art and science, and for graciously contributing the calligraphy shown on p. 5;
—to John Lennon, Gordon Onslow-Ford, and Gary Snyder for inspiring me through their art and their lives, and to Bob Dylan for two decades of powerful music and poetry;
—to Daniel Cohn-Bendit, Angela Davis, Victor Jara, Herbert Marcuse, and Adrienne Rich for raising my political consciousness;
—to Charlene Spretnak and Miriam Monasch for their friendship and support, and for sharpening my feminist awareness in theory and in practice;
—to my brother, Bernt Capra, my English publisher, Oliver Caldecott, and my friend Lenore Weiss for reading the entire manuscript and giving me their valuable advice and guidance;
—to all the people who came to my lectures, seminars, and workshops for providing the stimulating environment that led me to write this book;
—to the Esalen community, and particularly to Rick Tarnas, for their continuing support and generous hospitality, and for allowing me to discuss many tentative ideas in an informal setting;
—to the President and faculty of Macalester College for their hospitality and for giving me the opportunity, as a Visiting Professor, to present an early version of my thesis in a series of public lectures;
—to Susan Corrente, Howard Kornfeld, Ken Meter, and Annelies Rainer for research and advice;
—to my secretaries, Murray Lamp and Jake Walter, for helping me with innumerable chores with efficiency, imagination, and good humor, and to Alma Taylor for superb typing and proofreading;
—and to my editors at Simon and Schuster, Alice Mayhew and John Cox, for their patience, support, and encouragement, and for helping me turn a huge manuscript into a well-proportioned book.

CONTENTS

PREFACE

My main professional interest during the 1970s has been in the dramatic change of concepts and ideas that has occurred in physics during the first three decades of the century, and that is still being elaborated in our current theories of matter. The new concepts in physics have brought about a profound change in our world view; from the mechanistic conception of Descartes and Newton to a holistic and ecological view, a view which I have found to be similar to the views of mystics of all ages and traditions.

The new view of the physical universe was by no means easy for scientists at the beginning of the century to accept. The exploration of the atomic and subatomic world brought them in contact with a strange and unexpected reality that seemed to defy any coherent description. In their struggle to grasp this new reality, scientists became painfully aware that their basic concepts, their language, and their whole way of thinking were inadequate to describe atomic phenomena. Their problems were not merely intellectual but amounted to an intense emotional and, one could say, even existential crisis. It took them a long time to overcome this crisis, but in the end they were rewarded with deep insights into the nature of matter and its relation to the human mind.

I have come to believe that today our society as a whole finds itself in a similar crisis. We can read about its numerous manifestations every day in the newspapers. We have high inflation and unemployment, we have an energy crisis, a crisis in health care, pollution and other environmental disasters, a rising wave of violence and crime, and so on. The basic thesis of this book is that these are all different facets of one and the same crisis, and that this crisis is essentially a crisis of perception. Like the crisis in physics in the 1920s, it derives from the fact that we are trying to apply the concepts of an outdated world view—the mechanistic world view of Cartesian-Newtonian science—

to a reality that can no longer be understood in terms of these concepts. We live today in a globally interconnected world, in which biological, psychological, social, and environmental phenomena are all interdependent. To describe this world appropriately we need an ecological perspective which the Cartesian world view does not offer.

What we need, then, is a new "paradigm"—a new vision of reality; a fundamental change in our thoughts, perceptions, and values. The beginnings of this change, of the shift from the mechanistic to the holistic conception of reality, are already visible in all fields and are likely to dominate the present decade. The various manifestations and implications of this "paradigm shift" are the subject of this book. The sixties and seventies have generated a whole series of social movements that all seem to go in the same direction, emphasizing different aspects of the new vision of reality. So far, most of these movements still operate separately and have not yet recognized how their intentions interrelate. The purpose of this book is to provide a coherent conceptual framework that will help them recognize the communality of their aims. Once this happens, we can expect the various movements to flow together and form a powerful force for social change. The gravity and global extent of our current crisis indicate that this change is likely to result in a transformation of unprecedented dimensions, a turning point for the planet as a whole.

My discussion of the paradigm shift falls into four parts. The first part introduces the main themes of the book. The second part describes the historical development of the Cartesian world view and the dramatic shift of basic concepts that has occurred in modern physics. In the third part I discuss the profound influence of Cartesian-Newtonian thought on biology, medicine, psychology, and economics, and present my critique of the mechanistic paradigm in these disciplines. In doing so, I emphasize especially how the limitations of the Cartesian world view and of the value system which lies at its basis are now seriously affecting our individual and social health.

This critique is followed, in the fourth part of the book, by a detailed discussion of the new vision of reality. This new vision includes the emerging systems view of life, mind, consciousness, and evolution; the corresponding holistic approach to health and healing; the integration of Western and Eastern approaches to psychology and psychotherapy; a new conceptual framework for economics and technology;

and an ecological and feminist perspective which is spiritual in its ultimate nature and will lead to profound changes in our social and political structures.

The entire discussion covers a very broad range of ideas and phenomena, and I am well aware that my presentation of detailed developments in various fields is bound to be superficial, given the limitations of space and of my time and knowledge. However, as I wrote the book, I came to feel very strongly that the systems view I advocate in it also applies to the book itself. None of its elements is really original, and several of them may be represented in somewhat simplistic fashion. But the ways in which the various parts are integrated into the whole are more important than the parts themselves. The interconnections and interdependencies between the numerous concepts represent the essence of my own contribution. The resulting whole, I hope, will be more than the sum of its parts.

This book is for the general reader. All technical terms are defined in footnotes on the pages where they first appear. However, I hope that it will also be of interest to professionals in the various fields I have discussed. Although some may find my critique disturbing, I hope they will take none of it personally. My intent has never been to criticize particular professional groups as such, but rather to show how the dominant concepts and attitudes in various fields reflect the same unbalanced world view, a world view that is still shared by the majority of our culture but is now rapidly changing.

Much of what I say in this book is a reflection of my personal development. My life was decisively influenced by the two revolutionary trends of the 1960s, one operating in the social sphere, the other in the spiritual domain. In my first book, *The Tao of Physics,* I was able to make a connection between the spiritual revolution and my work as a physicist. At the same time, I believed that the conceptual shift in modern physics also had important social implications. Indeed, at the end of the book I wrote:

> I believe that the world-view implied by modern physics is inconsistent with our present society, which does not reflect the harmonious interrelatedness we observe in nature. To achieve such a state of dynamic balance, a radically different social and economic structure will be needed: a cultural revolution in the true sense of the word. The survival of our

> whole civilization may depend on whether we can bring about such a change.

Over the past six years, this statement evolved into the present book.

Berkeley,
April, 1981

—Fritjof Capra

I

CRISIS AND TRANSFORMATION

1·The Turning of the Tide

At the beginning of the last two decades of our century, we find ourselves in a state of profound, world-wide crisis. It is a complex, multidimensional crisis whose facets touch every aspect of our lives—our health and livelihood, the quality of our environment and our social relationships, our economy, technology, and politics. It is a crisis of intellectual, moral, and spiritual dimensions; a crisis of a scale and urgency unprecedented in recorded human history. For the first time we have to face the very real threat of extinction of the human race and of all life on this planet.

We have stockpiled tens of thousands of nuclear weapons, enough to destroy the entire world several times over, and the arms race continues at undiminished speed. In November 1978, while the United States and the Soviet Union were completing their second round of talks on the Strategic Arms Limitation Treaties, the Pentagon launched its most ambitious nuclear weapons production program in two decades; two years later this culminated in the biggest military boom in history: a five-year defense budget of 1,000 billion dollars.[1] Since then, American bomb factories have been running at full capacity. At Pantex, the Texas factory where every nuclear weapon owned by the United States is assembled, additional workers were hired and second and third shifts were added to increase the production of weapons of unprecedented destructive power.[2]

The costs of this collective nuclear madness are staggering. In 1978, before the latest escalation of costs, world military spending was about 425 billion dollars—over one billion dollars a day. More than a hundred countries, most of them in the Third World, are in the business of buying arms, and sales of military equipment for both nuclear and conventional wars are larger than the national incomes of all but ten nations in the world.[3]

In the meantime more than fifteen million people—most of them children—die of starvation each year; another 500 million are seriously undernourished. Almost 40 percent of the world's population has no access to professional health services; yet developing countries spend more than three times as much on armaments as on health care. Thirty-five percent of humanity lacks safe drinking water, while half of its scientists and engineers are engaged in the technology of making weapons.

In the United States, where the military-industrial complex has become an integral part of government, the Pentagon tries to persuade us that building more and better weapons will make the country safer. In fact, the opposite is true—more nuclear weapons mean more danger. Over the past few years an alarming change in American defense policy has been noticeable, a trend toward a nuclear arsenal aimed not at retaliation but at a first strike. There is increasing evidence that first-strike strategies are no longer a military option but have become central to American defense policy.[4] In such a situation each new missile makes nuclear war more likely. Nuclear weapons do not increase our security, as the military establishment would have us believe; they merely increase the likelihood of global destruction.

The threat of nuclear war is the greatest danger humanity is facing today, but it is by no means the only one. While the military powers increase their lethal arsenal of nuclear weapons, the industrial world is busy building equally dangerous nuclear power plants that threaten to extinguish life on our planet. Twenty-five years ago world leaders decided to use "atoms for peace" and presented nuclear power as the reliable, clean, and cheap energy source of the future. Today we are becoming painfully aware that nuclear power is neither safe, nor clean, nor cheap. The 360 nuclear reactors now operating world-wide, and the hundreds more planned, have become a major threat to our well-being.[5] The radioactive elements released by nuclear reactors are the same as those making up the fallout of atomic bombs. Thousands of tons of these toxic materials have already been discharged into the environment by nuclear explosions and reactor spills. As they continue to accumulate in the air we breathe, the food we eat, and the water we drink, our risk of developing cancer and genetic diseases continues to increase. The most toxic of these radioactive poisons, plutonium, is itself fissionable, which means that it can be used to build atomic bombs. Thus nuclear power and nuclear weapons are inextricably linked, being but different aspects of the same threat to humankind. With their continuing proliferation, the likelihood of global extinction becomes greater every day.

Even discounting the threat of a nuclear catastrophe, the global ecosystem and the further evolution of life on earth are seriously endangered and may well end in a large-scale ecological disaster. Overpopulation and industrial technology have contributed in various ways to a severe degradation of the natural environment upon which we are completely dependent for life. As a result, our health and well-being are seriously endangered. Our major cities are covered by blankets of choking, mustard-colored smog. Those of us who live in cities can see it every day; we feel it when it burns our eyes and irritates our lungs. In Los Angeles, according to a statement by sixty faculty members of the University of California Medical School,[6] "air pollution has now become a major health hazard to most of this community during much of the year." But smog is not confined to the big metropolitan areas of the United States. It is equally irritating, if not worse, in Mexico City, Athens, and Istanbul. This continual pollution of the air not only affects humans but also upsets ecological systems. It injures and kills plants, and these changes in plant life can induce drastic changes in animal populations that depend on the plants. In today's world, smog is not only found in the vicinity of large cities but disperses throughout the earth's atmosphere and may severely affect the global climate. Meteorologists speak of a nebulous veil of air pollution encircling the entire planet.

In addition to air pollution, our health is also threatened by the water we drink and the food we eat, both contaminated by a wide variety of toxic chemicals. In the United States synthetic food additives, pesticides, plastics, and other chemicals are marketed at a rate currently estimated at a thousand new chemical compounds a year. As a result, chemical poisoning has become an increasing part of our affluent life. Moreover, the threats to our health through the pollution of air, water, and food are merely the most obvious, direct effects of human technology on the natural environment. Less obvious but possibly far more dangerous effects have been recognized only recently and are still not fully understood.[7] However, it has become clear that our technology is severely disturbing, and may even be destroying, the ecological systems upon which our very existence depends.

The deterioration of our natural environment has been accompanied by a corresponding increase in health problems of individuals. Whereas nutritional and infectious diseases are the greatest killers in the Third World, the industrialized countries are plagued by the chronic and degenerative diseases appropriately called "diseases of civilization," the principal killers being heart disease, cancer, and strokes. On the psychological side, severe depression, schizophrenia, and other psychiatric disorders appear to spring from a parallel deterioration of our social environment. There are numerous signs of social disintegration, including a rise in violent crimes, accidents, and suicides; increased alcoholism and drug abuse; and growing numbers of children with learning disabilities and behavioral disorders. The rise in violent crimes and suicides by young people is so dramatic that it has been called an epidemic of violent deaths. At the same time, the loss of young lives from accidents, especially motor accidents, is twenty times higher than the death rate from polio when it was at its worst. According to health economist Victor Fuchs, " 'epidemic' is almost too weak a word to describe this situation."[8]

Along with these social pathologies we have been witnessing economic anomalies that seem to confound all our leading economists and politicians. Rampant inflation, massive unemployment, and a gross maldistribution of income and wealth have become structural features of most national economies. The resulting dismay among the general public and its appointed leaders is aggravated by the perception that energy and natural resources—the basic ingredients of all industrial activity—are rapidly being depleted.

Faced with the triple threat of energy depletion, inflation, and unemployment, our politicians no longer know where to turn first to minimize the damage. They, and the media, argue about priorities—should we deal with the energy crisis first or should we first fight inflation?—without realizing that both these problems, as well as all the others mentioned here, are but different facets of a single crisis. Whether we talk about cancer, crime, pollution, nuclear power, inflation, or energy shortage, the dynamics underlying these problems are the same. The central purpose of this book is to clarify these dynamics and to point to directions for change.

It is a striking sign of our time that the people who are supposed to be experts in various fields can no longer deal with the urgent problems that have arisen in their areas of expertise. Economists are unable to understand inflation, oncologists are totally confused about the causes of cancer, psychiatrists are mystified by schizophrenia, police are helpless in the face of rising crime, and the list goes on. In the United States it has been traditional for presidents to turn to academic people for counsel, either directly or through "brain trusts" and "think tanks" set up explicitly to advise government on various policy matters. This intellectual elite has formulated the "mainstream academic view" and generally agreed on the basic conceptual framework underlying its advice. Today this consensus no longer exists. In 1979 the *Washington Post* ran a story under the heading "The Cupboard of Ideas is Bare," in which prominent thinkers admitted they were unable to solve the nation's most urgent policy problems.[9] According to the *Post*, "Talks with noted intellectuals in Cambridge, Mass., and New York, in fact, not only confirm that the mainstream of ideas has split into dozens of rivulets, but that in some areas it has dried up altogether." One of the academics interviewed was Irving Kristol, Henry R. Luce professor of urban values at New York University, who said that he was resigning his chair because "I don't have anything to say anymore. I don't think anybody does. When a problem becomes too difficult, you lose interest."

As sources of their confusion or retreat the intellectuals cited "new circumstances" or "the course of events"— Vietnam, Watergate, and the persistence of slums, poverty, and crime. None of them, however, identified the real problem that underlies our crisis of ideas: the fact that most academics subscribe to narrow perceptions of reality which are inadequate for dealing with the major problems of our time. These problems, as we shall see in detail, are systemic problems, which means that they are closely interconnected and interdependent. They cannot be understood within the fragmented methodology characteristic of our academic disciplines and government agencies. Such an approach will never resolve any of our difficulties but will merely shift them around in the complex web of social and ecological relations. A resolution can be found only if the structure of the web itself is changed, and this will involve profound transformations of our social institutions, values, and ideas. As we examine the sources of our cultural crisis it will become apparent that most of our leading thinkers use outdated conceptual models and irrelevant variables. It will also become evident that a significant aspect of our conceptual impasse is that all of the prominent intellectuals interviewed by the *Washington Post* were men.

To understand our multifaceted cultural crisis we need to adopt an extremely broad view and see our situation in the context of human cultural evolution. We have to shift our perspective from the end of the twentieth century to a time span encompassing thousands of years; from the notion of static social structures to the perception of dynamic patterns of change. Seen from this perspective, crisis appears as an aspect of transformation. The Chinese, who have always had a thoroughly dynamic world view and a keen sense of history, seem to have been well aware of this profound connection between crisis and change. The term they use for "crisis"—*wei-ji*—is composed of the characters for "danger" and "opportunity."

Western sociologists have confirmed this ancient intuition. Studies of periods of cultural transformation in various societies have shown that these transformations are typically preceded by a variety of social indicators, many of them identical to the symptoms of our current crisis. They include a sense of alienation and an increase in mental illness, violent crime, and social disruption, as well as an increased interest in religious cultism—all of which have been observed in our society during the past decade. In times of historic cultural change these indicators have tended to appear one to three decades before the central transformation, rising in frequency and intensity as the transformation is approaching, and falling again after it has occurred.[10]

Cultural transformations of this kind are essential steps in the development of civilizations. The forces underlying this development are complex, and historians are far from having a comprehensive theory of cultural dynamics, but it seems that all civilizations go through similar cyclical processes of genesis, growth, breakdown, and disintegration.

The following graph shows this striking pattern for the major civilizations around the Mediterranean.[11]

Among the foremost if more conjectural studies of these patterns in the rise and fall of civilizations is Arnold Toynbee's *A Study of History*.[12] According to Toynbee, the genesis of a civilization consists of a transition from a static condition to dynamic activity. This transition may occur spontaneously, through the influence of some civilization that is already in existence or through the disintegration of one or more civilizations of an older generation. Toynbee sees the basic pattern in the genesis of civilizations as a pattern of interaction which he calls "challenge-and-response." A challenge from the natural or social environment provokes a creative response in a society, or a social group, which induces that society to enter the process of civilization.

The civilization continues to grow when its successful response to the initial challenge generates cultural momentum that carries the society beyond a state of equilibrium into an overbalance that presents itself as a fresh challenge. In this way the initial pattern of challenge-

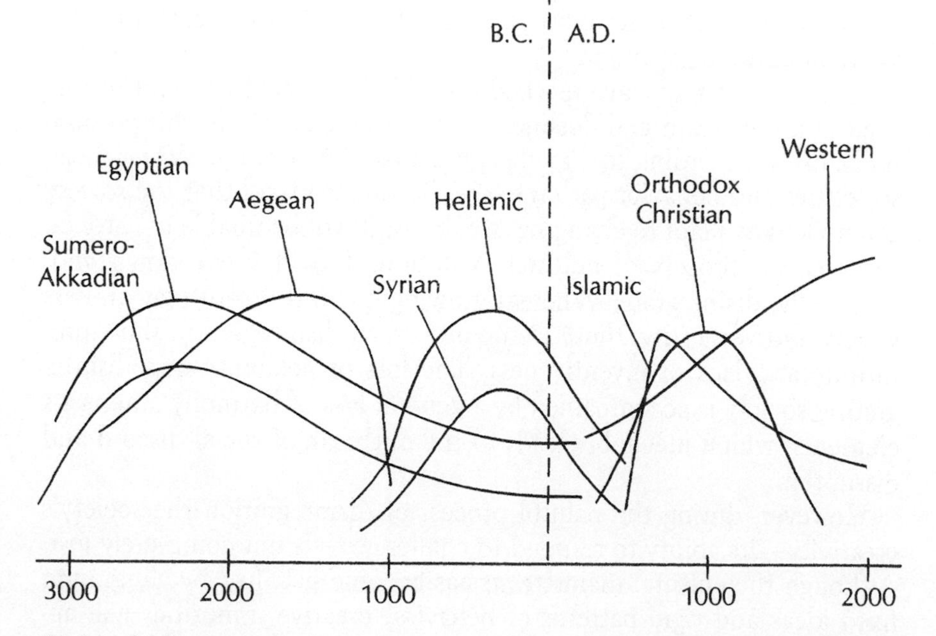

Rise-and-fall patterns of the major civilizations around the Mediterranean.

and-response is repeated in successive phases of growth, each successful response producing a disequilibrium that requires new creative adjustments.

The recurrent rhythm in cultural growth seems to be related to processes of fluctuation that have been observed throughout the ages and were always regarded as part of the fundamental dynamics of the universe. Ancient Chinese philosophers believed that all manifestations of reality are generated by the dynamic interplay between two polar forces which they called the yin and the yang. Heraclitus, in ancient Greece, compared the world order to an ever living fire, "kindling in measures and going out in measures." Empedocles attributed the changes in the universe to the ebb and flow of two complementary forces, which he called "love" and "hate."

The idea of a fundamental universal rhythm has also been expressed by numerous philosophers of modern times.[13] Saint-Simon saw the histories of civilizations as a series of alternating "organic"and "critical" periods; Herbert Spencer viewed the universe as moving through a series of "integrations" and "differentiations"; and Hegel saw human history as a spiral development from one form of unity through a phase of disunity, and on to reintegration on a higher plane. Indeed, the notion of fluctuating patterns seems always to be extremely useful for the study of cultural evolution.

After civilizations have reached a peak of vitality, they tend to lose their cultural steam and decline. An essential element in this cultural breakdown, according to Toynbee, is a loss of flexibility. When social structures and behavior patterns have become so rigid that the society can no longer adapt to changing situations, it will be unable to carry on the creative process of cultural evolution. It will break down and, eventually, disintegrate. Whereas growing civilizations display endless variety and versatility, those in the process of disintegration show uniformity and lack of inventiveness. The loss of flexibility in a disintegrating society is accompanied by a general loss of harmony among its elements, which inevitably leads to the outbreak of social discord and disruption.

However, during the painful process of disintegration the society's creativity—its ability to respond to challenges—is not completely lost. Although the cultural mainstream has become petrified by clinging to fixed ideas and rigid patterns of behavior, creative minorities will appear on the scene and carry on the process of challenge-and-response.

The dominant social institutions will refuse to hand over their leading roles to these new cultural forces, but they will inevitably go on to decline and disintegrate, and the creative minorities may be able to transform some of the old elements into a new configuration. The process of cultural evolution will then continue, but in new circumstances and with new protagonists.

The cultural patterns Toynbee described seem to fit our current situation very well. Looking at the nature of our challenges—not at the various symptoms of crisis but at the underlying changes in our natural and social environments—we can recognize the confluence of several transitions.[14] Some of them are connected with natural resources, others with cultural values and ideas; some are parts of periodic fluctuations, others occur within patterns of rise-and-fall. Each of these processes has a distinct time span, or periodicity, but all of them involve periods of transitions that happen to coincide at the present moment. Among these transitions are three that will shake the very foundations of our lives and will deeply affect our social, economic, and political system.

The first and perhaps most profound transition is due to the slow and reluctant but inevitable decline of patriarchy.[15] The time span associated with patriarchy is at least three thousand years, a period so long that we cannot say whether we are dealing with a cyclical process because the information we have about prepatriarchal eras is far too tenuous. What we do know is that for the past three thousand years Western civilization and its precursors, as well as most other cultures, have been based on philosophical, social, and political systems "in which men—by force, direct pressure, or through ritual, tradition, law and language, customs, etiquette, education, and the division of labor—determine what part women shall or shall not play, and in which the female is everywhere subsumed under the male."[16]

The power of patriarchy has been extremely difficult to understand because it is all-pervasive. It has influenced our most basic ideas about human nature and about our relation to the universe—"man's" nature and "his" relation to the universe, in patriarchal language. It is the one system which, until recently, had never in recorded history been openly challenged, and whose doctrines were so universally accepted that they seemed to be laws of nature; indeed, they were usually presented as such. Today, however, the disintegration of patriarchy is in sight. The feminist movement is one of the strongest cultural currents of our time and will have a profound effect on our further evolution.

The second transition that will have a profound impact on our lives is forced upon us by the decline of the fossil-fuel age. Fossil fuels*—coal, oil, and natural gas—have been the principal sources of energy for the modern industrial era, and as we run out of them this era will come to an end. From the broad historical perspective of cultural evolution, the fossil-fuel age and the industrial era are but a brief episode, a thin peak around the year 2000 on our graph. Fossil fuels will be exhausted by the year 2300, but the economic and political effects of this decline are already being felt. This decade will be marked by the transition from the fossil-fuel age to a solar age, powered by renewable energy from the sun; a shift that will involve radical changes in our economic and political systems.

The third transition is again connected with cultural values. It involves what is now often called a "paradigm † shift"—a profound change in the thoughts, perceptions, and values that form a particular vision of reality.[17] The paradigm that is now shifting has dominated our culture for several hundred years, during which it has shaped our modern Western society and has significantly influenced the rest of the

* Fossil fuels are residues of fossilized plants, plants that were buried in the earth's crust and transformed into their present state by chemical reactions over long periods of time.

† From the Greek *paradeigma* ("pattern").

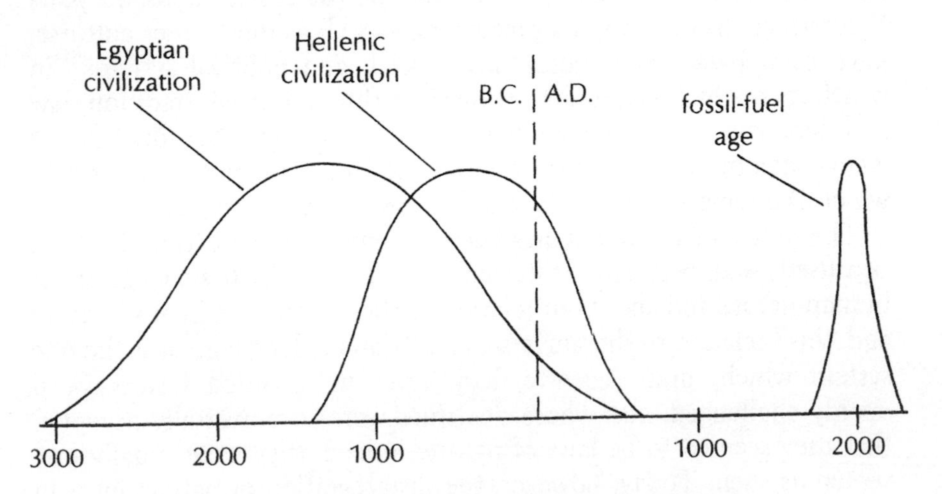

Fossil-fuel age in the context of cultural evolution.

world. This paradigm comprises a number of ideas and values that differ sharply from those of the Middle Ages; values that have been associated with various streams of Western culture, among them the Scientific Revolution, the Enlightenment, and the Industrial Revolution. They include the belief in the scientific method as the only valid approach to knowledge; the view of the universe as a mechanical system composed of elementary material building blocks; the view of life in society as a competitive struggle for existence; and the belief in unlimited material progress to be achieved through economic and technological growth. During the past decades all these ideas and values have been found severely limited and in need of radical revision.

From our broad perspective of cultural evolution, the current paradigm shift is part of a larger process, a strikingly regular fluctuation of value systems that can be traced throughout Western civilization and most other cultures. These fluctuating changes of values and their effects on all aspects of society, at least in the West, have been mapped out by the sociologist Pitirim Sorokin in a monumental four-volume work written between 1937 and 1941.[18] Sorokin's grand scheme for the synthesis of Western history is based on the cyclical waxing and waning of three basic value systems that underlie all manifestations of a culture.

Sorokin calls these three value systems the sensate, the ideational, and the idealistic. The sensate value system holds that matter alone is the ultimate reality, and that spiritual phenomena are but a manifestation of matter. It professes that all ethical values are relative and that sensory perception is the only source of knowledge and truth. The ideational value system is profoundly different. It holds that true reality lies beyond the material world, in the spiritual realm, and that knowledge can be obtained through inner experience. It subscribes to absolute ethical values and superhuman standards of justice, truth, and beauty. Western representations of the ideational concept of spiritual reality include Platonic ideas, the soul, and Judeo-Christian images of God, but Sorokin points out that similar ideas are expressed in the East, in different form, in Hindu, Buddhist, and Taoist cultures.

Sorokin contends that the cyclical rhythms of interplay between sensate and ideational expressions of human culture also produce an intermediate, synthesizing stage—the idealistic—which represents their harmonious blending. According to idealistic beliefs, true reality has both sensory and supersensory aspects which coexist within an all-embracing unity. Idealistic cultural periods thus tend to attain the highest and noblest expressions of both ideational and sensate styles, producing balance, integration, and esthetic fulfillment in art, philosophy, science, and technology. Examples of such idealistic periods are the Greek flowering of the fifth and fourth centuries, B.C., and the European Renaissance.

These three basic patterns of human cultural expression have, according to Sorokin, produced identifiable cycles in Western civilization, which he has plotted on dozens of charts for belief systems, wars and internal conflicts, scientific and technological development, and law and various other social institutions. He has also charted fluctuations of styles in architecture, painting, sculpture, and literature. In Sorokin's model the current paradigm shift and the decline of the Industrial Age are another period of maturation and decline of sensate culture. The rise of our current sensate era was preceded by the ascendancy of ideational culture during the rise of Christianity and the Middle Ages, and by the subsequent flowering of an idealistic stage during the European Renaissance. It was the slow decline of these ideational and idealistic epochs in the fifteenth and sixteenth centuries that gave way to the rise of a new sensate period in the seventeenth, eighteenth, and nineteenth centuries, an era marked by the value system of the Enlightenment, the scientific views of Descartes and Newton, and the technology of the Industrial Revolution. In the twentieth century these sensate values and ideas are on the decline again, and thus in 1937, with great foresight, Sorokin predicted as the twilight of sensate culture the paradigm shift and social upheavals we are witnessing today.[19]

Sorokin's analysis suggests very forcefully that the crisis we are facing today is no ordinary crisis but one of the great transition phases that have occurred in previous cycles of human history. These profound cultural transformations do not take place very often. According to Lewis Mumford, there may have been fewer than half a dozen in the entire history of Western civilization, among them the rise of civilization with the invention of agriculture at the beginning of the neolithic period, the rise of Christianity at the fall of the Roman Empire, and the transition from the Middle Ages to the Scientific Age.[20]

The transformation we are experiencing now may well be more dramatic than any of the preceding ones, because the rate of change in our age is faster than ever before, because the changes are more extensive, involving the entire globe, and because several major transitions are coinciding. The rhythmic recurrences and patterns of rise and decline that seem to dominate human cultural evolution have somehow conspired to reach their points of reversal at the same time. The decline of patriarchy, the end of the fossil-fuel age, and the paradigm shift occurring in the twilight of the sensate culture are all contributing to the same global process. The current crisis, therefore, is not just a crisis of individuals, governments, or social institutions; it is a transition of planetary dimensions. As individuals, as a society, as a civilization, and as a planetary ecosystem, we are reaching the turning point.

Cultural transformations of this magnitude and depth cannot be prevented. They should not be opposed but, on the contrary, should be welcomed as the only escape from agony, collapse, or mummification. What we need, to prepare ourselves for the great transition we are about to enter, is a deep reexamination of the main premises and values of our culture, a rejection of those conceptual models that have outlived their usefulness, and a new recognition of some of the values discarded in previous periods of our cultural history. Such a thorough change in the mentality of Western culture must naturally be accompanied by a profound modification of most social relationships and forms of social organization—by changes that will go far beyond the superficial measures of economic and political readjustment being considered by today's political leaders.

During this phase of revaluation and cultural rebirth it will be important to minimize the hardship, discord, and disruption that are inevitably involved in periods of great social change, and to make the transition as painless as possible. It will therefore be crucial to go beyond attacking particular social groups or institutions, and to show how their attitudes and behavior reflect a value system that underlies our whole culture and that has now become outdated. It will be necessary to recognize and widely communicate the fact that our current social changes are manifestations of a much broader, and inevitable, cultural transformation. Only then will we be able to approach the kind of harmonious, peaceful cultural transition described in one of humanity's oldest books of wisdom, the Chinese *I Ching,* or Book of Changes: "The movement is natural, arising spontaneously. For this reason the transformation of the old becomes easy. The old is discarded and the new is introduced. Both measures accord with the time; therefore no harm results."[21]

The model of cultural dynamics that will be used in our discussion of the current social transformation is based in part on Toynbee's ideas about the rise and fall of civilizations; on the age-old notion of a fundamental universal rhythm resulting in fluctuating cultural patterns; on Sorokin's analysis of the fluctuation of value systems; and on the ideal of harmonious cultural transitions portrayed in the *I Ching*.

The major alternative to this model, which is related to it but different in several aspects, is the Marxist view of history known as dialectic or historical materialism. According to Marx, the roots of social evolution lie not in a change of ideas or values but in economic and technological developments. The dynamics of change is that of a "dialectic" interplay of opposites arising from contradictions that are intrinsic to all things. Marx took this idea from the philosophy of Hegel and adapted it to his analysis of social change, asserting that all changes in society arise from the development of its internal contradictions. He saw the contradictory principles of social organization as being embodied in society's classes, and class struggle as a consequence of their dialectic interaction.

The Marxist view of cultural dynamics, being based on the Hegelian notion of recurrent rhythmic change, is not unlike the models of Toynbee, Sorokin, and the *I Ching* in that respect.[22] However, it differs significantly from those models in its emphasis on conflict and struggle. Class struggle was the driving force of history for Marx, who held that all important historical progress was born in conflict, struggle, and violent revolution. Human suffering and sacrifice was a necessary price that had to be paid for social change.

The emphasis on struggle in Marx's theory of historical evolution paralleled Darwin's emphasis on struggle in biological evolution. In fact Marx's favorite image of himself is said to have been that of "the Darwin of sociology." The idea of life as an ongoing struggle for existence, which both Darwin and Marx owed to the economist Thomas Malthus, was vigorously promoted in the nineteenth century by the Social Darwinists, who influenced, if not Marx, certainly many of his followers.[23] I believe their view of social evolution overemphasizes the role of struggle and conflict, overlooking the fact that all struggle in nature takes place within a wider context of cooperation. Although conflict and struggle have brought about important social progress in our past and will often be an essential part of the dynamics of change, this does not mean that they are the source of this dynamics. Therefore, following the philosophy of the *I Ching* rather than the Marxist view, I believe that conflict should be minimized in times of social transition.

In our discussion of cultural values and attitudes throughout this book we will make extensive use of a framework that is developed in great detail in the *I Ching,* and that lies at the very basis of Chinese thought. Like Sorokin's framework, it is based on the idea of continuous cyclical fluctuation, but it involves the much broader notion of two archetypal poles—yin and yang—underlying the fundamental rhythm of the universe.

The Chinese philosophers saw reality, whose ultimate essence they called Tao, as a process of continual flow and change. In their view all phenomena we observe participate in this cosmic process and are thus intrinsically dynamic. The principal characteristic of the Tao is the cyclical nature of its ceaseless motion; all developments in nature—those in the physical world as well as those in the psychological and social realms—show cyclical patterns. The Chinese gave this idea of cyclical patterns a definite structure by introducing the polar opposites yin and yang, the two poles that set the limits for the cycles of change: "The yang having reached its climax retreats in favor of the yin; the yin having reached its climax retreats in favor of the yang."[24]

In the Chinese view, all manifestations of the Tao are generated by the dynamic interplay of these two archetypal poles, which are associated with many images of opposites taken from nature and from social life. It is important, and very difficult for us Westerners, to understand that these opposites do not belong to different categories but are extreme poles of a single whole. Nothing is only yin or only yang. All natural phenomena are manifestations of a continuous oscillation between the two poles, all transitions taking place gradually and in unbroken progression. The natural order is one of dynamic balance between yin and yang.

The terms yin and yang have recently become quite popular in the West, but they are rarely used in our culture in the Chinese sense. Most Western usage reflects cultural preconceptions that severely distort the original meanings. One of the best interpretations is given by Manfred Porkert in his comprehensive study of Chinese medicine.[25]

According to Porkert, yin corresponds to all that is contractive, responsive, and conservative, whereas yang implies all that is expansive, aggressive, and demanding. Further associations include, among many others:

YIN	YANG
EARTH	HEAVEN
MOON	SUN
NIGHT	DAY
WINTER	SUMMER
MOISTURE	DRYNESS
COOLNESS	WARMTH
INTERIOR	SURFACE

In Chinese culture yin and yang have never been associated with moral values. What is good is not yin or yang but the dynamic balance between the two; what is bad or harmful is imbalance.

From the earliest times of Chinese culture, yin was associated with the feminine and yang with the masculine. This ancient association is extremely difficult to assess today because of its reinterpretation and distortion in subsequent patriarchal eras. In human biology masculine and feminine characteristics are not neatly separated but occur, in varying proportions, in both sexes.[26] Similarly, the Chinese ancients believed that all people, whether men or women, go through yin and yang phases. The personality of each man and each woman is not a static entity but a dynamic phenomenon resulting from the interplay between feminine and masculine elements. This view of human nature is in sharp contrast to that of our patriarchal culture, which has established a rigid order in which all men are supposed to be masculine and all women feminine, and has distorted the meaning of those terms by giving men the leading roles and most of society's privileges.

In view of this patriarchal bias, the frequent association of yin with passivity and yang with activity is particularly dangerous. In our culture women have traditionally been portrayed as passive and receptive, men as active and creative. This imagery goes back to Aristotle's theory of sexuality and has been used throughout the centuries as a "scientific" rationale for keeping women in a subordinate role, subservient to men.[27] The association of yin with passivity and yang with activity seems to be yet another expression of patriarchal stereotypes, a modern Western interpretation that is very unlikely to reflect the original meaning of the Chinese terms.

One of the most important insights of ancient Chinese culture was the recognition that activity—"the constant flow of transformation and change," as Chuang Tzu called it[28]—is an essential aspect of the universe. Change, in this view, does not occur as a consequence of some force but is a natural tendency, innate in all things and situations. The universe is engaged in ceaseless motion and activity, in a continual cosmic process that the Chinese called Tao—the Way. The notion of absolute rest, or inactivity, was almost entirely absent from Chinese philosophy. According to Hellmut Wilhelm, one of the leading Western interpreters of the *I Ching*. "The state of absolute immobility is such an abstraction that the Chinese . . . could not conceive it."[29]

The term *wu wei* is frequently used in Taoist philosophy and means literally "nonaction." In the West the term is usually interpreted as referring to passivity. This is quite wrong. What the Chinese mean by *wu wei* is not abstaining from activity but abstaining from a certain *kind* of activity, activity that is out of harmony with the ongoing cosmic process. The distinguished sinologist Joseph Needham defines *wu wei* as "refraining from action contrary to nature" and justifies his translation with a quotation from Chuang Tzu: "Nonaction does not mean doing nothing and keeping silent. Let everything be allowed to do what it naturally does, so that its nature will be satisfied."[30] If one refrains from acting contrary to nature or, as Needham says, from "going against the grain of things," one is in harmony with the Tao and thus one's actions will be successful. This is the meaning of Lao Tzu's seemingly puzzling statement: "By nonaction everything can be done."[31]

In the Chinese view, then, there seem to be two kinds of activity—activity in harmony with nature and activity against the natural flow of things. The idea of passivity, the complete absence of any action, is not entertained. Therefore the frequent Western association of yin and yang with passive and active behavior, respectively, does not seem to be consistent with Chinese thought. In view of the original imagery associated with the two archetypal poles, it would seem that yin can be interpreted as corresponding to responsive, consolidating, cooperative activity; yang as referring to aggressive, expanding, competitive activity. Yin action is conscious of the environment, yang action is conscious of the self. In modern terminology one could call the former "eco-action" and the latter "ego-action."

These two kinds of activity are closely related to two kinds of knowledge, or two modes of consciousness, which have been recognized as characteristic properties of the human mind throughout the ages. They are usually called the intuitive and the rational and have traditionally been associated with religion or mysticism and with science. Although the association of yin and yang with these two modes of consciousness is not part of the original Chinese terminology, it seems to be a natural extension of the ancient imagery and will be so regarded in our discussion.

The rational and the intuitive are complementary modes of functioning of the human mind. Rational thinking is linear, focused, and analytic. It belongs to the realm of the intellect, whose function it is to discriminate, measure, and categorize. Thus rational knowledge tends to be fragmented. Intuitive knowledge, on the other hand, is based on a direct, nonintellectual experience of reality arising in an expanded state of awareness. It tends to be synthesizing, holistic,* and nonlinear. From this it is apparent that rational knowledge is likely to generate self-centered, or yang, activity, whereas intuitive wisdom is the basis of ecological, or yin, activity.

This, then, is the framework for our exploration of cultural values and attitudes. For our purposes these associations of yin and yang will be most useful:

YIN	YANG
FEMININE	MASCULINE
CONTRACTIVE	EXPANSIVE
CONSERVATIVE	DEMANDING
RESPONSIVE	AGGRESSIVE
COOPERATIVE	COMPETITIVE
INTUITIVE	RATIONAL
SYNTHESIZING	ANALYTIC

* The term "holistic," from the Greek *holos* ("whole"), refers to an understanding of reality in terms of integrated wholes whose properties cannot be reduced to those of smaller units.

Looking at this list of opposites, it is easy to see that our society has consistently favored the yang over the yin—rational knowledge over intuitive wisdom, science over religion, competition over cooperation, exploitation of natural resources over conservation, and so on. This emphasis, supported by the patriarchal system and further encouraged by the dominance of sensate culture during the past three centuries, has led to a profound cultural imbalance which lies at the very root of our current crisis—an imbalance in our thoughts and feelings, our values and attitudes, and our social and political structures. In describing the various manifestations of this cultural imbalance, I shall pay particular attention to their effects on health, and want to use the concept of health in a very broad sense, including in it not only individual health but also social and ecological health. These three levels of health are closely interrelated and our current crisis constitutes a serious threat to all three of them. It threatens the health of individuals, of the society, and of the ecosystems of which we are a part.

Throughout this book I will attempt to show how the strikingly consistent preference for yang values, attitudes, and behavior patterns has resulted in a system of academic, political, and economic institutions that are mutually supportive and have become all but blind to the dangerous imbalance of the value system that motivates their activities. According to Chinese wisdom, none of the values pursued by our culture is intrinsically bad, but by isolating them from their polar opposites, by focusing on the yang and investing it with moral virtue and political power, we have brought about the current sad state of affairs. Our culture takes pride in being scientific; our time is referred to as the Scientific Age. It is dominated by rational thought, and scientific knowledge is often considered the only acceptable kind of knowledge. That there can be intuitive knowledge, or awareness, which is just as valid and reliable, is generally not recognized. This attitude, known as scientism, is widespread, pervading our educational system and all other social and political institutions. When President Lyndon Johnson needed advice about warfare in Vietnam, his administration turned to theoretical physicists—not because they were specialists in the methods of electronic warfare, but because they were considered the high priests of science, guardians of supreme knowledge. We can now say, with hindsight, that Johnson might have been much better served had he sought his advice from some of the poets. But that, of course, was—and still is—unthinkable.

The emphasis on rational thought in our culture is epitomized in Descartes' celebrated statement "*Cogito, ergo sum*"—"I think, therefore I exist"—which forcefully encouraged Western individuals to equate their identity with their rational mind rather than with their whole organism. We shall see that the effects of this division between mind and body are felt throughout our culture. Retreating into our minds, we have forgotten how to "think" with our bodies, how to use them as agents of knowing. In doing so we have also cut ourselves off from our natural environment and have forgotten how to commune and cooperate with its rich variety of living organisms.

The division between mind and matter led to a view of the universe as a mechanical system consisting of separate objects, which in turn were reduced to fundamental material building blocks whose properties and interactions were thought to completely determine all natural phenomena. This Cartesian view of nature was further extended to living organisms, which were regarded as machines constructed from separate parts. We shall see that such a mechanistic conception of the world is still at the basis of most of our sciences and continues to have a tremendous influence on many aspects of our lives. It has led to the well-known fragmentation in our academic disciplines and government agencies and has served as a rationale for treating the natural environment as if it consisted of separate parts, to be exploited by different interest groups.

Exploitation of nature has gone hand in hand with that of women, who have been identified with nature throughout the ages. From the earliest times, nature—and especially the earth—was seen as a kind and nurturing mother, but also as a wild and uncontrollable female. In prepatriarchal eras her many aspects were identified with the numerous manifestations of the Goddess. Under patriarchy the benign image of nature changed into one of passivity, whereas the view of nature as wild and dangerous gave rise to the idea that she was to be dominated by man. At the same time women were portrayed as passive and subservient to men. With the rise of Newtonian science, finally, nature became a mechanical system that could be manipulated and exploited, together with the manipulation and exploitation of women. The ancient association of woman and nature thus interlinks women's history and the history of the environment, and is the source of a natural kinship between feminism and ecology which is manifesting itself increasingly. In the words of Carolyn Merchant, historian of science at the University of California, Berkeley:

> In investigating the roots of our current environmental dilemma and its connections to science, technology and the economy, we must re-examine the formation of a world-view and a science which, by reconceptualizing reality as a machine rather than a living organism, sanctioned the domination of both nature and women. The contributions of such founding "fathers" of modern science as Francis Bacon, William Harvey, René Descartes, Thomas Hobbes and Isaac Newton must be reevaluated.[32]

The view of man as dominating nature and woman, and the belief in the superior role of the rational mind, have been supported and encouraged by the Judeo-Christian tradition, which adheres to the image of a male god, personification of supreme reason and source of ultimate power, who rules the world from above by imposing his divine law on it. The laws of nature searched for by the scientists were seen as reflections of this divine law, originating in the mind of God.

It is now becoming apparent that overemphasis on the scientific method and on rational, analytic thinking has led to attitudes that are profoundly antiecological. In truth, the understanding of ecosystems is hindered by the very nature of the rational mind. Rational thinking is linear, whereas ecological awareness arises from an intuition of nonlinear systems. One of the most difficult things for people in our culture to understand is the fact that if you do something that is good, then more of the same will not necessarily be better. This, to me, is the essence of ecological thinking. Ecosystems sustain themselves in a dynamic balance based on cycles and fluctuations, which are nonlinear processes. Linear enterprises, such as indefinite economic and technological growth—or, to give a more specific example, the storage of radioactive waste over enormous time spans—will necessarily interfere with the natural balance and, sooner or later, will cause severe damage.

Ecological awareness, then, will arise only when we combine our rational knowledge with an intuition for the nonlinear nature of our environment. Such intuitive wisdom is characteristic of traditional, nonliterate cultures, especially of American Indian cultures, in which life was organized around a highly refined awareness of the environment.

In the mainstream of our culture, on the other hand, the cultivation of intuitive wisdom has been neglected. This may be related to the fact that, in our evolution, there has been an increasing separation between the biological and cultural aspects of human nature. Biological evolution of the human species stopped some fifty thousand years ago. From then on, evolution proceeded no longer genetically but socially and culturally, while the human body and brain remained essentially the same in structure and size.[33] In our civilization we have modified our environment to such an extent during this cultural evolution that we have lost touch with our biological and ecological base more than any other culture and any other civilization in the past. This separation manifests itself in a striking disparity between the development of intellectual power, scientific knowledge, and technological skills, on the one hand, and of wisdom, spirituality, and ethics on the other. Scientific and technological knowledge has grown enormously since the Greeks embarked on the scientific venture in the sixth century B.C. But during these twenty-five centuries there has been hardly any progress in the conduct of social affairs. The spirituality and moral standards of Lao Tzu and Buddha, who also lived in the sixth century B.C., were clearly not inferior to ours.

Our progress, then, has been largely a rational and intellectual affair, and this one-sided evolution has now reached a highly alarming stage, a situation so paradoxical that it borders insanity. We can control the soft landings of space craft on distant planets, but we are unable to control the polluting fumes emanating from our cars and factories. We propose Utopian communities in gigantic space colonies, but cannot manage our cities. The business world makes us believe that huge industries producing pet foods and cosmetics are a sign of our high standard of living, while economists try to tell us that we cannot "afford" adequate health care, education, or public transport. Medical science and pharmacology are endangering our health, and the Defense Department has become the greatest threat to our national security. Those are the results of overemphasizing our yang, or masculine side—rational knowledge, analysis, expansion—and neglecting our yin, or feminine side—intuitive wisdom, synthesis, and ecological awareness.

The yin/yang terminology is especially useful in an analysis of cultural imbalance that adopts a broad ecological view, a view that could also be called a systems view, in the sense of general systems theory.[34]

Systems theory looks at the world in terms of the interrelatedness and interdependence of all phenomena, and in this framework an integrated whole whose properties cannot be reduced to those of its parts is called a system. Living organisms, societies, and ecosystems are all systems. It is fascinating to see that the ancient Chinese idea of yin and yang is related to an essential property of natural systems that has only recently been studied in Western science.

Living systems are organized in such a way that they form multileveled structures, each level consisting of subsystems which are wholes in regard to their parts, and parts with respect to the larger wholes. Thus molecules combine to form organelles, which in turn combine to form cells. The cells form tissues and organs, which themselves form larger systems, like the digestive system or the nervous system. These, finally, combine to form the living woman or man; and the "stratified order"* does not end there. People form families, tribes, societies, nations. All these entities—from molecules to human beings, and on to social systems—can be regarded as wholes in the sense of being integrated structures, and also as parts of larger wholes at higher levels of complexity. In fact, we shall see that parts and wholes in an absolute sense do not exist at all.

Arthur Koestler has coined the word "holons" for these subsystems which are both wholes and parts, and he has emphasized that each holon has two opposite tendencies: an integrative tendency to function as part of the larger whole, and a self-assertive tendency to preserve its individual autonomy.[35] In a biological or social system each holon must assert its individuality in order to maintain the system's stratified order, but it must also submit to the demands of the whole in order to make the system viable. These two tendencies are opposite but complementary. In a healthy system—an individual, a society, or an ecosystem—there is a balance between integration and self-assertion. This balance is not static but consists of a dynamic interplay between the two complementary tendencies, which makes the whole system flexible and open to change.

The relation between modern systems theory and ancient Chinese thought now becomes apparent. The Chinese sages seem to have recognized the basic polarity that is characteristic of living systems. Self-assertion is achieved by displaying yang behavior; by being demanding,

* See Chapter 9.

aggressive, competitive, expanding, and—as far as human behavior is concerned—by using linear, analytic thinking. Integration is furthered by yin behavior; by being responsive, cooperative, intuitive, and aware of one's environment. Both yin and yang, integrative and self-assertive tendencies, are necessary for harmonious social and ecological relationships.

Excessive self-assertion manifests itself as power, control, and domination of others by force; and these are, indeed, the patterns prevalent in our society. Political and economic power is exerted by a dominant corporate class; social hierarchies are maintained along racist and sexist lines, and rape has become a central metaphor of our culture—rape of women, of minority groups, and of the earth herself. Our science and technology are based on the seventeenth-century belief that an understanding of nature implies domination of nature by "man." Combined with the mechanistic model of the universe, which also originated in the seventeenth century, and with excessive emphasis on linear thinking, this attitude has produced a technology that is unhealthy and inhuman; a technology in which the natural, organic habitat of complex human beings is replaced by a simplified, synthetic, and prefabricated environment.[36]

This technology is aimed at control, mass production, and standardization, and is subjected, most of the time, to centralized management that pursues the illusion of indefinite growth. Thus the self-assertive tendency keeps increasing, and with it the requirement of submission, which is not the complement to self-assertion but the reverse side of the same phenomenon. While self-assertive behavior is presented as the ideal for men, submissive behavior is expected from women, but also from employees and executives who are required to deny their personal identities and to adopt the corporate identity and behavior patterns. A similar situation exists in our educational system, in which self-assertiveness is rewarded as far as competitive behavior is concerned, but is discouraged when expressed in terms of original ideas and questioning of authority.

Promotion of competitive behavior over cooperation is one of the principal manifestations of the self-assertive tendency in our society. It is rooted in the erroneous view of nature held by the Social Darwinists of the nineteenth century, who believed that all life in society had to be a struggle for existence ruled by "survival of the fittest." Accordingly, competition has been seen as the driving force of the economy, the "aggressive approach" has become the ideal of the business world, and this behavior has been combined with the exploitation of natural resources to create patterns of competitive consumption.

Aggressive, competitive behavior alone, of course, would make life impossible. Even the most ambitious, goal-oriented individuals need sympathetic support, human contact, and times of carefree spontaneity and relaxation. In our culture women are expected, and often forced, to fulfill these needs. They are the secretaries, receptionists, hostesses, nurses, and homemakers who perform the services that make life more comfortable and create the atmosphere in which the competitors can succeed. They cheer up their bosses and make coffee for them; they help smooth out conflicts in the office; they are the first to receive visitors and entertain them with small talk. In doctors' offices and hospitals women provide most of the human contact with patients that initiates the healing process. In physics departments women make the tea and serve the cookies over which the men discuss their theories. All these services involve yin, or integrative, activities, and since they rank lower in our value system than the yang, or self-assertive, activities, those who perform them get paid less. Indeed, many of them, such as mothers and housewives, are not paid at all.

From this short survey of cultural attitudes and values we can see that our culture has consistently promoted and rewarded the yang, the masculine or self-assertive elements of human nature, and has disregarded its yin, the feminine or intuitive aspects. Today, however, we are witnessing the beginning of a tremendous evolutionary movement. The turning point we are about to reach marks, among many other things, a reversal in the fluctuation between yin and yang. As the Chinese text says, "The yang, having reached its climax, retreats in favor of the yin." Our 1960s and 1970s have generated a whole series of philosophical, spiritual, and political movements that seem to go in the same direction. They all counteract the overemphasis on yang attitudes and values, and try to reestablish a balance between the masculine and feminine sides of human nature.

There is a rising concern with ecology, expressed by citizen movements that are forming around social and environmental issues, pointing out the limits to growth, advocating a new ecological ethic, and developing appropriate "soft" technologies. In the political arena the antinuclear movement is fighting the most extreme outgrowth of our self-assertive "macho" technology and, in doing so, is likely to become one of the most powerful political forces of this decade. At the same time there is the beginning of a significant shift in values—from the admiration of large-scale enterprises and institutions to the notion of "small is beautiful," from material consumption to voluntary simplicity, from economic and technological growth to inner growth and development. These new values are being promoted by the "human potential" movement, the "holistic-health" movement, and various spiritual movements. Perhaps most important, the old value system is being challenged and profoundly changed by the rise of feminist awareness originating in the women's movement.

These various movements form what cultural historian Theodore Roszak has called the counter culture.[37] So far, many of them still operate separately and have not yet seen how much their purposes interrelate. Thus the human potential movement and the holistic health movement often lack a social perspective, while spiritual movements tend to lack ecological awareness, with Eastern gurus displaying Western capitalist status symbols and spending considerable time building their economic empires. However, some movements have recently begun to form coalitions. As would be expected, the ecology movement and the feminist movement are joining forces on several issues, notably nuclear power, and environmental groups, consumer groups, and ethnic liberation movements are beginning to make contacts. We can anticipate that, once they have recognized the commonality of their aims, all these movements will flow together and form a powerful force of social transformation. I shall call this force the rising culture, following Toynbee's persuasive model of cultural dynamics:

> During the disintegration of a civilization, two separate plays with different plots are being performed simultaneously side by side. While an unchanging dominant minority is perpetually rehearsing its own defeat, fresh challenges are perpetually evoking fresh creative responses from newly recruited minorities, which proclaim their own creative power by rising, each time, to the occasion. The drama of challenge-and-response continues to be performed, but in new circumstances and with new actors.[38]

From this broad historical perspective cultures are seen to come and go in rhythms, and preserving cultural traditions may not always be the most desirable aim. What we have to do to minimize the hardship of inevitable change is recognize the changing conditions as clearly as possible and transform our lives and our social institutions accordingly. I shall argue that physicists can play an important role in this process. Since the seventeenth century physics has been the shining example of an "exact" science, and has served as the model for all the other sciences. For two and a half centuries physicists have used a mechanistic view of the world to develop and refine the conceptual framework known as classical physics. They have based their ideas on the mathematical theory of Isaac Newton, the philosophy of René Descartes, and the scientific methodology advocated by Francis Bacon, and developed them in accordance with the general conception of reality prevalent during the seventeenth, eighteenth, and nineteenth centuries. Matter was thought to be the basis of all existence, and the material world was seen as a multitude of separate objects assembled into a huge machine. Like human-made machines, the cosmic machine was thought to consist of elementary parts. Consequently it was believed that complex phenomena could always be understood by reducing them to their basic building blocks and by looking for the mechanisms through which these interacted. This attitude, known as reductionism, has become so deeply ingrained in our culture that it has often been identified with the scientific method. The other sciences accepted the mechanistic and reductionistic views of classical physics as the correct description of reality and modeled their own theories accordingly. Whenever psychologists, sociologists, or economists wanted to be scientific, they naturally turned toward the basic concepts of Newtonian physics.

In the twentieth century, however, physics has gone through several conceptual revolutions that clearly reveal the limitations of the mechanistic world view and lead to an organic, ecological view of the world which shows great similarities to the views of mystics of all ages and traditions. The universe is no longer seen as a machine, made up of a multitude of separate objects, but appears as a harmonious indivisible whole; a network of dynamic relationships that include the human observer and his or her consciousness in an essential way. The fact that modern physics, the manifestation of an extreme specialization of the rational mind, is now making contact with mysticism, the essence of religion and manifestation of an extreme specialization of the intuitive mind, shows very beautifully the unity and complementary nature of the rational and intuitive modes of consciousness; of the yang and the yin. Physicists, therefore, can provide the scientific background to the changes in attitudes and values that our society so urgently needs. In a culture dominated by science, it will be much easier to convince our social institutions that fundamental changes are necessary if we can give our arguments a scientific basis. This is what physicists can now provide. Modern physics can show the other sciences that scientific thinking does not necessarily have to be reductionist and mechanistic, that holistic and ecological views are also scientifically sound.

One of the main lessons that physicists have had to learn in this century has been the fact that all the concepts and theories we use to describe nature are limited. Because of the essential limitations of the rational mind, we have to accept the fact that, as Werner Heisenberg phrases it, "every word or concept, clear as it may seem to be, has only a limited range of applicability."[39] Scientific theories can never provide a complete and definitive description of reality. They will always be approximations to the true nature of things. To put it bluntly, scientists do not deal with truth; they deal with limited and approximate descriptions of reality.

At the beginning of the century, when physicists extended the range of their investigations into the realms of atomic and subatomic phenomena, they suddenly became aware of the limitations of their classical ideas and had to radically revise many of their basic concepts about reality. The experience of questioning the very basis of their conceptual framework and of being forced to accept profound modifications of their most cherished ideas was dramatic and often painful for those scientists, especially during the first three decades of the century, but it was rewarded by deep insights into the nature of matter and the human mind.

I believe this experience can serve as a useful lesson for other scientists, many of whom have now reached the limits of the Cartesian world view in their fields. Like the physicists, they will have to accept the fact that we must modify or even abandon some of our concepts when we expand the realm of our experience or field of study. The following chapters will show how the natural sciences, as well as the humanities and social sciences, have modeled themselves after classical Newtonian physics. Now that physicists have gone far beyond this model, it is time for the other sciences to expand their underlying philosophies.

Among the sciences that have been influenced by the Cartesian world view and by Newtonian physics, and will have to change to be consistent with the views of modern physics, I shall concentrate on those dealing with health in the broadest ecological sense: from biology and medical science to psychology and psychotherapy, sociology, economics, and political science. In all these fields the limitations of the classical, Cartesian world view are now becoming apparent. To transcend the classical models scientists will have to go beyond the mechanistic and reductionist approach as we have done in physics, and develop holistic and ecological views. Although their theories will need to be consistent with those of modern physics, the concepts of physics will generally not be appropriate as a model for the other sciences. However, they may still be very helpful. Scientists will not need to be reluctant to adopt a holistic framework, as they often are today, for fear of being unscientific. Modern physics can show them that such a framework is not only scientific but is in agreement with the most advanced scientific theories of physical reality.

II

The Two Paradigms

2·The Newtonian World-Machine

The world view and value system that lie at the basis of our culture and that have to be carefully reexamined were formulated in their essential outlines in the sixteenth and seventeenth centuries. Between 1500 and 1700 there was a dramatic shift in the way people pictured the world and in their whole way of thinking. The new mentality and the new perception of the cosmos gave our Western civilization the features that are characteristic of the modern era. They became the basis of the paradigm that has dominated our culture for the past three hundred years and is now about to change.

Before 1500 the dominant world view in Europe, as well as in most other civilizations, was organic. People lived in small, cohesive communities and experienced nature in terms of organic relationships, characterized by the interdependence of spiritual and material phenomena and the subordination of individual needs to those of the community. The scientific framework of this organic world view rested on two authorities—Aristotle and the Church. In the thirteenth century Thomas Aquinas combined Aristotle's comprehensive system of nature with Christian theology and ethics and, in doing so, established the conceptual framework that remained unquestioned throughout the Middle Ages. The nature of medieval science was very different from that of contemporary science. It was based on both reason and faith and its main goal was to understand the meaning and significance of things, rather than prediction and control. Medieval scientists, looking for the purposes underlying various natural phenomena, considered questions relating to God, the human soul, and ethics to be of the highest significance.

The medieval outlook changed radically in the sixteenth and seventeenth centuries. The notion of an organic, living, and spiritual universe was replaced by that of the world as a machine, and the world-machine became the dominant metaphor of the modern era. This development was brought about by revolutionary changes in physics and astronomy, culminating in the achievements of Copernicus, Galileo, and Newton. The science of the seventeenth century was based on a new method of inquiry, advocated forcefully by Francis Bacon, which involved the mathematical description of nature and the analytic method of reasoning conceived by the genius of Descartes. Acknowledging the crucial role of science in bringing about these far-reaching changes, historians have called the sixteenth and seventeenth centuries the Age of the Scientific Revolution.

The Scientific Revolution began with Nicolas Copernicus, who overthrew the geocentric view of Ptolemy and the Bible that had been accepted dogma for more than a thousand years. After Copernicus, the earth was no longer the center of the universe but merely one of the many planets circling a minor star at the edge of the galaxy, and man was robbed of his proud position as the central figure of God's creation. Copernicus was fully aware that his view would deeply offend the religious consciousness of his time; he delayed its publication until 1543, the year of his death, and even then he presented the heliocentric view merely as a hypothesis.

Copernicus was followed by Johannes Kepler, a scientist and mystic who searched for the harmony of the spheres and was able, through painstaking work with astronomical tables, to formulate his celebrated empirical laws of planetary motion, which gave further support to the Copernican system. But the real change in scientific opinion was brought about by Galileo Galilei, who was already famous for discovering the laws of falling bodies when he turned his attention to astronomy. Directing the newly invented telescope to the skies and applying his extraordinary gift for scientific observation to celestial phenomena, Galileo was able to discredit the old cosmology beyond any doubt and to establish the Copernican hypothesis as a valid scientific theory.

The role of Galileo in the Scientific Revolution goes far beyond his achievements in astronomy, although these are most widely known because of his clash with the Church. Galileo was the first to combine scientific experimentation with the use of mathematical language to formulate the laws of nature he discovered, and is therefore considered the father of modern science. "Philosophy,"* he believed, "is written in that great book which ever lies before our eyes; but we cannot understand it if we do not first learn the language and characters in which it is written. This language is mathematics, and the characters are triangles, circles, and other geometrical figures."[1] The two aspects of Galileo's pioneering work—his empirical approach and his use of a mathematical description of nature—became the dominant features of science in the seventeenth century and have remained important criteria of scientific theories up to the present day.

To make it possible for scientists to describe nature mathematically, Galileo postulated that they should restrict themselves to studying the essential properties of material bodies—shapes, numbers, and movement—which could be measured and quantified. Other properties, like color, sound, taste, or smell, were merely subjective mental projections which should be excluded from the domain of science.[2] Galileo's strategy of directing the scientist's attention to the quantifiable properties of matter has proved extremely successful throughout modern science, but it has also exacted a heavy toll, as the psychiatrist R. D. Laing emphatically reminds us: "Out go sight, sound, taste, touch and smell and along with them has since gone aesthetics and ethical sensibility, values, quality, form; all feelings, motives, intentions, soul, consciousness, spirit. Experience as such is cast out of the realm of scientific discourse."[3] According to Laing, hardly anything has changed our world more during the past four hundred years than the obsession of scientists with measurement and quantification.

While Galileo devised ingenious experiments in Italy, Francis Bacon set forth the empirical method of science explicitly in England. Bacon was the first to formulate a clear theory of the inductive procedure—to make experiments and to draw general conclusions from them, to be tested in further experiments—and he became extremely influential by vigorously advocating the new method. He boldly attacked traditional schools of thought and developed a veritable passion for scientific experimentation.

The "Baconian spirit" profoundly changed the nature and purpose of the scientific quest. From the time of the ancients the goals of sci-

* From the Middle Ages to the nineteenth century the term "philosophy" was used in a very broad sense and included what we now call "science."

ence had been wisdom, understanding the natural order and living in harmony with it. Science was pursued "for the glory of God," or, as the Chinese put it, to "follow the natural order" and "flow in the current of the Tao."[4] These were yin, or integrative, purposes; the basic attitude of scientists was ecological, as we would say in today's language. In the seventeenth century this attitude changed into its polar opposite; from yin to yang, from integration to self-assertion. Since Bacon, the goal of science has been knowledge that can be used to dominate and control nature, and today both science and technology are used predominantly for purposes that are profoundly antiecological.

The terms in which Bacon advocated his new empirical method of investigation were not only passionate but often outright vicious. Nature, in his view, had to be "hounded in her wanderings," "bound into service," and made a "slave." She was to be "put in constraint," and the aim of the scientist was to "torture nature's secrets from her."[5] Much of this violent imagery seems to have been inspired by the witch trials that were held frequently in Bacon's time. As attorney general for King James I, Bacon was intimately familiar with such prosecutions, and because nature was commonly seen as female, it is not surprising that he should carry over the metaphors used in the courtroom into his scientific writings. Indeed, his view of nature as a female whose secrets have to be tortured from her with the help of mechanical devices is strongly suggestive of the widespread torture of women in the witch trials of the early seventeenth century.[6] Bacon's work thus represents an outstanding example of the influence of patriarchal attitudes on scientific thought.

The ancient concept of the earth as nurturing mother was radically transformed in Bacon's writings, and it disappeared completely as the Scientific Revolution proceeded to replace the organic view of nature with the metaphor of the world as a machine. This shift, which was to become of overwhelming importance for the further development of Western civilization, was initiated and completed by two towering figures of the seventeenth century, Descartes and Newton.

René Descartes is usually regarded as the founder of modern philosophy. He was a brilliant mathematician and his philosophical outlook was profoundly affected by the new physics and astronomy. He did not accept any traditional knowledge, but set out to build a whole new system of thought. According to Bertrand Russell, "This had not happened since Aristotle, and is a sign of the new self-confidence that resulted from the progress of science. There is a freshness about his work that is not to be found in any eminent previous philosopher since Plato."[7]

At the age of twenty-three, Descartes experienced an illuminating vision that was to shape his entire life.[8] After several hours of intense concentration, during which he reviewed systematically all the knowledge he had accumulated, he perceived, in a sudden flash of intuition, the "foundations of a marvellous science" which promised the unification of all knowledge. This intuition had been foreshadowed in a letter to a friend in which Descartes announced his ambitious aim: "And so as to not hide anything from you about the nature of my work, I would like to give the public . . . a completely new science which would resolve generally all questions of quantity, continuous or discontinuous."[9] In his vision Descartes perceived how he could realize this plan. He saw a method that would allow him to construct a complete science of nature about which he could have absolute certainty; a science based, like mathematics, on self-evident first principles. Descartes was overwhelmed by this revelation. He felt that he had made the supreme discovery of his life and had no doubt that his vision came from divine inspiration. This conviction was enforced by an extraordinary dream the following night in which the new science was presented to him in symbolic form. Descartes was now certain that God had shown him his mission, and he set out to build a new scientific philosophy.

Descartes' vision had implanted in him the firm belief in the certainty of scientific knowledge, and his vocation in life was to distinguish truth from error in all fields of learning. "All science is certain, evident knowledge," he wrote. "We reject all knowledge which is merely probable and judge that only those things should be believed which are perfectly known and about which there can be no doubts."[10]

The belief in the certainty of scientific knowledge lies at the very basis of Cartesian philosophy and of the world view derived from it, and it was here, at the very outset, that Descartes went wrong. Twentieth-century physics has shown us very forcefully that there is no absolute truth in science, that all our concepts and theories are limited and approximate.The Cartesian belief in scientific truth is still widespread today and is reflected in the scientism that has become typical of our Western culture. Many people in our society, scientists as well as nonscientists, are convinced that the scientific method is the only valid way of understanding the universe. Descartes' method of thought and his view of nature have influenced all branches of modern science and can still be very useful today. But they will be useful only if their limitations are recognized. The acceptance of the Cartesian view as absolute truth and of Descartes' method as the only valid way to knowledge has played an important role in bringing about our current cultural imbalance.

Cartesian certainty is mathematical in its essential nature. Descartes believed that the key to the universe was its mathematical structure, and in his mind science was synonymous with mathematics. Thus he wrote, regarding the properties of physical objects, "I admit nothing as true of them that is not deduced, with the clarity of a mathematical demonstration, from common notions whose truth we cannot doubt. Because all the phenomena of nature can be explained in this way, I think that no other principles of physics need be admitted, nor are to be desired."[11]

Like Galileo, Descartes believed that the language of nature—"that great book which ever lies before our eyes"—was mathematics, and his desire to describe nature in mathematical terms led him to his most celebrated discovery. By applying numerical relations to geometrical figures, he was able to correlate algebra and geometry and, in doing so, founded a new branch of mathematics, now known as analytic geometry. This included the representation of curves by algebraic equations whose solutions he studied in a systematic way. His new method allowed Descartes to apply a very general type of mathematical analysis to the study of moving bodies, in accordance with his grand scheme of reducing all physical phenomena to exact mathematical relationships. Thus he could say, with great pride, "My entire physics is nothing other than geometry."[12]

Descartes' genius was that of a mathematician, and this is apparent also in his philosophy. To carry out his plan of building a complete and exact natural science, he developed a new method of reasoning which he presented in his most famous book, *Discourse on Method.* Although this text has become one of the great philosophical classics, its original purpose was not to teach philosophy but to serve as an introduction to science. Descartes' method was designed to reach scientific truth, as is evident from the book's full title, *Discourse on the*

Method of Rightly Conducting One's Reason and Searching the Truth in the Sciences.

The crux of Descartes' method is radical doubt. He doubts everything he can manage to doubt—all traditional knowledge, the impressions of his senses, and even the fact that he has a body—until he reaches one thing he cannot doubt, the existence of himself as a thinker. Thus he arrives at his celebrated statement, "*Cogito, ergo sum,*" "I think, therefore I exist." From this Descartes deduces that the essence of human nature lies in thought, and that all the things we conceive clearly and distinctly are true. Such clear and distinct conception—"the conception of the pure and attentive mind"[13]—he calls "intuition," and he affirms that "there are no paths to the certain knowledge of truth open to man except evident intuition and necessary deduction."[14] Certain knowledge, then, is achieved through intuition and deduction, and these are the tools Descartes uses in his attempt to rebuild the edifice of knowledge on firm foundations.

Descartes' method is analytic. It consists in breaking up thoughts and problems into pieces and in arranging these in their logical order. This analytic method of reasoning is probably Descartes' greatest contribution to science. It has become an essential characteristic of modern scientific thought and has proved extremely useful in the development of scientific theories and the realization of complex technological projects. It was Descartes' method that made it possible for NASA to put a man on the moon. On the other hand, overemphasis on the Cartesian method has led to the fragmentation that is characteristic of both our general thinking and our academic disciplines, and to the widespread attitude of reductionism in science—the belief that all aspects of complex phenomena can be understood by reducing them to their constituent parts.

Descartes' *cogito,* as it has come to be called, made mind more certain for him than matter and led him to the conclusion that the two were separate and fundamentally different. Thus he asserted that "there is nothing included in the concept of body that belongs to the mind; and nothing in that of mind that belongs to the body"[15] The Cartesian division between mind and matter has had a profound effect on Western thought. It has taught us to be aware of ourselves as isolated egos existing "inside" our bodies; it has led us to set a higher value on mental than manual work; it has enabled huge industries to sell products—especially to women—that would make us owners of the "ideal body"; it has kept doctors from seriously considering the psychological dimensions of illness, and psychotherapists from dealing with their patients' bodies. In the life sciences, the Cartesian division has led to endless confusion about the relation between mind and brain, and in physics it made it extremely difficult for the founders of quantum theory to interpret their observations of atomic phenomena. According to Heisenberg, who struggled with the problem for many years, "This partition has penetrated deeply into the human mind during the three centuries following Descartes and it will take a long time for it to be replaced by a really different attitude toward the problem of reality."[16]

Descartes based his whole view of nature on this fundamental division between two independent and separate realms; that of mind, or *res cogitans*, the "thinking thing," and that of matter, or *res extensa*, the "extended thing." Both mind and matter were the creations of God, who represented their common point of reference, being the source of the exact natural order and of the light of reason that enabled the human mind to recognize this order. For Descartes, the existence of God was essential to his scientific philosophy, but in subsequent centuries scientists omitted any explicit reference to God and developed their theories according to the Cartesian division, the humanities concentrating on the *res cogitans* and the natural sciences on the *res extensa*.

To Descartes the material universe was a machine and nothing but a machine. There was no purpose, life, or spirituality in matter. Nature worked according to mechanical laws, and everything in the material world could be explained in terms of the arrangement and movement of its parts. This mechanical picture of nature became the dominant paradigm of science in the period following Descartes. It guided all scientific observation and the formulation of all theories of natural phenomena until twentieth-century physics brought about radical change. The whole elaboration of mechanistic science in the seventeenth, eighteenth and nineteenth centuries, including Newton's grand synthesis, was but the development of the Cartesian idea. Descartes gave scientific thought its general framework—the view of nature as a perfect machine, governed by exact mathematical laws.

The drastic change in the image of nature from organism to machine had a strong effect on people's attitudes toward the natural environment. The organic world view of the Middle Ages had implied a value system conducive to ecological behavior. In the words of Carolyn Merchant:

> The image of the earth as a living organism and nurturing mother served as a cultural constraint restricting the actions of human beings. One does not readily slay a mother, dig into her entrails for gold, or mutilate her body . . . As long as the earth was considered to be alive and sensitive, it could be considered a breach of human ethical behavior to carry out destructive acts against it.[17]

These cultural constraints disappeared as the mechanization of science took place. The Cartesian view of the universe as a mechanical system provided a "scientific" sanction for the manipulation and exploitation of nature that has become typical of Western culture. In fact, Descartes himself shared Bacon's view that the aim of science was the domination and control of nature, affirming that scientific knowledge could be used to "render ourselves the masters and possessors of nature."[18]

In his attempt to build a complete natural science, Descartes extended his mechanistic view of matter to living organisms. Plants and animals were considered simply machines; human beings were inhabited by a rational soul that was connected with the body through the pineal gland in the center of the brain. As far as the human body was concerned, it was indistinguishable from an animal-machine. Descartes explained at great length how the motions and various biological functions of the body could be reduced to mechanical operations, in order to show that living organisms were nothing but automata. In doing so he was strongly influenced by the preoccupation of the baroque seventeenth century with artful, "lifelike" machinery that delighted people with the magic of its seemingly spontaneous movements. Like most of his contemporaries, Descartes was fascinated by these automata and even constructed a few of them himself. Inevitably, he compared their functioning to that of living organisms: "We see clocks, artificial fountains, mills and other similar machines which, though merely man-made, have nonetheless the power to move by themselves in several different ways . . . I do not recognize any difference between the machines made by craftsmen and the various bodies that nature alone composes."[19]

Clockmaking in particular had attained a high degree of perfection by Descartes' time, and the clock was thus a privileged model for other automatic machines. Descartes compared animals to a "clock . . . composed . . . of wheels and springs," and he extended this comparison to the human body: "I consider the human body as a machine . . . My thought . . . compares a sick man and an ill-made clock with my idea of a healthy man and a well-made clock."[20]

Descartes' view of living organisms has had a decisive influence on the development of the life sciences. The careful description of the mechanisms that make up living organisms has been the major task of biologists, physicians, and psychologists for the past three hundred years. The Cartesian approach has been very successful, especially in biology, but it has also limited the directions of scientific research. The problem is that scientists, encouraged by their success in treating living organisms as machines, tend to believe that they are *nothing but* machines. The adverse consequences of this reductionist fallacy have become especially apparent in medicine, where the adherence to the Cartesian model of the human body as a clockwork has prevented doctors from understanding many of today's major illnesses.

This, then, was Descartes' "marvellous science." Using his method of analytic thought, he attempted to give a precise account of all natural phenomena in one single system of mechanical principles. His science was to be complete, and the knowledge it gave was to provide absolute mathematical certainty. Descartes, of course, was not able to carry out this ambitious plan, and he himself recognized that his science was incomplete. But his method of reasoning and the general outline of the theory of natural phenomena he provided have shaped Western scientific thought for three centuries.

Today, although the severe limitations of the Cartesian world view are becoming apparent in all the sciences, Descartes' general method of approaching intellectual problems and his clarity of thought remain immensely valuable. I was vividly reminded of this after a lecture on modern physics in which I emphasized the limitations of the mechanistic world view in quantum theory and the necessity of overcoming this view in other fields, when a Frenchwoman complimented me on my "Cartesian clarity." As Montesquieu wrote in the eighteenth century, "Descartes has taught those who came after him how to discover his own errors."[21]

• • •

Descartes created the conceptual framework for seventeenth-century science, but his view of nature as a perfect machine, governed by exact mathematical laws, had to remain a vision during his lifetime. He could not do more than sketch the outlines of his theory of natural phenomena. The man who realized the Cartesian dream and completed the Scientific Revolution was Isaac Newton, born in England in 1642, the year of Galileo's death. Newton developed a complete mathematical formulation of the mechanistic view of nature, and thus accomplished a grand synthesis of the works of Copernicus and Kepler, Bacon, Galileo, and Descartes. Newtonian physics, the crowning achievement of seventeenth-century science, provided a consistent mathematical theory of the world that remained the solid foundation of scientific thought well into the twentieth century. Newton's grasp of mathematics was far more powerful than that of his contemporaries. He invented a completely new method, known today as differential calculus, to describe the motion of solid bodies; a method that went far beyond the mathematical techniques of Galileo and Descartes. This tremendous intellectual achievement has been praised by Einstein as "perhaps the greatest advance in thought that a single individual was ever privileged to make."[22]

Kepler had derived empirical laws of planetary motion by studying astronomical tables, and Galileo had performed ingenious experiments to discover the laws of falling bodies. Newton combined those two discoveries by formulating the general laws of motion governing all objects in the solar system, from stones to planets.

According to legend, the decisive insight occurred to Newton in a sudden flash of inspiration when he saw an apple fall from a tree. He realized that the apple was pulled toward the earth by the same force that pulled the planets toward the sun, and thus found the key to his grand synthesis. He then used his new mathematical method to formulate the exact laws of motion for all bodies under the influence of the force of gravity. The significance of these laws lay in their universal application. They were found to be valid throughout the solar system and thus seemed to confirm the Cartesian view of nature. The Newtonian universe was, indeed, one huge mechanical system, operating according to exact mathematical laws.

Newton presented his theory of the world in great detail in his *Mathematical Principles of Natural Philosophy.* The *Principia,* as the work is usually called for short after its original Latin title, comprises a comprehensive system of definitions, propositions, and proofs which scientists regarded as the correct description of nature for more than two hundred years. It also contains an explicit discussion of Newton's experimental method, which he saw as a systematic procedure whereby the mathematical description is based, at every step, on critical evaluation of experimental evidence:

> Whatever is not deduced from the phenomena is to be called a hypothesis; and hypotheses, whether metaphysical or physical, whether of occult qualities or mechanical, have no place in experimental philosophy. In this philosophy, particular propositions are inferred from the phenomena, and afterwards rendered general by induction.[23]

Before Newton there had been two opposing trends in seventeenth-century science; the empirical, inductive method represented by Bacon and the rational, deductive method represented by Descartes. Newton, in his *Principia,* introduced the proper mixture of both methods, emphasizing that neither experiments without systematic interpretation nor deduction from first principles without experimental evidence will lead to a reliable theory. Going beyond Bacon in his systematic experimentation and beyond Descartes in his mathematical analysis, Newton unified the two trends and developed the methodology upon which natural science has been based ever since.

Isaac Newton was a much more complex personality than one would think from a reading of his scientific writings. He excelled not only as a scientist and mathematician but also, at various stages of his life, as a lawyer, historian, and theologian, and he was deeply involved in research into occult and esoteric knowledge. He looked at the world as a riddle and believed that its clues could be found not only through scientific experiments but also in the cryptic revelations of esoteric traditions. Newton was tempted to think, like Descartes, that his powerful mind could unravel all the secrets of the universe, and he applied it with equal intensity to the study of natural and esoteric science. While working at Trinity College, Cambridge, on the *Principia,* he accumulated, during the very same years, voluminous notes on alchemy, apocalyptic texts, unorthodox theological theories, and various occult matters. Most of these esoteric writings have never been published, but what is known of them indicates that Newton, the great genius of the

Scientific Revolution, was at the same time the "last of the magicians."[24]

The stage of the Newtonian universe, on which all physical phenomena took place, was the three-dimensional space of classical Euclidean geometry. It was an absolute space, an empty container that was independent of the physical phenomena occurring in it. In Newton's own words, "Absolute space, in its own nature, without regard to anything external, remains always similar and immovable."[25] All changes in the physical world were described in terms of a separate dimension, time, which again was absolute, having no connection with the material world and flowing smoothly from the past through the present to the future. "Absolute, true, and mathematical time," wrote Newton, "of itself and by its own nature, flows uniformly, without regard to anything external."[26]

The elements of the Newtonian world which moved in this absolute space and absolute time were material particles; small, solid and indestructible objects out of which all matter was made. The Newtonian model of matter was atomistic, but it differed from the modern notion of atoms in that the Newtonian particles were all thought to be made of the same material substance. Newton assumed matter to be homogeneous; he explained the difference between one type of matter and another not in terms of atoms of different weights or densities but in terms of more or less dense packing of atoms. The basic building blocks of matter could be of different sizes but consisted of the same "stuff," and the total amount of material substance in an object was given by the object's mass.

The motion of the particles was caused by the force of gravity, which, in Newton's view, acted instantaneously over a distance. The material particles and the forces between them were of a fundamentally different nature, the inner constitution of the particles being independent of their mutual interaction. Newton saw both the particles and the force of gravity as created by God and thus not subject to further analysis. In his *Opticks,* Newton gave a clear picture of how he imagined God's creation of the material world:

> It seems probable to me that God in the beginning formed matter in solid, massy, hard, impenetrable, movable particles, of such sizes and

> figures, and with such other properties, and in such proportion to space, as most conduced to the end for which he formed them; and that these primitive particles being solids, are incomparably harder than any porous bodies compounded of them; even so very hard, as never to wear or break in pieces; no ordinary power being able to divide what God himself made one in the first creation.[27]

In Newtonian mechanics all physical phenomena are reduced to the motion of material particles, caused by their mutual attraction, that is, by the force of gravity. The effect of this force on a particle or any other material object is described mathematically by Newton's equations of motion, which form the basis of classical mechanics. These were considered fixed laws according to which material objects moved, and were thought to account for all changes observed in the physical world. In the Newtonian view, God created in the beginning the material particles, the forces between them, and the fundamental laws of motion. In this way the whole universe was set in motion, and it has continued to run ever since, like a machine, governed by immutable laws. The mechanistic view of nature is thus closely related to a rigorous determinism, with the giant cosmic machine completely causal and determinate. All that happened had a definite cause and gave rise to a definite effect, and the future of any part of the system could—in principle—be predicted with absolute certainty if its state at any time was known in all details.

This picture of a perfect world-machine implied an external creator; a monarchical god who ruled the world from above by imposing his divine law on it. The physical phenomena themselves were not thought to be divine in any sense, and when science made it more and more difficult to believe in such a god, the divine disappeared completely from the scientific world view, leaving behind the spiritual vacuum that has become characteristic of the mainstream of our culture. The philosophical basis of this secularization of nature was the Cartesian division between spirit and matter. As a consequence of this division, the world was believed to be a mechanical system that could be described objectively, without ever mentioning the human observer, and such an objective description of nature became the ideal of all science.

The eighteenth and nineteenth centuries used Newtonian mechanics with tremendous success. The Newtonian theory was able to explain the motion of the planets, moons, and comets down to the smallest details, as well as the flow of the tides and various other phenomena related to gravity. Newton's mathematical system of the world established itself quickly as the correct theory of reality and generated enormous enthusiasm among scientists and the lay public alike. The picture of the world as a perfect machine, which had been introduced by Descartes, was now considered a proved fact and Newton became its symbol. During the last twenty years of his life Sir Isaac Newton reigned in eighteenth-century London as the most famous man of his time, the great white-haired sage of the Scientific Revolution. Accounts of this period of Newton's life sound quite familiar to us because of our memories and photographs of Albert Einstein, who played a very similar role in our century.

Encouraged by the brilliant success of Newtonian mechanics in astronomy, physicists extended it to the continuous motion of fluids and to the vibrations of elastic bodies, and again it worked. Finally, even the theory of heat could be reduced to mechanics when it was realized that heat was the energy generated by a complicated "jiggling" motion of atoms and molecules. Thus many thermal phenomena, such as the evaporation of a liquid, or the temperature and pressure of a gas, could be understood quite well from a purely mechanistic point of view.

The study of the physical behavior of gases led John Dalton to the formulation of his celebrated atomic hypothesis, which was probably the most important step in the entire history of chemistry. Dalton had a vivid pictorial imagination and tried to explain the properties of gas mixtures with the help of elaborate drawings of geometric and mechanical models of atoms. His main assumptions were that all chemical elements are made up of atoms, and that the atoms of a given element are all alike but differ from those of every other element in mass, size, and properties.Using Dalton's hypothesis, chemists of the nineteenth century developed a precise atomic theory of chemistry which paved the way for the conceptual unification of physics and chemistry in the twentieth century. Thus Newtonian mechanics was extended far beyond the description of macroscopic bodies. The behavior of solids, liquids, and gases, including the phenomena of heat and sound, was explained successfully in terms of the motion of elementary material particles. For the scientists of the eighteenth and nineteenth centuries this tremendous success of the mechanistic model confirmed their belief that the universe was indeed a huge mechanical system, running according to the Newtonian laws of motion, and that Newton's mechanics was the ultimate theory of natural phenomena.

Although the properties of atoms were studied by chemists rather than physicists throughout the nineteenth century, classical physics was based on the Newtonian idea of atoms as hard and solid building blocks of matter. This image no doubt contributed to the reputation of physics as a "hard science," and to the development of the "hard technology" based upon it. The overwhelming success of Newtonian physics and the Cartesian belief in the certainty of scientific knowledge led directly to the emphasis on hard science and hard technology in our culture. Not until the mid-twentieth century would it become clear that the idea of a hard science was part of the Cartesian-Newtonian paradigm, a paradigm that would be transcended.

With the firm establishment of the mechanistic world view in the eighteenth century, physics naturally became the basis of all the sciences. If the world is really a machine, the best way to find out how it works is to turn to Newtonian mechanics. It was thus an inevitable consequence of the Cartesian world view that the sciences of the eighteenth and nineteenth centuries modeled themselves after Newtonian physics. In fact, Descartes was well aware of the basic role of physics in his view of nature. "All philosophy," he wrote, "is like a tree. The roots are metaphysics, the trunk is physics, and the branches are all the other sciences."[28]

Descartes himself had sketched the outlines of a mechanistic approach to physics, astronomy, biology, psychology, and medicine. The thinkers of the eighteenth century carried this program further by applying the principles of Newtonian mechanics to the sciences of human nature and human society. The newly created social sciences generated great enthusiasm, and some of their proponents even claimed to have discovered a "social physics." The Newtonian theory of the universe and the belief in the rational approach to human problems spread so rapidly among the middle classes of the eighteenth century that the whole era became the "Age of Enlightenment." The dominant figure in this development was the philosopher John Locke, whose most important writings were published late in the seventeenth century. Strongly influenced by Descartes and Newton, Locke's work had a decisive impact on eighteenth-century thought.

Following Newtonian physics, Locke developed an atomistic view of society, describing it in terms of its basic building block, the human being. As physicists reduced the properties of gases to the motion of their atoms, or molecules, so Locke attempted to reduce the patterns observed in society to the behavior of its individuals. Thus he proceeded to study first the nature of the individual human being, and then tried to apply the principles of human nature to economic and political problems. Locke's analysis of human nature was based on that of an earlier philosopher, Thomas Hobbes, who had declared that all knowledge was based on sensory perception. Locke adopted this theory of knowledge and, in a famous metaphor, compared the human mind at birth to a *tabula rasa*, a completely blank tablet on which knowledge is imprinted once it is acquired through sensory experience. This image was to have a strong influence on two major schools of classical psychology, behaviorism and psychoanalysis, as well as on political philosophy. According to Locke, all human beings—"all men," as he would say—were equal at birth and depended in their development entirely on their environment. Their actions, Locke believed, were always motivated by what they assumed to be their own interest.

When Locke applied his theory of human nature to social phenomena, he was guided by the belief that there were laws of nature governing human society similar to those governing the physical universe. As the atoms in a gas would establish a balanced state, so human individuals would settle down in a society in a "state of nature." Thus the function of government was not to impose its laws on the people, but rather to discover and enforce the natural laws that existed before any government was formed. According to Locke, these natural laws included the freedom and equality of all individuals as well as the right to property, which represented the fruits of one's labor.

Locke's ideas became the basis for the value system of the Enlightenment and had a strong influence on the development of modern economic and political thought. The ideals of individualism, property rights, free markets, and representative government, all of which can be traced back to Locke, contributed significantly to the thinking of Thomas Jefferson and are reflected in the Declaration of Independence and the American Constitution.

During the nineteenth century scientists continued to elaborate the mechanistic model of the universe in physics, chemistry, biology, psychology, and the social sciences. As a result the Newtonian world-machine became a much more complex and subtle structure. At the same time, new discoveries and new ways of thinking made the limitations of the Newtonian model apparent and prepared the way for the scientific revolutions of the twentieth century.

One of these nineteenth-century developments was the discovery and investigation of electric and magnetic phenomena that involved a new type of force and could not be described appropriately by the mechanistic model. The important step was taken by Michael Faraday and completed by Clerk Maxwell—the first one of the greatest experimenters in the history of science, the second a brilliant theorist. Faraday and Maxwell not only studied the effects of the electric and magnetic forces, but made the forces themselves the primary object of their investigation. By replacing the concept of a force with the much subtler concept of a force field they were the first to go beyond Newtonian physics,[29] showing that the fields had their own reality and could be studied without any reference to material bodies. This theory, called electrodynamics, culminated in the realization that light was in fact a rapidly alternating electromagnetic field traveling through space in the form of waves.

In spite of these far-reaching changes, Newtonian mechanics still held its position as the basis of all physics. Maxwell himself tried to explain his results in mechanical terms, interpreting the fields as states of mechanical stress in a very light, all-pervasive medium, called ether, and the electromagnetic waves as elastic waves of this ether. However, he used several mechanical interpretations of his theory at the same time and apparently took none of them really seriously, knowing intuitively that the fundamental entities in his theory were the fields and not the mechanical models. It remained for Einstein to clearly recognize this fact in our century, when he declared that no ether existed, and that the electromagnetic fields were physical entities in their own right which could travel through empty space and could not be explained mechanically.

While electromagnetism dethroned Newtonian mechanics as the ultimate theory of natural phenomena, a new trend of thinking arose that went beyond the image of the Newtonian world-machine and was to dominate not only the nineteenth century but all future scientific thinking. It involved the idea of evolution; of change, growth, and development. The notion of evolution had arisen in geology, where careful studies of fossils led scientists to the idea that the present state of the earth was the result of a continuous development caused by the action of natural forces over immense periods of time. But geologists were not the only ones who thought in those terms. The theory of the solar system proposed by both Immanuel Kant and Pierre Laplace was based on evolutionary, or developmental, thinking; evolutionary concepts were crucial to the political philosophies of Hegel and Engels; poets and philosophers alike, throughout the nineteenth century, were deeply concerned with the problem of becoming.

These ideas formed the intellectual background to the most precise and most far-reaching formulation of evolutionary thought—the theory of the evolution of species in biology. Ever since antiquity natural philosophers had entertained the idea of a "great chain of being." This chain, however, was conceived as a static hierarchy, starting with God at the top and descending through angels, human beings, and animals, to ever lower forms of life. The number of species was fixed; it had not changed since the day of their creation. As Linnaeus, the great botanist and classifier, put it: "We reckon as many species as issued in pairs from the hands of the Creator."[30] This view of biological species was in complete agreement with Judeo-Christian doctrine and was well suited for the Newtonian world.

The decisive change came with Jean Baptiste Lamarck, at the beginning of the nineteenth century; a change that was so dramatic that Gregory Bateson, one of the deepest and broadest thinkers of our time, has compared it to the Copernican Revolution:

> Lamarck, probably the greatest biologist in history, turned that ladder of explanation upside down. He was the man who said it starts with the infusoria and that there were changes leading up to man. His turning the taxonomy upside down is one of the most astonishing feats that has ever happened. It was the equivalent in biology of the Copernican revolution in astronomy.[31]

Lamarck was the first to propose a coherent theory of evolution, according to which all living beings have evolved from earlier, simpler forms under the pressure of their environment. Although the details of the Lamarckian theory had to be abandoned later on, it was nevertheless the important first step.

Several decades later Charles Darwin presented an overwhelming mass of evidence in favor of biological evolution, establishing the phenomenon for scientists beyond any doubt. He also proposed an explanation, based on the concepts of chance variation—now known as random mutation—and natural selection, which were to remain the cornerstones of modern evolutionary thought. Darwin's monumental *Origin of Species* synthesized the ideas of previous thinkers and has shaped all subsequent biological thought. Its role in the life sciences was similar to that of Newton's *Principia* in physics and astronomy two centuries earlier.

The discovery of evolution in biology forced scientists to abandon the Cartesian conception of the world as a machine that had emerged fully constructed from the hands of its Creator. Instead, the universe had to be pictured as an evolving and ever changing system in which complex structures developed from simpler forms. While this new way of thinking was elaborated in the life sciences, evolutionary concepts also emerged in physics. However, whereas in biology evolution meant a movement toward increasing order and complexity, in physics it came to mean just the opposite—a movement toward increasing disorder.

The application of Newtonian mechanics to the study of thermal phenomena, which involved treating liquids and gases as complicated mechanical systems, led physicists to the formulation of thermodynamics, the "science of complexity." The first great achievement of this new science was the discovery of one of the most fundamental laws of physics, the law of the conservation of energy. It states that the total energy involved in a process is always conserved. It may change its form in the most complicated way, but none of it is lost. This law, which physicists discovered in their study of steam engines and other heat-producing machines, is also known as the first law of thermodynamics.

It was followed by the second law of thermodynamics, that of the dissipation of energy. While the total energy involved in a process is always constant, the amount of useful energy is diminishing, dissipating into heat, friction, and so on. The second law was formulated first by Sadi Carnot in terms of the technology of thermal engines, but was soon recognized to be of much broader significance. It introduced into physics the idea of irreversible processes, of an "arrow of time." According to the second law, there is a certain trend in physical phenomena. Mechanical energy is dissipated into heat and cannot be completely recovered; when hot and cold water are brought together, the result will be lukewarm water and the two liquids will not separate. Similarly, when a bag of white sand and a bag of black sand are mixed, the result will be gray sand, and the more we shake the mixture the more uniform the gray will be; we will not see the two kinds of sand separate spontaneously.

What all these processes have in common is that they proceed in a certain direction—from order to disorder—and this is the most general formulation of the second law of thermodynamics: Any isolated physical system will proceed spontaneously in the direction of ever increasing disorder. In mid-century, to express this direction in the evolution of physical systems in precise mathematical form, Rudolf Clausius introduced a new quantity which he called "entropy." The term represents a combination of "energy" and "tropos," the Greek word for transformation, or evolution. Thus entropy is a quantity that measures the degree of evolution of a physical system. According to the second law, the entropy of an isolated physical system will keep increasing, and because this evolution is accompanied by increasing disorder, entropy can also be seen as a measure of disorder.

The formulation of the concept of entropy and the second law of thermodynamics was one of the most important contributions to physics in the nineteenth century. The increase of entropy in physical systems, which marks the direction of time, could not be explained by the laws of Newtonian mechanics and remained mysterious until Ludwig Boltzmann clarified the situation by introducing an additional idea, the concept of probability. With the help of probability theory, the behavior of complex mechanical systems could be described in terms of statistical laws, and thermodynamics could be put on a solid Newtonian basis, known as statistical mechanics.

Boltzmann showed that the second law of thermodynamics is a statistical law. Its affirmation that certain processes do not occur—for example, the spontaneous conversion of heat energy into mechanical energy—does not mean that they are impossible but merely that they are extremely unlikely. In microscopic systems, consisting of only a few molecules, the second law is violated regularly, but in macroscopic systems, which consist of vast numbers of molecules,* the probability that the total entropy of the system will increase becomes virtual certainty. Thus in any isolated system, made up of a large number of mol-

* For example, every cubic centimeter of air contains some ten billion billion (10^{19}) molecules.

ecules, the entropy—or disorder—will keep increasing until, eventually, the system reaches a state of maximum entropy, also known as "heat death"; in this state all activity has ceased, all material being evenly distributed and at the same temperature. According to classical physics, the universe as a whole is going toward such a state of maximum entropy; it is running down and will eventually grind to a halt.

This grim picture of cosmic evolution is in sharp contrast to the evolutionary idea held by biologists, who observe that the living universe evolves from disorder to order, toward states of ever increasing complexity. The emergence of the concept of evolution in physics thus brought to light another limitation of the Newtonian theory. The mechanistic conception of the universe as a system of small billiard balls in random motion is far too simplistic to deal with the evolution of life.

At the end of the nineteenth century Newtonian mechanics had lost its role as the fundamental theory of natural phenomena. Maxwell's electrodynamics and Darwin's theory of evolution involved concepts that clearly went beyond the Newtonian model and indicated that the universe was far more complex than Descartes and Newton had imagined. Nevertheless, the basic ideas underlying Newtonian physics, though insufficient to explain all natural phenomena, were still believed to be correct. The first three decades of our century changed this situation radically. Two developments in physics, culminating in relativity theory and in quantum theory, shattered all the principal concepts of the Cartesian world view and Newtonian mechanics. The notion of absolute space and time, the elementary solid particles, the fundamental material substance, the strictly causal nature of physical phenomena, and the objective description of nature—none of these concepts could be extended to the new domains into which physics was now penetrating.

3·The New Physics

At the beginning of modern physics stands the extraordinary intellectual feat of one man—Albert Einstein. In two articles, both published in 1905, Einstein initiated two revolutionary trends in scientific thought. One was his special theory of relativity; the other was a new way of looking at electromagnetic radiation which was to become characteristic of quantum theory, the theory of atomic phenomena. The complete quantum theory was worked out twenty years later by a whole team of physicists. Relativity theory, however, was constructed in its complete form almost entirely by Einstein himself. Einstein's scientific papers are intellectual monuments that mark the beginning of twentieth-century thought.

Einstein strongly believed in nature's inherent harmony, and throughout his scientific life his deepest concern was to find a unified foundation of physics. He began to move toward this goal by constructing a common framework for electrodynamics and mechanics, the two separate theories of classical physics. This framework is known as the special theory of relativity. It unified and completed the structure of classical physics, but at the same time it involved radical changes in the traditional concepts of space and time and thus undermined one of the foundations of the Newtonian world view. Ten years later Einstein proposed his general theory of relativity, in which the framework of the special theory is extended to include gravity. This is achieved by further drastic modifications of the concepts of space and time.

The other major development in twentieth-century physics was a consequence of the experimental investigation of atoms. At the turn of the century physicists discovered several phenomena connected with the structure of atoms, such as X-rays and radioactivity, which were inexplicable in terms of classical physics. Besides being objects of intense study, these phenomena were used, in most ingenious ways, as new tools to probe deeper into matter than had ever been possible before. For example, the so-called alpha particles emanating from radioactive substances were perceived to be high-speed projectiles of subatomic size that could be used to explore the interior of the atom. They could be fired at atoms, and from the way they were deflected one could draw conclusions about the atoms' structure.

This exploration of the atomic and subatomic world brought scientists in contact with a strange and unexpected reality that shattered the foundations of their world view and forced them to think in entirely new ways. Nothing like that had ever happened before in science. Revolutions like those of Copernicus and Darwin had introduced profound changes in the general conception of the universe, changes that were shocking to many people, but the new concepts themselves were not difficult to grasp. In the twentieth century, however, physicists faced, for the first time, a serious challenge to their ability to understand the universe. Every time they asked nature a question in an atomic experiment, nature answered with a paradox, and the more they tried to clarify the situation, the sharper the paradoxes became. In their struggle to grasp this new reality, scientists became painfully aware that their basic concepts, their language, and their whole way of thinking were inadequate to describe atomic phenomena. Their problem was not only intellectual but involved an intense emotional and existential experience, as vividly described by Werner Heisenberg: "I remember discussions with Bohr which went through many hours till very late at night and ended almost in despair; and when at the end of the discussion I went alone for a walk in the neighboring park I repeated to myself again and again the question: Can nature possibly be so absurd as it seemed to us in these atomic experiments?"[1]

It took these physicists a long time to accept the fact that the paradoxes they encountered are an essential aspect of atomic physics, and to realize that they arise whenever one tries to describe atomic phenomena in terms of classical concepts. Once this was perceived, the physicists began to learn to ask the right questions and to avoid contradictions. As Heisenberg says, "They somehow got into the spirit of the quantum theory,"[2] and finally they found the precise and consistent mathematical formulation of that theory. Quantum theory, or quantum mechanics as it is also called, was formulated during the first three decades of the century by an international group of physicists including Max Planck, Albert Einstein, Niels Bohr, Louis De Broglie, Erwin Schrödinger, Wolfgang Pauli, Werner Heisenberg, and Paul Dirac. These men joined forces across national borders to shape one of the most exciting periods of modern science, one that saw not only brilliant intellectual exchanges but also dramatic human conflicts, as well as deep personal friendships, among the scientists.

Even after the mathematical formulation of quantum theory was completed, its conceptual framework was by no means easy to accept. Its effect on the physicists' view of reality was truly shattering. The new physics necessitated profound changes in concepts of space, time, matter, object, and cause and effect; and because these concepts are so fundamental to our way of experiencing the world, their transformation came as a great shock. To quote Heisenberg again, "The violent reaction to the recent development of modern physics can only be understood when one realizes that here the foundations of physics have started moving; and that this motion has caused the feeling that the ground would be cut from science."[3]

Einstein experienced the same shock when he was confronted with the new concepts of physics, and he described his feelings in terms very similar to Heisenberg's: "All my attempts to adapt the theoretical foundation of physics to this [new type of] knowledge failed completely. It was as if the ground had been pulled out from under one, with no firm foundation to be seen anywhere, upon which one could have built."[4]

Out of the revolutionary changes in our concepts of reality that were brought about by modern physics, a consistent world view is now emerging. This view is not shared by the entire physics community, but is being discussed and elaborated by many leading physicists whose interest in their science goes beyond the technical aspects of their research. These scientists are deeply interested in the philosophical implications of modern physics and are trying in an open-minded way to improve their understanding of the nature of reality.

In contrast to the mechanistic Cartesian view of the world, the world view emerging from modern physics can be characterized by words like organic, holistic, and ecological. It might also be called a systems view, in the sense of general systems theory.[5] The universe is no longer seen as a machine, made up of a multitude of objects, but has to be pictured as one indivisible, dynamic whole whose parts are essentially interrelated and can be understood only as patterns of a cosmic process.

The basic concepts underlying this world view of modern physics are discussed in the following pages. I described this world view in detail in *The Tao of Physics,* showing how it is related to the views held in mystical traditions, especially those of Eastern mysticism. Many physicists, brought up, as I was, in a tradition that associates mysticism with things vague, mysterious, and highly unscientific, were shocked at having their ideas compared to those of mystics.[6] Fortunately, this attitude is now changing. As Eastern thought has begun to interest a significant number of people, and meditation is no longer viewed with ridicule or suspicion, mysticism is being taken seriously even within the scientific community. An increasing number of scientists are aware that mystical thought provides a consistent and relevant philosophical background to the theories of contemporary science, a conception of the world in which the scientific discoveries of men and women can be in perfect harmony with their spiritual aims and religious beliefs.

The experimental investigation of atoms at the beginning of the century yielded sensational and totally unexpected results. Far from being the hard, solid particles of time-honored theory, atoms turned out to consist of vast regions of space in which extremely small particles—the electrons—moved around the nucleus. A few years later quantum theory made it clear that even the subatomic particles—the electrons and the protons and neutrons in the nucleus—were nothing like the solid objects of classical physics. These subatomic units of matter are very abstract entities which have a dual aspect. Depending on how we look at them, they appear sometimes as particles, sometimes as waves; and this dual nature is also exhibited by light, which can take the form of electromagnetic waves or particles. The particles of light were first called "quanta" by Einstein—hence the origin of the term "quantum theory"—and are now known as photons.

This dual nature of matter and of light is very strange. It seems impossible to accept that something can be, at the same time, a particle, an entity confined to a very small volume, and a wave, which is spread out over a large region of space. And yet this is exactly what physicists had to accept. The situation seemed hopelessly paradoxical until it was realized that the terms "particle" and "wave" refer to classical concepts which are not fully adequate to describe atomic phenomena. An electron is neither a particle nor a wave, but it may show particle-like aspects in some situations and wave-like aspects in others. While it acts like a particle, it is capable of developing its wave nature at the expense of its particle nature, and vice versa, thus undergoing continual transformations from particle to wave and from wave to particle. This means that neither the electron nor any other atomic "object" has any intrinsic properties independent of its environment. The properties it shows—particle-like or wave-like—will depend on the experimental situation, that is, on the apparatus it is forced to interact with.[7]

It was Heisenberg's great achievement to express the limitations of classical concepts in a precise mathematical form, which is known as the uncertainty principle. It consists of a set of mathematical relations that determine the extent to which classical concepts can be applied to atomic phenomena; these relations stake out the limits of human imagination in the atomic world. Whenever we use classical terms—particle, wave, position, velocity—to describe atomic phenomena, we find that there are pairs of concepts, or aspects, which are interrelated and cannot be defined simultaneously in a precise way. The more we emphasize one aspect in our description the more the other aspect becomes uncertain, and the precise relation between the two is given by the uncertainty principle.

For a better understanding of this relation between pairs of classical concepts, Niels Bohr introduced the notion of complementarity. He considered the particle picture and the wave picture two complementary descriptions of the same reality, each of them only partly correct and having a limited range of application. Both pictures are needed to give a full account of the atomic reality, and both are to be applied within the limitations set by the uncertainty principle. The notion of complementarity has become an essential part of the way physicists think about nature, and Bohr has often suggested that it might also be a useful concept outside the field of physics. Indeed, this seems to be true, and we shall come back to it in discussions of biological and psychological phenomena. Complementarity has already been used extensively in our survey of the Chinese yin/yang terminology, since the yin and yang opposites are interrelated in a polar, or complementary, way. Clearly the modern concept of complementarity is reflected in ancient Chinese thought, a fact that made a deep impression on Niels Bohr.[8]

The resolution of the particle/wave paradox forced physicists to accept an aspect of reality that called into question the very foundation of the mechanistic world view—the concept of the reality of matter. At the subatomic level, matter does not exist with certainty at definite places, but rather shows "tendencies to exist," and atomic events do not occur with certainty at definite times and in definite ways, but rather show "tendencies to occur." In the formalism of quantum mechanics, these tendencies are expressed as probabilities and are associated with quantities that take the form of waves; they are similar to the mathematical forms used to describe, say, a vibrating guitar string, or sound wave. This is how particles can be waves at the same time. They are not "real" three-dimensional waves like water waves or sound waves. They are "probability waves"—abstract mathematical quantities with all the characteristic properties of waves—that are related to the probabilities of finding the particles at particular points in space and at particular times. All the laws of atomic physics are expressed in terms of these probabilities. We can never predict an atomic event with certainty; we can only predict the likelihood of its happening.

The discovery of the dual aspect of matter and of the fundamental role of probability has demolished the classical notion of solid objects. At the subatomic level, the solid material objects of classical physics dissolve into wave-like patterns of probabilities. These patterns, furthermore, do not represent probabilities of things, but rather probabilities of interconnections. A careful analysis of the process of observation in atomic physics shows that the subatomic particles have no meaning as isolated entities but can be understood only as interconnections, or correlations, between various processes of observation and measurement. As Niels Bohr wrote, "Isolated material particles are abstractions, their properties being definable and observable only through their interaction with other systems."[9]

Subatomic particles, then, are not "things" but are interconnections between "things," and these "things," in turn, are interconnections between other "things," and so on. In quantum theory you never end up with "things"; you always deal with interconnections.

This is how modern physics reveals the basic oneness of the universe. It shows that we cannot decompose the world into independently existing smallest units. As we penetrate into matter, nature does not show us any isolated basic building blocks, but rather appears as a complicated web of relations between the various parts of a unified whole. As Heisenberg expresses it, "The world thus appears as a complicated tissue of events, in which connections of different kinds alternate or overlap or combine and thereby determine the texture of the whole."[10]

The universe, then, is a unified whole that can to some extent be divided into separate parts, into objects made of molecules and atoms, themselves made of particles. But here, at the level of particles, the notion of separate parts breaks down. The subatomic particles—and therefore, ultimately, all parts of the universe—cannot be understood as isolated entities but must be defined through their interrelations. Henry Stapp, of the University of California, writes, "An elementary particle is not an independently existing unanalyzable entity. It is, in essence, a set of relationships that reach outward to other things."[11]

This shift from objects to relationships has far-reaching implications for science as a whole. Gregory Bateson even argued that relationships should be used as a basis for *all* definitions, and that this should be taught to our children in elementary school.[12] Any thing, he believed, should be defined not by what it is in itself, but by its relations to other things.

In quantum theory the fact that atomic phenomena are determined by their connections to the whole is closely related to the fundamental role of probability.[13] In classical physics, probability is used whenever the mechanical details involved in an event are unknown. For example, when we throw a die, we could—in principle—predict the outcome if we knew all the details of the objects involved: the exact composition of the die, of the surface on which it falls, and so on. These details are called local variables because they reside within the objects involved. Local variables are important in atomic and subatomic physics too. Here they are represented by connections between spatially separated events through signals—particles and networks of particles—that respect the usual laws of spatial separation. For example, no signal can be transmitted faster than the speed of light. But beyond these local connections are other, nonlocal connections that are instantaneous and cannot be predicted, at present, in a precise mathematical way. These nonlocal connections are the essence of quantum reality. Each event is influenced by the whole universe, and although we cannot describe this influence in detail, we recognize some order that can be expressed in terms of statistical laws.

Thus probability is used in classical and quantum physics for similar reasons. In both cases there are "hidden" variables, unknown to us, and this ignorance prevents us from making exact predictions. There is a crucial difference, however. Whereas the hidden variables in classical physics are local mechanisms, those in quantum physics are nonlocal; they are instantaneous connections to the universe as a whole. In the ordinary, macroscopic world nonlocal connections are relatively unimportant, and thus we can speak of separate objects and formulate the laws of physics in terms of certainties. But as we go to smaller dimensions, the influence of nonlocal connections becomes stronger; here the laws of physics can be formulated only in terms of probabilities, and it becomes more and more difficult to separate any part of the universe from the whole.

Einstein could never accept the existence of nonlocal connections and the resulting fundamental nature of probability. This was the subject of the historic debate in the 1920s with Bohr, in which Einstein expressed his opposition to Bohr's interpretation of quantum theory in the famous metaphor "God does not play dice."[14] At the end of the debate, Einstein had to admit that quantum theory, as interpreted by Bohr and Heisenberg, formed a consistent system of thought, but he remained convinced that a deterministic interpretation in terms of local hidden variables would be found some time in the future.

Einstein's unwillingness to accept the consequences of the theory that his earlier work had helped to establish is one of the most fascinating episodes in the history of science. The essence of his disagreement with Bohr was his firm belief in some external reality, consisting of independent spatially separated elements. This shows that Einstein's philosophy was essentially Cartesian. Although he initiated the revolution of twentieth-century science and went far beyond Newton in his theory of relativity, it seems that Einstein, somehow, could not bring himself to go beyond Descartes. This kinship between Einstein and Descartes is even more intriguing in view of Einstein's attempts, toward the end of his life, to construct a unified field theory by geometrizing physics along the lines of his general theory of relativity. Had these attempts been successful, Einstein could well have said, like Descartes, that his entire physics was nothing other than geometry.

In his attempt to show that Bohr's interpretation of quantum theory was inconsistent, Einstein devised a thought experiment that has become known as the Einstein-Podolsky-Rosen (EPR) experiment.[15] Three decades later John Bell derived a theorem, based on the EPR experiment, which proves that the existence of local hidden variables is inconsistent with the statistical predictions of quantum mechanics.[16] Bell's theorem dealt a shattering blow to Einstein's position by showing that the Cartesian conception of reality as consisting of separate parts, joined by local connections, is incompatible with quantum theory.

The EPR experiment provides a fine example of a situation in which a quantum phenomenon clashes with our deepest intuition of reality. It is thus ideally suited to show the difference between classical and quantum concepts. A simplified version of the experiment involves two spinning electrons, and, if we are to grasp the essence of the situation, it is necessary to understand some properties of electron spin.[17] The classical image of a spinning tennis ball is not fully adequate to describe a spinning subatomic particle. Particle spin is in a sense a rotation about the particle's own axis, but, as always in subatomic physics, this classical concept is limited. In the case of an electron, the particle's spin is restricted to two values: the amount of spin is always the same, but the particle can spin in one or the other direction, for a given axis of rotation. Physicists often denote these two values of spin by "up" and "down," assuming the electron's axis of rotation, in this case, to be vertical.

The crucial property of a spinning electron, which cannot be understood in terms of classical ideas, is the fact that its axis of rotation can-

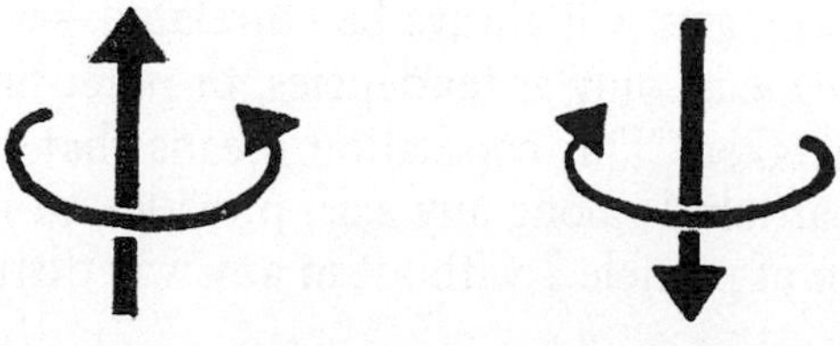

"SPIN UP" "SPIN DOWN"

not always be defined with certainty. Just as electrons show tendencies to exist in certain places, they also show tendencies to spin about certain axes. Yet whenever a measurement is performed for any axis of rotation, the electron will be found to spin in one or the other direction about that axis. In other words, the particle acquires a definite axis of rotation in the process of measurement, but before the measurement is taken, it cannot generally be said to spin about a definite axis; it merely has a certain tendency, or potentiality, to do so.

With this understanding of electron spin we can now examine the EPR experiment and Bell's theorem. To set up the experiment, any one of several methods is used to put two electrons in a state in which their total spin is zero, that is, they are spinning in opposite directions. Now suppose the two particles in this system of total spin zero are made to drift apart by some process that does not affect their spins. As they go off in opposite directions, their combined spin will still be zero, and once they are separated by a large distance, their individual spins are measured. An important aspect of the experiment is the fact that the distance between the two particles at the time of the measurement is macroscopic. It can be arbitrarily large; one particle may be in Los Angeles and the other in New York, or one on the earth and the other on the moon.

Suppose now that the spin of particle 1 is measured along a vertical axis and is found to be "up." Because the combined spin of the two particles is zero, this measurement tells us that the spin of particle 2 must be "down." Similarly, if we choose to measure the spin of particle 1 along a horizontal axis and find it to be "right," we know that in that case the spin of particle 2 must be "left." Quantum theory tells us that in a system of two particles having total spin zero, the spins of the particles about any axis will always be correlated—will be opposite—even though they exist only as tendencies, or potentialities, before the measurement is taken. This correlation means that the measurement of the spin of particle 1, along any axis, provides an indirect measurement of the spin of particle 2 without in any way disturbing that particle.

The paradoxical aspect of the EPR experiment arises from the fact that the observer is free to choose the axis of measurement. Once this choice is made, the measurement transforms the tendencies of the particles to spin about various axes into certainties. The crucial point is that we can choose our axis of measurement at the last minute, when the particles are already far apart. At the instant we perform our measurement on particle 1, particle 2, which may be thousands of miles away, will acquire a definite spin—"up" or "down" if we have chosen a vertical axis, "left" or "right" if we have chosen a horizontal axis. How does particle 2 know which axis we have chosen? There is no time for it to receive that information by any conventional signal.

This is the crux of the EPR experiment, and this is where Einstein disagreed with Bohr. According to Einstein, since no signal can travel faster than the speed of light, it is therefore impossible that the measurement performed on one particle will instantly determine the direction of the other particle's spin, thousands of miles away. According to Bohr, the two-particle system is an indivisible whole, even if the particles are separated by a great distance; the system cannot be analyzed in terms of independent parts. In other words, the Cartesian view of reality cannot be applied to the two electrons. Even though they are far apart in space, they are nevertheless linked by instantaneous, nonlocal connections. These connections are not signals in the Einsteinian sense; they transcend our conventional notions of information transfer. Bell's theorem supports Bohr's interpretation of the two particles as an indivisible whole and proves rigorously that Einstein's Cartesian view is incompatible with the laws of quantum theory. As Stapp sums up the situation, "The theorem of Bell proves, in effect, the profound truth that the world is either fundamentally lawless or fundamentally inseparable."[18]

The fundamental role of nonlocal connections and of probability in atomic physics implies a new notion of causality that is likely to have profound implications for all fields of science. Classical science was constructed by the Cartesian method of analyzing the world into parts and arranging those parts according to causal laws. The resulting deterministic picture of the universe was closely related to the image of nature as a clockwork. In atomic physics, such a mechanical and deterministic picture is no longer possible. Quantum theory has shown us that the world cannot be analyzed into independently existing isolated elements. The notion of separate parts—like atoms, or subatomic particles—is an idealization with only approximate validity; these parts are not connected by causal laws in the classical sense.

In quantum theory individual events do not always have a well-defined cause. For example, the jump of an electron from one atomic orbit to another, or the disintegration of a subatomic particle, may occur spontaneously without any single event causing it. We can never predict when and how such a phenomenon is going to happen; we can only predict its probability. This does not mean that atomic events occur in completely arbitrary fashion; it means only that they are not brought about by local causes. The behavior of any part is determined by its nonlocal connections to the whole, and since we do not know these connections precisely, we have to replace the narrow classical notion of cause and effect by the wider concept of statistical causality. The laws of atomic physics are statistical laws, according to which the probabilities for atomic events are determined by the dynamics of the whole system. Whereas in classical mechanics the properties and behavior of the parts determine those of the whole, the situation is reversed in quantum mechanics: it is the whole that determines the behavior of the parts.

The concepts of nonlocality and statistical causality imply quite clearly that the structure of matter is not mechanical. Hence the term "quantum mechanics" is very much a misnomer, as David Bohm has pointed out.[19] In his 1951 textbook on quantum theory Bohm offered some interesting speculations on the analogies between quantum processes and thought processes,[20] thus carrying further the celebrated statement made by James Jeans two decades earlier: "Today there is a wide measure of agreement . . . that the stream of knowledge is heading towards a non-mechanical reality; the universe begins to look more like a great thought than like a great machine."[21]

The apparent similarities between the structure of matter and the structure of mind should not surprise us too much, since human consciousness plays a crucial role in the process of observation, and in atomic physics determines to a large extent the properties of the observed phenomena. This is another important insight of quantum theory that is likely to have far-reaching consequences. In atomic physics the observed phenomena can be understood only as correlations between various processes of observation and measurement, and the end of this chain of processes lies always in the consciousness of the human observer. The crucial feature of quantum theory is that the observer is not only necessary to observe the properties of an atomic phenomenon, but is necessary even to bring about these properties. My conscious decision about how to observe, say, an electron will determine the electron's properties to some extent. If I ask it a particle question, it will give me a particle answer; if I ask it a wave question, it will give me a wave answer. The electron does not *have* objective properties independent of my mind. In atomic physics the sharp Cartesian division between mind and matter, between the observer and the observed, can no longer be maintained. We can never speak about nature without, at the same time, speaking about ourselves.

In transcending the Cartesian division, modern physics has not only invalidated the classical ideal of an objective description of nature but has also challenged the myth of a value-free science. The patterns scientists observe in nature are intimately connected with the patterns of their minds; with their concepts, thoughts, and values. Thus the scientific results they obtain and the technological applications they investigate will be conditioned by their frame of mind. Although much of their detailed research will not depend explicitly on their value system, the larger paradigm within which this research is pursued will never be value-free. Scientists, therefore, are responsible for their research not only intellectually but also morally. This responsibility has become an important issue in many of today's sciences, but especially so in physics, in which the results of quantum mechanics and relativity theory have opened up two very different paths for physicists to pursue. They may lead us—to put it in extreme terms—to the Buddha or to the Bomb, and it is up to each of us to decide which path to take.

The conception of the universe as an interconnected web of relations is one of two major themes that recur throughout modern physics. The other theme is the realization that the cosmic web is intrinsically dynamic. The dynamic aspect of matter arises in quantum theory as a consequence of the wave nature of subatomic particles, and is even more central in relativity theory, which has shown us that the being of matter cannot be separated from its activity. The properties of its basic patterns, the subatomic particles, can be understood only in a dynamic context, in terms of movement, interaction, and transformation.

The fact that particles are not isolated entities but wave-like probability patterns implies that they behave in a very peculiar way. Whenever a subatomic particle is confined to a small region of space, it reacts to this confinement by moving around. The smaller the region of confinement, the faster the particle will "jiggle" around in it. This behavior is a typical "quantum effect," a feature of the subatomic world which has no analogy in macroscopic physics: the more a particle is confined, the faster it will move around.[22] This tendency of particles to react to confinement with motion implies a fundamental "restlessness" of matter which is characteristic of the subatomic world. In this world most of the material particles *are* confined; they are bound to the molecular, atomic, and nuclear structures, and therefore are not at rest but have an inherent tendency to move about. According to quantum theory, matter is always restless, never quiescent. To the extent that things can be pictured to be made of smaller constituents—molecules, atoms, and particles—these constituents are in a state of continual motion. Macroscopically, the material objects around us may seem passive and inert, but when we magnify such a "dead" piece of stone or metal, we see that it is full of activity. The closer we look at it, the more alive it appears. All the material objects in our environment are made of atoms that link up with each other in various ways to form an enormous variety of molecular structures which are not rigid and motionless but vibrate according to their temperature and in harmony with the thermal vibrations of their environment. Inside the vibrating atoms the electrons are bound to the atomic nuclei by electric forces that try to keep them as close as possible, and they respond to this confinement by whirling around extremely fast. In the nuclei, finally, protons and neutrons are pressed into a minute volume by the strong nuclear forces, and consequently race about at unimaginable velocities.

Modern physics thus pictures matter not at all as passive and inert but as being in a continuous dancing and vibrating motion whose rhythmic patterns are determined by the molecular, atomic, and nuclear configurations. We have come to realize that there are no static structures in nature. There is stability, but this stability is one of dynamic balance, and the further we penetrate into matter the more we need to understand its dynamic nature to understand its patterns.

In this penetration into the world of submicroscopic dimensions, a decisive point is reached in the study of atomic nuclei in which the velocities of protons and neutrons are often so high that they come close to the speed of light. This fact is crucial for the description of their interactions, because any description of natural phenomena involving such high velocities has to take the theory of relativity into account. To understand the properties and interactions of subatomic particles we need a framework that incorporates not only quantum theory but also relativity theory; and it is relativity theory that reveals the dynamic nature of matter to its fullest extent.

Einstein's theory of relativity has brought about a drastic change in our concepts of space and time. It has forced us to abandon the classical ideas of an absolute space as the stage of physical phenomena and absolute time as a dimension separate from space. According to Einstein's theory, both space and time are relative concepts, reduced to the subjective role of elements of the language a particular observer uses to describe natural phenomena. To provide an accurate description of phenomena involving velocities close to the speed of light, a "relativistic" framework has to be used, one that incorporates time with the three space coordinates, making it a fourth coordinate to be specified relative to the observer. In such a framework space and time are intimately and inseparably connected and form a four-dimensional continuum called "space-time." In relativistic physics, we can never talk about space without talking about time, and vice versa.

Physicists have now lived with relativity theory for many years and have become thoroughly familiar with its mathematical formalism. Nevertheless, this has not helped our intuition very much. We have no direct sensory experience of the four-dimensional space-time, and whenever this relativistic reality manifests itself—that is, in all situations where high velocities are involved—we find it very hard to deal with it at the level of intuition and ordinary language. An extreme example of such a situation occurs in quantum electrodynamics, one of the most successful relativistic theories of particle physics, in which antiparticles may be interpreted as particles moving backward in time. According to this theory, the same mathematical expression describes either a positron—the antiparticle of the electron—moving from the past to the future, or an electron moving from the future to the past. Particle interactions can stretch in any direction of four-dimensional space-time, moving backward and forward in time just as they move left and right in space. To picture these interactions we need four-dimensional maps covering the whole span of time as well as the whole region of space. These maps, known as space-time diagrams, have no definite direction of time attached to them. Consequently there is no "before" and "after" in the processes they picture, and thus no linear relation of cause and effect. All events are interconnected, but the connections are not causal in the classical sense.

Mathematically there are no problems with this interpretation of particle interactions, but when we want to express it in ordinary language we run into serious difficulties, since all our words refer to the conventional notions of time and are inappropriate to describe relativistic phenomena. Thus relativity theory has taught us the same lesson as quantum mechanics. It has shown us that our common notions of reality are limited to our ordinary experience of the physical world and have to be abandoned whenever we extend this experience.

The concepts of space and time are so basic for our description of natural phenomena that their radical modification in relativity theory entailed a modification of the whole framework we use in physics to describe nature. The most important consequence of the new relativistic framework has been the realization that mass is nothing but a form of energy. Even an object at rest has energy stored in its mass, and the relation between the two is given by Einstein's famous equation $E = m c^2$, c being the speed of light.

Once it is seen to be a form of energy, mass is no longer required to be indestructible, but can be transformed into other forms of energy. This happens continually in the collision processes of high-energy physics, in which material particles are created and destroyed, their masses being transformed into energy of motion and vice versa. The collisions of subatomic particles are our main tool for studying their properties, and the relation between mass and energy is essential for their description. The equivalence of mass and energy has been verified innumerable times and physicists have become completely familiar with it—so familiar, in fact, that they measure the masses of particles in the corresponding energy units.

The discovery that mass is a form of energy has had a profound influence on our picture of matter and has forced us to modify our concept of a particle in an essential way. In modern physics, mass is no longer associated with a material substance, and hence particles are not seen as consisting of any basic "stuff," but as bundles of energy. Energy, however, is associated with activity, with processes, and this implies that the nature of subatomic particles is intrinsically dynamic. To understand this better we must remember that these particles can be conceived only in relativistic terms, that is, in terms of a framework where space and time are fused into a four-dimensional continuum. In such a framework the particles can no longer be pictured as small billiard balls, or small grains of sand. These images are inappropriate not only because they represent particles as separate objects, but also because they are static, three-dimensional images. Subatomic particles must be conceived as four-dimensional entities in space-time. Their forms have to be understood dynamically, as forms in space and time. Particles are dynamic patterns, patterns of activity which have a space aspect and a time aspect. Their space aspect makes them appear as objects with a certain mass, their time aspect as processes involving the equivalent energy. Thus the being of matter and its activity cannot be separated; they are but different aspects of the same space-time reality.

The relativistic view of matter has drastically affected not only our conception of particles, but also our picture of the forces between these particles. In a relativistic description of particle interactions, the forces between the particles—their mutual attraction or repulsion—are pictured as the exchange of other particles. This concept is very difficult to visualize, but it is needed for an understanding of subatomic phenomena. It links the forces between constituents of matter to the properties of other constituents of matter, and thus unifies the two concepts, force and matter, which had seemed to be fundamentally different in Newtonian physics. Both force and matter are now seen to have their common origin in the dynamic patterns that we call particles. These energy patterns of the subatomic world form the stable nuclear, atomic, and molecular structures which build up matter and give it its macroscopic solid aspect, thus making us believe that it is made of some material substance. At the macroscopic level this notion of substance is a useful approximation, but at the atomic level it no longer makes sense. Atoms consist of particles, and these particles are not made of any material stuff. When we observe them we never see any substance; what we observe are dynamic patterns continually changing into one another—the continuous dance of energy.

The two basic theories of modern physics have thus transcended the principal aspects of the Cartesian world view and of Newtonian physics. Quantum theory has shown that subatomic particles are not isolated grains of matter but are probability patterns, interconnections in an inseparable cosmic web that includes the human observer and her*

* The feminine pronoun is used here as a general reference to a person who may be a woman or a man. Similarly, I shall occasionally use the masculine pronoun as a general reference, including both men and women. I think this the best way to avoid being either sexist or awkward.

consciousness. Relativity theory has made the cosmic web come alive, so to speak, by revealing its intrinsically dynamic character; by showing that its activity is the very essence of its being. In modern physics, the image of the universe as a machine has been transcended by a view of it as one indivisible, dynamic whole whose parts are essentially interrelated and can be understood only as patterns of a cosmic process. At the subatomic level the interrelations and interactions between the parts of the whole are more fundamental than the parts themselves. There is motion but there are, ultimately, no moving objects; there is activity but there are no actors; there are no dancers, there is only the dance.

Current research in physics aims at unifying quantum mechanics and relativity theory in a complete theory of subatomic particles. We have not yet been able to formulate such a complete theory, but we do have several partial theories, or models, which describe certain aspects of subatomic phenomena very well. At present there are two different kinds of "quantum-relativistic" theories in particle physics that have been successful in different areas. The first are a group of quantum field theories which apply to electromagnetic and weak interactions; the second is the theory known as S-matrix theory, which has been successful in describing the strong interactions.[23] Of these two approaches, S-matrix theory is more relevant to the theme of this book, since it has deep implications for science as a whole.[24]

The philosophical foundation of S-matrix theory is known as the bootstrap approach. Geoffrey Chew proposed it in the early 1960s, and he and other physicists have used it to develop a comprehensive theory of strongly interacting particles, together with a more general philosophy of nature. According to this bootstrap philosophy, nature cannot be reduced to fundamental entities, like fundamental building blocks of matter, but has to be understood entirely through self-consistency. All of physics has to follow uniquely from the requirement that its components be consistent with one another and with themselves. This idea constitutes a radical departure from the traditional spirit of basic research in physics which had always been bent on finding the fundamental constituents of matter. At the same time it is the culmination of the conception of the material world as an interconnected web of relations that emerged from quantum theory. The bootstrap philosophy not only abandons the idea of fundamental building blocks of matter, but accepts no fundamental entities whatsoever—no fundamental constants, laws, or equations. The universe is seen as a dynamic web of interrelated events. None of the properties of any part of this web is fundamental; they all follow from the properties of the other parts, and the overall consistency of their interrelations determines the structure of the entire web.

The fact that the bootstrap approach does not accept any fundamental entities makes it, in my opinion, one of the most profound systems of Western thought, raising it to the level of Buddhist or Taoist philosophy.[25] At the same time it is a very difficult approach to physics, one that has been pursued by only a small minority of physicists. The bootstrap philosophy is too foreign to traditional ways of thinking to be seriously appreciated yet, and this lack of appreciation extends also to S-matrix theory. It is curious that although the basic concepts of the theory are used by all particle physicists whenever they analyze the results of particle collisions and compare them to their theoretical predictions, not a single Nobel prize has so far been awarded to any of the outstanding physicists who contributed to the development of S-matrix theory over the past two decades.

In the framework of S-matrix theory, the bootstrap approach attempts to derive all properties of particles and their interactions uniquely from the requirement of self-consistency. The only "fundamental" laws accepted are a few very general principles that are required by the methods of observation and are essential parts of the scientific framework. All other aspects of particle physics are expected to emerge as a necessary consequence of self-consistency. If this approach can be carried out successfully, the philosophical implications will be very profound. The fact that all the properties of particles are determined by principles closely related to the methods of observation would mean that the basic structures of the material world are determined, ultimately, by the way we look at this world; that the observed patterns of matter are reflections of patterns of mind.

The phenomena of the subatomic world are so complex that it is by no means certain whether a complete, self-consistent theory will ever be constructed, but one can envisage a series of partly successful models of smaller scope. Each of them would be intended to cover only a part of the observed phenomena and would contain some unexplained aspects, or parameters, but the parameters of one model might be explained by another. Thus more and more phenomena could gradually be covered with ever increasing accuracy by a mosaic of interlocking models whose net number of unexplained parameters keeps decreasing. The adjective "bootstrap" is thus never appropriate for any individual model, but can be applied only to a combination of mutually consistent models, none of which is any more fundamental than the others. Chew explains succinctly: "A physicist who is able to view any number of different partially successful models without favoritism is automatically a bootstrapper."[26]

Progress in S-matrix theory was steady but slow until several important developments of recent years resulted in a major breakthrough, which made it quite likely that the bootstrap program for the strong interactions will be completed in the near future, and that it may also be extended successfully to the electromagnetic and weak interactions.[27] These results have generated great enthusiasm among S-matrix theorists and are likely to force the rest of the physics community to reevaluate its attitudes toward the bootstrap approach.

The key element of the new bootstrap theory of subatomic particles is the notion of order as a new and important aspect of particle physics. Order, in this context, means order in the interconnectedness of subatomic processes. Since there are various ways in which subatomic events can interconnect, one can define various categories of order. The language of topology—well known to mathematicians but never before applied to particle physics—is used to classify these categories of order. When this concept of order is incorporated into the mathematical framework of S-matrix theory, only a few special categories of ordered relationships turn out to be consistent with that framework. The resulting patterns of particle interactions are precisely those observed in nature.

The picture of subatomic particles that emerges from the bootstrap theory can be summed up in the provocative phrase "Every particle consists of all other particles." It must not be imagined, however, that each of them contains all the others in a classical, static sense. Subatomic particles are not separate entities but interrelated energy patterns in an ongoing dynamic process. These patterns do not "contain" one another but rather "involve" one another in a way that can be given a precise mathematical meaning but cannot easily be expressed in words.

The emergence of order as a new and central concept in particle physics has not only led to a major breakthrough in S-matrix theory, but may well have great implications for science as a whole. The significance of order in subatomic physics is still obscure, and the extent to which it can be incorporated into the S-matrix framework is not yet fully known, but it is intriguing to remind ourselves that the notion of order plays a very basic role in the scientific approach to reality and is a crucial aspect of all methods of observation. The ability to recognize order seems to be an essential aspect of the rational mind; every perception of a pattern is, in a sense, a perception of order. The clarification of the concept of order in a field of research where patterns of matter and patterns of mind are increasingly being recognized as reflections of one another promises to open fascinating frontiers of knowledge.

Further extensions of the bootstrap approach in subatomic physics will eventually have to go beyond the present framework of S-matrix theory, which has been developed specifically to describe the strong interactions. To enlarge the bootstrap program a more general framework will have to be found, in which some of the concepts that are now accepted without explanation will have to be "bootstrapped," derived from overall self-consistency. These may include our conception of macroscopic space-time and, perhaps, even our conception of human consciousness. Increased use of the bootstrap approach opens up the unprecedented possibility of being forced to include the study of human consciousness explicitly in future theories of matter. The question of consciousness has already arisen in quantum theory in connection with the problem of observation and measurement, but the pragmatic formulation of the theory scientists use in their research does not refer to consciousness explicitly. Some physicists argue that consciousness may be an essential aspect of the universe, and that we may be blocked from further understanding of natural phenomena if we insist on excluding it.

At present there are two approaches in physics that come very close to dealing with consciousness explicitly. One is the notion of order in Chew's S-matrix theory; the other is a theory developed by David Bohm, who follows a much more general and more ambitious approach.[28] Bohm's starting point is the notion of "unbroken wholeness," and his aim is to explore the order he believes to be inherent in the cosmic web of relations at a deeper, "nonmanifest" level. He calls this order "implicate," or "enfolded," and describes it with the analogy of a hologram, in which each part, in some sense, contains the whole.[29] If any part of a hologram is illuminated, the entire image will be reconstructed, although it will show less detail than the image obtained from the complete hologram. In Bohm's view the real world is structured according to the same general principles, with the whole enfolded in each of its parts.

Bohm realizes that the hologram is too static to be used as a scientific model for the implicate order at the subatomic level. To express the essentially dynamic nature of reality at this level he has coined the term "holomovement." In his view the holomovement is a dynamic phenomenon out of which all forms of the material universe flow. The aim of his approach is to study the order enfolded in this holomovement, not by dealing with the structure of objects, but rather with the structure of movement, thus taking into account both the unity and the dynamic nature of the universe. To understand the implicate order Bohm has found it necessary to regard consciousness as an essential feature of the holomovement and to take it into account explicitly in his theory. He sees mind and matter as being interdependent and correlated, but not causally connected. They are mutually enfolding projections of a higher reality which is neither matter nor consciousness.

Bohm's theory is still tentative, but there seems to be an intriguing kinship, even at this preliminary stage, between his theory of the implicate order and Chew's S-matrix theory. Both approaches are based on a view of the world as a dynamic web of relations; both attribute a central role to the notion of order; both use matrices to represent change and transformation, and topology to classify categories of order. Finally, both theories recognize that consciousness may well be an essential aspect of the universe that will have to be included in a future theory of physical phenomena. Such a future theory may well arise from the merging of Bohm's and Chew's theories, which represent two of the most imaginative and philosophically profound contemporary approaches to physical reality.

My presentation of modern physics in this chapter has been influenced by my personal beliefs and allegiances. I have emphasized certain concepts and theories that are not yet accepted by the majority of physicists, but that I consider significant philosophically, of great importance for the other sciences and for our culture as a whole. Every contemporary physicist, however, will accept the main theme of the presentation—that modern physics has transcended the mechanistic Cartesian view of the world and is leading us to a holistic and intrinsically dynamic conception of the universe.

This world view of modern physics is a systems view, and it is consistent with the systems approaches that are now emerging in other fields, although the phenomena studied by these disciplines are generally of a different nature and require different concepts. In transcending the metaphor of the world as a machine, we also have to abandon the idea of physics as the basis of all science. According to the bootstrap or systems view of the world, different but mutually consistent concepts may be used to describe different aspects and levels of reality, without the need to reduce the phenomena of any level to those of another.

Before I describe the conceptual framework for such a multidisciplinary, holistic approach to reality, we may find it useful to see how the other sciences have adopted the Cartesian world view and have modeled their concepts and theories after those of classical physics. The limitations of the Cartesian paradigm in the natural and social sciences can also be brought to light, and their exposure is intended to help scientists and nonscientists change their underlying philosophies in order to participate in the current cultural transformation.

III

The Influence of Cartesian-Newtonian Thought

4·The Mechanistic View of Life

While the new physics was developing in the twentieth century, the mechanistic Cartesian world view and the principles of Newtonian physics maintained their strong influence on Western scientific thinking, and even today many scientists still hold to the mechanistic paradigm, although physicists themselves have gone beyond it.

However, the new conception of the universe that has emerged from modern physics does not mean that Newtonian physics is wrong, or that quantum theory, or relativity theory, is right. Modern science has come to realize that all scientific theories are approximations to the true nature of reality; and that each theory is valid for a certain range of phenomena. Beyond this range it no longer gives a satisfactory description of nature, and new theories have to be found to replace the old one, or, rather, to extend it by improving the approximation. Thus scientists construct a sequence of limited and approximate theories, or "models," each more accurate than the previous one but none of them representing a complete and final account of natural phenomena. Louis Pasteur said it beautifully: "Science advances through tentative answers to a series of more and more subtle questions which reach deeper and deeper into the essence of natural phenomena."[1]

The question, then, will be: How good an approximation is the Newtonian model as a basis for various sciences, and where are the limits of the Cartesian world view in those fields? In physics the mechanistic paradigm had to be abandoned at the level of the very small (in atomic and subatomic physics) and the level of the very large (in astrophysics and cosmology). In other fields the limitations may be of different kinds; they need not be connected with the dimensions of the phenomena to be described. What we are concerned with is not so much the application of Newtonian physics to other phenomena, but rather the application of the mechanistic world view on which Newtonian physics is based. Each science will need to find out the limitations of this world view in each context.

In biology the Cartesian view of living organisms as machines, constructed from separate parts, still provides the dominant conceptual framework. Although Descartes' simple mechanistic biology could not be carried very far and had to be modified considerably during the subsequent three hundred years, the belief that all aspects of living organisms can be understood by reducing them to their smallest constituents, and by studying the mechanisms through which these interact, lies at the very basis of most contemporary biological thinking. This passage from a current textbook on modern biology is a clear expression of the reductionist credo: "One of the acid tests of understanding an object is the ability to put it together from its component parts. Ultimately, molecular biologists will attempt to subject their understanding of cell structure and function to this sort of test by trying to synthesize a cell."[2]

Although the reductionist approach has been extremely successful in biology, culminating in the understanding of the chemical nature of genes, the basic units of heredity, and in the unraveling of the genetic code, it nevertheless has its severe limitations. As the eminent biologist Paul Weiss has observed,

> We can assert definitely . . . on the basis of strictly empirical investigations, that the sheer reversal of our prior analytic dissection of the universe by putting the pieces together again, whether in reality or just in our minds, can yield no complete explanation of the behavior of even the most elementary living system.[3]

This is what most contemporary biologists find hard to admit. Carried away by the successes of the reductionist method, most notable recently in the field of genetic engineering, they tend to believe that it is the only valid approach, and they have organized biological research accordingly. Students are not encouraged to develop integrative concepts, and research institutions direct their funds almost exclusively toward the solution of problems formulated within the Cartesian framework. Biological phenomena that cannot be explained in reductionist terms are deemed unworthy of scientific investigation. Consequently biologists have developed very curious ways of dealing with living organisms. As the distinguished biologist and human ecologist René Dubos has pointed out, they usually feel most at ease when the thing they are studying is no longer living.[4]

It is not easy to determine the precise limitations of the Cartesian approach to the study of living organisms. Most biologists, being fervent reductionists, are not even interested in discussing this question, and it has taken me a long time and considerable effort to find out where the Cartesian model breaks down.[5] The problems that biologists cannot solve today, apparently because of their narrow, fragmented approach, all seem to be related to the function of living systems as wholes and to their interactions with their environment. For example, the integrative action of the nervous system remains a profound mystery. Although neuroscientists have been able to clarify many aspects of brain functioning, they still do not understand how neurons* work together—how they integrate themselves into the functioning of the whole system. In fact such a question is hardly ever asked. Biologists are busy dissecting the human body down to its minute components, and in doing so are gathering an impressive amount of knowledge about its cellular and molecular mechanisms, but they still do not know how we breathe, regulate our body temperature, digest, or focus our attention. They know some of the nervous circuits, but most of the integrative actions remain to be understood. The same is true of the healing of wounds, and the nature and pathways of pain also remain largely mysterious.

An extreme case of integrative activity that has fascinated scientists throughout the ages but has, so far, eluded all explanation is the phenomenon of embryogenesis—the formation and development of the embryo—which involves an orderly series of processes through which cells specialize to form the different tissues and organs of the adult body. The interaction of each cell with its environment is crucial to

* Neurons are nerve cells that have the ability to receive and transmit nervous impulses.

these processes, and the whole phenomenon is a result of the integral coordinating activity of the entire organism—a process far too complex to lend itself to reductionist analysis. Thus embryogenesis is considered a highly interesting but quite unrewarding topic for biological research.

The reason why most biologists are not concerned with the limitations of the reductionist approach is understandable. The Cartesian method has brought spectacular progress in certain areas and continues to produce exciting results. The fact that it is inappropriate for solving other problems has left these problems neglected, if not outright shunned, even though the proportions of the field as a whole are thereby severely distorted.

How, then, is this situation going to change? I believe that the change will come through medicine. The functions of a living organism that do not lend themselves to a reductionist description—those representing the organism's integrative activities and its interactions with the environment—are precisely the functions that are crucial for the organism's health. Because Western medicine has adopted the reductionist approach of modern biology, adhering to the Cartesian division and neglecting to treat the patient as a whole person, physicians now find themselves unable to understand, or to cure, many of today's major illnesses. There is a growing awareness among them that many of the problems our medical system faces stem from the reductionist model of the human organism on which it is based. This is recognized not only by physicians but also, and even more so, by nurses and other health professionals, and by the public at large. There is already considerable pressure on physicians to go beyond the narrow, mechanistic framework of contemporary medicine and develop a broader, holistic approach to health.

Transcending the Cartesian model will amount to a major revolution in medical science, and since current medical research is closely linked to research in biology—both conceptually and in its organization—such a revolution is bound to have a strong impact on the further development of biology. To see where this development may lead, it is useful to review the evolution of the Cartesian model in the history of biology. Such a historical perspective also shows that the association of biology with medicine is not something new but goes back to ancient times and has been an important factor throughout its history.[6]

• • •

The two outstanding Greek physicians, Hippocrates and Galen, both contributed decisively to the biological knowledge of antiquity and remained authoritative figures for medicine and biology throughout the Middle Ages. During the medieval era, when the Arabs became the custodians of Western science and dominated all its disciplines, biology was again advanced by physicians, the most famous being Rhazes, Avicenna, and Averroës, all of whom were also outstanding philosophers. During that time Arab alchemists, whose science was traditionally associated with medicine, were the first to attempt chemical analyses of living matter and, in doing so, became the precursors of modern biochemists.

The close association between biology and medicine continued through the Renaissance and into the modern era, where decisive advances in the life sciences were achieved again and again by scientists with medical backgrounds. Thus Linnaeus, the great classifier of the eighteenth century, was not only a botanist and zoologist but also a physician, and in fact botany itself developed from the study of plants with healing powers. Pasteur, though not a physician himself, laid the foundations of microbiology that were to revolutionize medical science. Claude Bernard, the founder of modern physiology, was a physician; Matthias Schleiden and Theodor Schwann, the originators of cell theory, had medical degrees, and so did Rudolf Virchow, who formulated cell theory in its modern form. Lamarck had medical training, and Darwin also studied medicine, albeit with very little success. These are just a few examples of the constant interplay between biology and medicine which is continuing in our time, where a significant proportion of funds for biological research is provided by medical institutions. It is quite likely, therefore, that medicine and biology will once more be revolutionized together when biomedical researchers recognize the need to go beyond the Cartesian paradigm to make further progress in the understanding of health and illness.

The Cartesian model of biology has had many failures and many successes since the seventeenth century. Descartes created an uncompromising image of living organisms as mechanical systems and thus established a rigid conceptual framework for subsequent research in physiology, but he did not spend much time on physiological observation or experiments and left it to his followers to work out the details of the mechanistic view of life. The first to be successful in this attempt was Giovanni Borelli, a student of Galileo, who managed to explain some basic aspects of muscle action in mechanistic terms. But the great triumph of seventeenth-century physiology came when William Harvey applied the mechanistic model to the phenomenon of blood circulation and solved what had been the most fundamental and difficult problem in physiology since ancient times. His treatise, *On the Movement of the Heart,* gives a lucid description of all that could be known of the blood system in terms of anatomy and hydraulics without the aid of a microscope. It represents the crowning achievement of mechanistic physiology and was praised as such with great enthusiasm by Descartes himself.

Inspired by Harvey's success, the physiologists of his time tried to apply the mechanistic method to describe other bodily functions, such as digestion and metabolism, but all their attempts were dismal failures. The phenomena physiologists tried to explain—often with the help of grotesque mechanical analogies—involved chemical and electrical processes that were unknown at the time and could not be described in mechanical terms. Although chemistry did not advance very far in the seventeenth century, there was a school of thought, rooted in alchemical tradition, that tried to explain the functioning of living organisms in terms of chemical processes. The originator of this school was Paracelsus von Hohenheim, a sixteenth-century medical pioneer and extremely successful healer, half sorcerer and half scientist, and altogether a most extraordinary figure in the history of medicine and biology. Paracelsus, who practiced his medicine as an art and an occult science based on alchemical concepts, believed that life was a chemical process and that disease was the result of an imbalance in the body chemistry. Such a view of illness was far too revolutionary for the science of his time and had to wait several hundred years to gain broad acceptance.

In the seventeenth century physiology was divided into two opposing camps. On the one side were the followers of Paracelsus, who called themselves "iatrochemists"* and believed that physiological functions could be explained in chemical terms. On the other side were those known as "iatromechanists," who followed the Cartesian approach and held that mechanical principles were the basis of all bod-

* From the Greek *iatros* ("physician").

ily functions. The mechanists, of course, were in the majority and continued to construct elaborate mechanical models which were often patently false but adhered to the paradigm that dominated seventeenth-century scientific thought.

This situation changed considerably in the eighteenth century, which saw a series of important discoveries in chemistry, including the discovery of oxygen and Antoine Lavoisier's formulation of the modern theory of combustion. The "father of modern chemistry" also demonstrated that respiration is a special form of oxidation and thus confirmed the relevance of chemical processes to the functioning of living organisms. At the end of the eighteenth century a further dimension was added to physiology when Luigi Galvani demonstrated that the transmission of nerve impulses was associated with an electric current. This discovery led Alessandro Volta to the study of electricity and thus became the source of two new sciences, neurophysiology and electrodynamics.

All these developments raised physiology to a new level of sophistication. The simplistic mechanical models of living organisms were abandoned, but the essence of the Cartesian idea survived. Animals were still machines, although they were much more complicated than mechanical clockworks, involving chemical and electrical phenomena. Thus biology ceased to be Cartesian in the sense of Descartes' strictly mechanical image of living organisms, but it remained Cartesian in the wider sense of attempting to reduce all aspects of living organisms to the physical and chemical interactions of their smallest constituents. At the same time the strict mechanistic physiology found its most forceful and elaborate expression in the polemical treatise *Man a Machine*, by La Mettrie, which remained famous well beyond the eighteenth century. La Mettrie abandoned the mind-body dualism of Descartes, denying that humans were essentially different from animals and comparing the human organism, including its mind, to an intricate clockwork:

> Does one need more . . . to prove that Man is but an Animal, or an assemblage of springs which all wind up one another in such a way that one cannot say at which point of the human circle Nature has begun? . . . Indeed, I am not mistaken; the human body is a clock, but immense and constructed with such ingenuity and skill that if the wheel

whose function it is to mark the seconds comes to a halt, that of the minutes turns and continues its course.[7]

La Mettrie's extreme materialism generated many debates and controversies, some of which reached into the twentieth century. As a young biologist Joseph Needham wrote an essay in defense of La Mettrie, published in 1928 and entitled, like La Mettrie's original, *Man a Machine.*[8] Needham made it clear that for him—at least at that time—science was to be identified with the mechanistic Cartesian approach. "Mechanism and materialism," he wrote, "lie at the foundation of scientific thought,"[9] and he explicitly included the study of mental phenomena in such a science: "I by no means accept the opinion that the phenomena of the mind are not amenable to physico-chemical description. All that we shall ever know of them scientifically will be mechanistic . . ."[10]

Toward the end of his essay Needham summed up his position on the scientific view of human nature with the forceful statement: "In science, man is a machine; or if he is not, then he is nothing at all."[11] Nevertheless, Needham later left the field of biology to become one of the leading historians of Chinese science and, as such, an ardent advocate of the organismic world view that is the basis of Chinese thought.

It would be foolish to categorically deny Needham's claim that scientists will be able, some day, to describe all biological phenomena in terms of the laws of physics and chemistry, or rather, as we would say today, in terms of biophysics and biochemistry. But this does not mean that these laws will be based on the view of living organisms as machines. To say so would be restricting science to Newtonian science. To understand the essence of living systems, scientists—whether in biophysics, biochemistry, or any other discipline concerned with the study of life—will have to abandon the reductionist belief that complex organisms can be described completely, like machines, in terms of the properties and behavior of their constituents. This should be easier to do today than in the 1920s, since the reductionist approach has had to be abandoned even in the study of inorganic matter.

In the history of the Cartesian model in the life sciences, the nineteenth century brought impressive new developments because of the remarkable advances in many areas of biology. The nineteenth century is best known for the establishment of the theory of evolution, but it also saw the formulation of cell theory, the beginning of modern embryology, the rise of microbiology, and the discovery of the laws of heredity. Biology was now firmly grounded in physics and chemistry, and scientists devoted all their efforts to the search for physicochemical explanations of life.

One of the most powerful generalizations in all of biology was the recognition that all animals and plants are composed of cells. It marked a decisive turn in the biologists' understanding of body structure, inheritance, fertilization, development and differentiation, evolution, and many other characteristics of life. The term "cell" was coined by Robert Hooke in the seventeenth century to describe various minute structures he saw through the newly invented microscope, but the development of a proper cell theory was a slow and gradual process that involved the work of many researchers and culminated in the nineteenth century, when biologists thought that they had definitely found the fundamental units of life. This belief gave the Cartesian paradigm a new meaning. From now on, all functions of a living organism had to be understood in terms of its cells. Rather than reflecting the organization of the organism as a whole, biological functions were seen as the results of the interactions between the cellular building blocks.

Understanding the structure and functioning of cells involves a problem that has become characteristic of all modern biology. The organization of a cell has often been compared to that of a factory, where different parts are manufactured at different sites, stored in intermediate facilities, and transported to assembly plants to be combined into finished products that are either used up by the cell itself or exported to other cells. Cell biology has made enormous progress in understanding the structures and functions of many of the cell's subunits, but it has remained largely ignorant about the coordinating activities that integrate those operations into the functioning of the cell as a whole. The complexity of this problem is increased considerably by the fact that, unlike those of a human-made factory, the equipment and machinery of a cell are not permanent fixtures but are periodically disassembled and rebuilt, always according to specific patterns and in harmony with the overall dynamics of cell functioning. Biologists have come to realize that cells are organisms in their own right, and they are becoming increasingly aware that the integrative activities of these living systems—especially the balancing of their interdependent metabolic* pathways and cycles—cannot be understood within the reductionist framework.

The invention of the microscope in the seventeenth century had opened up a new dimension for biology, but the instrument was not fully exploited until the nineteenth century, when various technical problems with the old lens system were finally solved. The newly perfected microscope generated an entire new field of research, microbiology, which revealed an unsuspected richness and complexity of living organisms of microscopic dimensions. Research in this field was dominated by the genius of Louis Pasteur, whose penetrating insights and clear formulations made a lasting impact on chemistry, biology, and medicine.

With the use of ingenious experimental techniques, Pasteur was able to clarify a question that had agitated biologists throughout the eighteenth century, the question of the origin of life. Since ancient times it had been the common belief that life, at least in its lower forms, could arise spontaneously from nonliving matter. During the seventeenth and eighteenth centuries that idea—known as "spontaneous generation"—was questioned, but the argument could not be settled until Pasteur demonstrated conclusively, with a series of clearly designed and rigorous experiments, that any microorganisms which developed under suitable conditions came from other microorganisms. It was Pasteur who brought to light the immense variety of the organic world at the level of the very small. In particular he was able to establish the role of bacteria in certain chemical processes, such as fermentation, and thus helped to lay the foundations of the new science of biochemistry.

After twenty years of research on bacteria, Pasteur turned to the study of diseases in higher animals and achieved another major advance—the demonstration of a definite correlation between germs† and disease. Although this discovery had a tremendous impact on the development of medicine, the exact nature of the correlation between bacteria and disease is still widely misunderstood. Pasteur's "germ theory of disease," in its simplistic and reductionist interpretation, meant

* Metabolism, from the Greek *metabolē* ("change"), denotes the sum of chemical changes occurring in living organisms, and in particular in cells, that are necessary to sustain life.

† "Germ" and "microbe" are early synonyms for the now generally used term "microorganism"; "bacterium" denotes a large group of microorganisms and "bacillus" refers to a particular kind of bacterium.

that biomedical researchers tended to regard bacteria as the only cause of disease. Consequently, they became obsessed with the identification of microbes and with the illusory goal of designing "magic bullets," drugs that would destroy specific bacteria without damaging the rest of the organism.

The reductionist view of disease eclipsed an alternative theory that had been taught a few decades earlier by Claude Bernard, a celebrated physician who is generally considered the founder of modern physiology. Although Bernard, adhering to the paradigm of his time, saw the living organism as "a machine which necessarily works by virtue of the physico-chemical properties of its constituent elements,"[12] his view of physiological functions was much subtler than those of his contemporaries. He insisted on the close and intimate relation between an organism and its environment, and was the first to point out that there was also a *milieu intérieur,* an internal environment in which the organs and tissues of the organism lived. Bernard observed that in a healthy organism this *milieu intérieur* remains essentially constant, even when the external environment fluctuates considerably. This discovery led him to formulate the famous dictum: "The constancy of the internal environment is the essential condition of independent life."[13]

Claude Bernard's strong emphasis on internal balance as a condition for health could not hold its ground against the rapid spread of the reductionist view of disease among biologists and physicians. The importance of his theory was rediscovered only in the twentieth century, when researchers became more aware of the crucial role of the environment in biological phenomena. Bernard's concept of the constancy of the internal environment has now been further elaborated and has led to the important notion of homeostasis, a word coined by the neurologist Walter Cannon to denote the tendency of living organisms to maintain a state of internal balance.[14]

The theory of evolution was biology's major contribution to the history of ideas in the nineteenth century. It forced scientists to abandon the Newtonian picture of the world as a machine that had emerged fully constructed from the hands of its Creator, and to replace it with the concept of an evolving and ever changing system. This did not, however, lead biologists to modify the reductionist paradigm; on the contrary, they concentrated on fitting the Darwinian theory into the Cartesian framework. They were extremely successful in explaining many of the physical and chemical mechanisms of heredity, but were unable to understand the essential nature of development and evolution.[15]

The first theory of evolution was formulated by Jean Baptiste Lamarck, a self-taught scientist who invented the word "biology" and turned to the study of animal species at the age of almost fifty. Lamarck observed that animals changed under environmental pressure, and he believed that they could pass on these changes to their offspring. This passing on of acquired characteristics was for him the main mechanism of evolution. Although it turned out that Lamarck was wrong in that respect,[16] his recognition of the phenomenon of evolution—the emergence of new biological structures in the history of species—was a revolutionary insight that profoundly affected all subsequent scientific thought.

In particular, Lamarck had a strong influence on Charles Darwin, who started his scientific career as a geologist but became interested in biology during an expedition to the Galapagos Islands, where he observed the great richness and variety of island fauna. These observations stimulated Darwin to speculate about the effect of geographical isolation on the formation of species and led him, eventually, to the formulation of his theory of evolution. Other major influences on Darwin's thought were the evolutionary ideas of geologist Charles Lyell, and the economist Thomas Malthus' idea of a competitive struggle for survival. Out of these observations and studies emerged the twin concepts on which Darwin based his theory—the concept of chance variation, later to be called random mutation, and the idea of natural selection through the "survival of the fittest."

Darwin published his theory of evolution in 1859 in his monumental *On the Origin of Species* and completed it twelve years later with *The Descent of Man*, in which the concept of evolutionary transformation of one species into another is extended to include human beings. Here Darwin showed that his ideas about human traits were strongly colored by the patriarchal bias of his time, in spite of the revolutionary nature of his theory. He saw the typical male as strong, brave, and intelligent; the typical female was passive, weak in body, and deficient in brains. "Man," he wrote, "is more courageous, pugnacious, and energetic than woman, and has more inventive genius."[17]

Although Darwin's concepts of chance variation and natural selection were to remain the cornerstones of modern evolutionary theory, it soon became clear that chance variations, as envisaged by Darwin, could never explain the emergence of new characteristics in the evolution of species. Nineteenth-century views of heredity were based on the assumption that the biological characteristics of an individual represented a "blend" of those of its parents, with both parents contributing more or less equal parts to the mixture. This meant that an offspring of a parent with a useful chance variation would inherit only 50 percent of the new characteristic, and would be able to pass on only 25 percent of it to the next generation. Thus the new characteristic would be diluted rapidly, with very little chance of establishing itself through natural selection. Darwin himself recognized that this was a serious flaw in his theory for which he had no remedy.

It is ironic that the solution to Darwin's problem was discovered by Gregor Mendel only a few years after the publication of the Darwinian theory, but was ignored until the rediscovery of Mendel's work at the turn of the century. From his careful experiments with garden peas, Mendel deduced that there were "units of heredity"—later to be called genes—that did not blend in the process of reproduction and thus become diluted, but were transmitted from generation to generation without changing their identity. With this discovery it could be assumed that random mutations would not disappear within a few generations but would be preserved, to be either reinforced or eliminated by natural selection.

Mendel's discovery not only played a decisive role in establishing the Darwinian theory of evolution but also opened up a whole new field of research—the study of heredity through the investigation of the chemical and physical nature of genes. William Bateson, a fervent advocate and popularizer of Mendel's work, named this new field "genetics" at the beginning of the century and introduced many of the terms now used by geneticists. He also named his youngest son Gregory, in Mendel's honor.

In the twentieth century genetics became the most active area in biological research and provided a strong reinforcement of the Cartesian approach to living organisms. It became clear quite early that the material of heredity lay in the chromosomes, those threadlike bodies that are present in the nucleus of every cell. Soon thereafter it was recognized that the genes occupied specific positions within the chromosomes; to be precise, they are arranged along the chromosomes in linear order. With these discoveries geneticists believed that they had now pinned down the "atoms of heredity" and proceeded to explain the biological characteristics of living organisms in terms of their elementary units, the genes, with each gene corresponding to a definite hereditary trait. Soon, however, further research showed that a single gene may affect a wide range of traits and that, conversely, many separate genes often combine to produce a single trait. Obviously the study of the cooperation and integrative activity of genes is of first importance, but here too the Cartesian framework has made it difficult to deal with these questions. When scientists reduce an integral whole to fundamental building blocks—whether they are cells, genes, or elementary particles—and try to explain all phenomena in terms of these elements, they lose the ability to understand the coordinating activities of the whole system.

Another fallacy of the reductionist approach in genetics is the belief that the character traits of an organism are uniquely determined by its genetic makeup. This "genetic determinism" is a direct consequence of regarding living organisms as machines controlled by linear chains of cause and effect. It ignores the fact that the organisms are multileveled systems, the genes being embedded in the chromosomes, the chromosomes functioning within the nuclei of their cells, the cells incorporated in the tissues, and so on. All these levels are involved in mutual interactions that influence the organism's development and result in wide variations of the "genetic blueprint."

Similar arguments apply to the evolution of a species. The Darwinian concepts of chance variation and natural selection are only two aspects of a complex phenomenon that can be understood much better within a holistic, or systemic, framework.[18] Such a framework is much more subtle and useful than the dogmatic position of so-called neo-Darwinian theory, forcefully expressed by the geneticist and Nobel laureate Jacques Monod:

> Chance alone is at the source of every innovation, of all creation in the biosphere. Pure chance, absolutely free but blind, at the very root of the stupendous edifice of evolution: this central concept of modern biology is no longer one among other conceivable hypotheses. It is today the *sole* conceivable hypothesis, the only one that squares with observed and tested fact. And nothing warrants the supposition—or the hope—that on this score our position is likely ever to be revised.[19]

More recently the fallacy of genetic determinism has given rise to a widely discussed theory known as sociobiology, in which all social behavior is seen as predetermined by genetic structure.[20] Numerous critics have pointed out that this view is not only scientifically unsound but also quite dangerous. It encourages pseudoscientific justifications for racism and sexism by interpreting differences in human behavior as genetically preprogrammed and unchangeable.[21]

Although genetics was very successful in clarifying many aspects of heredity during the first half of the twentieth century, the exact chemical and physical nature of its central concept, the gene, remained a mystery. The complicated chemistry of the chromosomes was not understood until the 1950s and 1960s, a full century after Darwin and Mendel.

Meanwhile, the new science of biochemistry progressed steadily and established the firm belief among biologists that all properties and functions of living organisms would eventually be explained in chemical and physical terms. This belief was most clearly expressed by Jacques Loeb in *The Mechanistic Conception of Life,* which had a tremendous influence on the biological thinking of its time. "Living organisms are chemical machines," wrote Loeb,[22] "possessing the peculiarity of preserving and reproducing themselves." To explain the functioning of these machines completely in terms of their basic building blocks was for Loeb, as for all reductionists, the essence of the scientific approach: "The ultimate aim of the physical sciences is the visualization of all phenomena in terms of groupings and displacements of ultimate particles, and since there is no discontinuity between the matter constituting the living and the non-living world, the goal of biology can be expressed in the same way."[23]

An extremely unfortunate consequence of the view of living things as machines has been excessive use of vivisection* in biomedical and behavioral research.[24] Descartes himself defended vivisection, believing that animals do not suffer and asserting that their cries meant nothing more than the creaking of a wheel; today the inhuman practice of systematically torturing animals still exists in the life sciences.

* Vivisection, in a broad sense, includes all types of experiments on living animals, whether or not cutting is done, and especially those considered to cause distress to the subject.

• • •

In the twentieth century a significant shift occurred in biological research that may well turn out to be the last step in the reductionist approach to the phenomena of life, leading to its greatest triumph and, at the same time, to its end. Whereas cells were regarded as the basic building blocks of living organisms during the nineteenth century, the attention shifted from cells to molecules toward the middle of our century, when geneticists began to explore the molecular structure of the gene. Their research culminated in the elucidation of the physical structure of DNA—the molecular basis of chromosomes—which stands as one of the greatest achievements of twentieth-century science. This triumph of molecular biology has led biologists to believe that all biological functions can be explained in terms of molecular structures and mechanisms, which has considerably distorted research in the life sciences.

In a general sense the term "molecular biology" refers to the study of any biological phenomenon in terms of the molecular structures and interactions involved in it. More specifically, it has come to mean the study of the very large biological molecules known as macromolecules. During the first half of the century it became clear that the essential constituents of all living cells—the proteins and nucleic acids*—were highly complex, chainlike structures containing thousands of atoms. The investigation of the chemical properties and exact three-dimensional form of these large chain molecules became the principal task of molecular biology.[25]

The first important step toward a molecular genetics came with the discovery that cells contain agents, called enzymes, that can promote specific chemical reactions. During the first half of the century biochemists managed to specify most of the chemical reactions that occur in cells, and found out that the most important of these reactions are essentially the same in all living organisms. Each of them depends crucially on the presence of a particular enzyme, and thus the study of enzymes became of primary importance.

During the 1940s geneticists achieved another decisive insight when they discovered that the primary function of genes was to control the synthesis of enzymes. With this discovery the broad outlines of the hereditary process emerged: genes determine hereditary traits by di-

* Nucleic acids—the acids found in cell nuclei—are of two basically different kinds, known as DNA and RNA.

recting the synthesis of enzymes, which in turn promote the chemical reactions corresponding to those traits. Although these discoveries represented major advances in understanding heredity, the nature of the gene remained unknown during this period. Geneticists ignored its chemical structure and were unable to explain how it managed to carry out its essential functions: the synthesis of enzymes, its own faithful replication in the process of cell division, and the sudden permanent changes known as mutations. As far as the enzymes were concerned, it was known that they were proteins, but their precise chemical structure was unknown and so, as a consequence, was the process by which enzymes promote chemical reactions.

This situation changed dramatically over the next two decades, which brought the major breakthrough in modern genetics, often referred to as the breaking of the genetic code: the discovery of the precise chemical structure of genes and enzymes, of the molecular mechanisms of protein synthesis, and of the mechanisms of gene replication and mutation.[26] This revolutionary achievement involved tremendous struggle and fierce competition, as well as stimulating collaboration, among a group of outstanding and highly original men and women, the main protagonists being Francis Crick, James Watson, Maurice Wilkins, Rosalind Franklin, Linus Pauling, Salvador Luria, and Max Delbrück.

A crucial element in the breaking of the genetic code was the fact that physicists moved into biology. Max Delbrück, Francis Crick, Maurice Wilkins, and several others had backgrounds in physics before they joined the biochemists and geneticists in their study of heredity. These scientists brought with them a new rigor, a new perspective, and new methods that thoroughly transformed genetic research. The interest of physicists in biology had begun in the 1930s, when Niels Bohr speculated about the relevance of the uncertainty principle and the concept of complementarity to biological research.[27] Bohr's speculations were further elaborated by Delbrück, whose ideas about the physical nature of genes led Erwin Schrödinger to write a small book entitled *What Is Life?* This book became a major influence on biological thought in the 1940s and was the main reason for several scientists to leave physics and turn to genetics.

The fascination of *What Is Life?* came from the clear and compelling way in which Schrödinger treated the gene not as an abstract unit but as a concrete physical substance, advancing definite hypotheses about its molecular structure that stimulated scientists to think about genetics in a new way. He was the first to suggest that the gene could be viewed as an information carrier whose physical structure corresponds to a succession of elements in a hereditary code script. Schrödinger's enthusiasm convinced physicists, biochemists, and geneticists that a new frontier of science had opened where great discoveries were imminent. From now on these scientists began to refer to themselves as "molecular biologists."

The basic structure of the biological molecules was discovered in the early 1950s through the confluence of three powerful methods of observation—chemical analysis, electron microscopy, and X-ray crystallography.* The first breakthrough came when Linus Pauling determined the structure of the protein molecule. Proteins were known to be long chain molecules, consisting of a sequence of different compounds, known as amino acids, linked end-to-end. Pauling showed that the backbone of the protein structure is coiled in a left-handed or right-handed helix, and that the rest of the structure is determined by the exact linear sequence of amino acids along this helical path. Subsequent further studies of the protein molecule showed how the specific structure of enzymes allows them to bind the molecules whose chemical reactions they promote.

Pauling's great success inspired James Watson and Francis Crick to concentrate all their efforts on elucidating the structure of DNA, which by then had been recognized as the genetic material in the chromosomes. After two years of strenuous work, of many false starts and great disappointments, Watson and Crick were finally rewarded with success. Using the X-ray data of Rosalind Franklin and Maurice Wilkins, they were able to determine the precise architecture of DNA, which is called the Watson-Crick structure. It is a double helix made up of two intertwined, structurally complementary chains. The compounds arranged on these chains in linear order are complex structures, known as nucleotides and existing in four different kinds.

It took another decade to understand the basic mechanism through which the DNA carries out its two fundamental functions: self-replication and protein synthesis. This research, again led by Watson and Crick, revealed explicitly how genetic information is encoded in the

* X-ray crystallography, invented in 1912 by Lawrence Bragg, is the method of determining the orderly array of atoms in molecular structures—originally crystals—by analyzing the ways in which X-rays are scattered by those structures.

chromosomes. To put it in greatly simplified terms, chromosomes are made of DNA molecules exhibiting the Watson-Crick structure. A gene is that length of a DNA double helix which specifies the structure of a particular enzyme. The synthesis of this enzyme occurs through a complicated two-step process involving RNA, the second nucleic acid. The elements of the hereditary code script are the four nucleotides which embody the genetic information in their aperiodic sequence along the chain. This linear sequence of nucleotides in the gene determines the linear sequence of amino acids in the corresponding enzyme. In the process of chromosome division, the two chains of the double helix separate, and each of them serves as a template for the construction of a new complementary chain. Gene mutation is caused by a chance error in this duplication process by which one nucleotide is substituted for another, resulting in a permanent change in the information carried by the gene.

These, then, are the basic elements of what has been hailed as the greatest discovery in biology since Darwin's theory of evolution. Advancing to ever smaller levels in their exploration of the phenomena of life, biologists found that the characteristics of all living organisms—from bacteria to humans—were encoded in their chromosomes in the same chemical substance, using the same code script. After two decades of intensive research, the precise details of this code had been unraveled. Biologists had discovered the alphabet of a truly universal language of life.

The spectacular success of molecular biology in the field of genetics led scientists to apply its methods to all areas of biology in an attempt to solve all problems by reducing them to their molecular level. Thus most biologists became fervent reductionists, concerned with molecular details. Molecular biology, originally a small branch of the life sciences, has now become a general, and exclusive, way of thinking that has led to a severe distortion of biological research. Funds are directed toward quick solutions and fashionable topics, while important theoretical problems that do not lend themselves to the reductionist approach are ignored. As Sidney Brenner, one of the leading researchers in the field, has noted, "Nobody publishes theory in biology—with few exceptions. Instead, they get out the structure of still another protein."[28]

The problems that resisted the reductionist approach of molecular biology became apparent around 1970, when the structure of DNA and the molecular mechanisms of heredity were well understood for simple single-cell organisms, such as bacteria, but had still to be worked out for multicellular organisms. This brought biologists face to face with the problems of cell development and differentiation that had been eclipsed during the unraveling of the genetic code. In the very early stages of the development of higher organisms, the number of their cells goes from one to two, to four, to eight, to sixteen, and so on. As the genetic information is thought to be identical in each cell, how can it happen that cells specialize in different ways, becoming muscle cells, blood cells, bone cells, nerve cells, and so on? This basic problem of development, which appears in many variations throughout biology, clearly shows the limitations of the reductionist approach. Today's biologists know the precise structure of a few genes, but they know very little of the ways in which genes communicate and cooperate in the development of an organism—how they interact, how they are grouped together, when they are switched on and off and in what order. Biologists know the alphabet of the genetic code but have almost no idea of its syntax. It is now apparent that only a small percentage of the DNA—less than 5 percent—is used to specify proteins; all the rest may well be used for integrative activities about which biologists are likely to remain ignorant as long as they adhere to their reductionist models.

The other area in which the limitations of the reductionist approach are quite apparent is the field of neurobiology. The higher nervous system is a holistic system par excellence whose integrative activities cannot be understood by reducing them to molecular mechanisms. At the same time, nerve cells are the largest cells and thus easiest to study. Neuroscientists may therefore be the first to propose holistic models of brain functioning to explain phenomena like perception, memory, and pain, which cannot be understood within the current reductionist framework. We shall see that some attempts in this direction have already been made, and promise exciting new perspectives. To go beyond the current reductionist approach, biologists will need to acknowledge that, as Paul Weiss has put it, "there is no phenomenon in a living system that is *not* molecular, but there is none that is *only* molecular either."[29] This will require a much broader conceptual framework than the one biology uses today. The biologists' spectacular advances have not broadened their basic philosophy; the Cartesian paradigm still dominates the life sciences.

A comparison between biology and physics is appropriate here. In the study of heredity, the period before 1940 is often called "classical genetics," as distinguished from the "modern genetics" of the subsequent decades. These terms probably derive from an analogy with the transition from classical to modern physics at the turn of the century.[30] As the atom was an indivisible unit of unknown structure in classical physics, so was the gene in classical genetics. But this analogy breaks down in a significant respect. The exploration of the atom has forced physicists to revise their basic concepts about the nature of physical reality in a radical way. The result of this revision is a coherent dynamic theory, quantum mechanics, which transcends the principal concepts of Cartesian-Newtonian science. In biology, on the other hand, the exploration of the gene has not led to a comparable revision of basic concepts, nor has it resulted in a universal dynamic theory. There is no unifying framework that would enable biologists to overcome the fragmentation of their science by evaluating the relative importance of research problems and recognizing how they interrelate. The only framework used for such an evaluation is still the Cartesian, in which living organisms are seen as physical and biochemical machines, to be explained completely in terms of their molecular mechanisms.

However, a few leading biologists of our time have expressed the feeling that molecular biology may be reaching the end of its usefulness. Francis Crick, who has dominated the field from the very beginning, acknowledges the severe limitations of the molecular approach in trying to understand basic biological phenomena:

> In one way, you could say all the genetic and molecular biological work of the last sixty years could be considered as a long interlude . . . Now that that program has been completed, we have come full circle—back to the problems . . . left behind unsolved. How does a wounded organism regenerate to exactly the same structure it had before? How does the egg form the organism?[31]

What is needed, to solve these problems, is a new paradigm; a new dimension of concepts transcending the Cartesian view. It is likely that the systems view of life will form the conceptual background of this new biology, as Sidney Brenner seems to indicate, without saying so explicitly, in some recent speculations about the future of his science:

> I think in the next twenty-five years we are going to have to teach biologists another language . . . I don't know what it's called yet; nobody knows. But what one is aiming at, I think, is the fundamental problem of the theory of elaborate systems . . . And here there is a grave problem of levels: it may be wrong to believe that all the logic is at the molecular level. We may need to get beyond the clock mechanisms.[32]

5 · The Biomedical Model

Throughout the history of Western science the development of biology has gone hand in hand with that of medicine. Naturally then, the mechanistic view of life, once firmly established in biology, has also dominated the attitudes of physicians toward health and illness. The influence of the Cartesian paradigm on medical thought resulted in the so-called biomedical model,* which constitutes the conceptual foundation of modern scientific medicine. The human body is regarded as a machine that can be analyzed in terms of its parts; disease is seen as the malfunctioning of biological mechanisms which are studied from the point of view of cellular and molecular biology; the doctor's role is to intervene, either physically or chemically, to correct the malfunctioning of a specific mechanism. Three centuries after Descartes, the science of medicine is still based, as George Engel writes, on "the notion of the body as a machine, of disease as the consequence of breakdown of the machine, and of the doctor's task as repair of the machine."[1]

By concentrating on smaller and smaller fragments of the body, modern medicine often loses sight of the patient as a human being, and by reducing health to mechanical functioning, it is no longer able to deal with the phenomenon of healing. This is perhaps the most seri-

* The biomedical model is often simply called the medical model. However, I shall use the term "biomedical" to distinguish it from conceptual models of other medical systems, such as the Chinese.

ous shortcoming of the biomedical approach. Although every practicing physician knows that healing is an essential aspect of all medicine, the phenomenon is considered outside the scientific framework; the term "healer" is viewed with suspicion, and the concepts of health and healing are generally not discussed in medical schools.

The reason for the exclusion of the phenomenon of healing from biomedical science is evident. It is a phenomenon that cannot be understood in reductionist terms. This applies to the healing of wounds, and even more to the healing of illnesses, which generally involve a complex interplay among the physical, psychological, social, and environmental aspects of the human condition. To reincorporate the notion of healing into the theory and practice of medicine, medical science will have to transcend its narrow view of health and illness. This does not mean that it will have to be less scientific. On the contrary, by broadening its conceptual basis it will become more consistent with recent developments in modern science.

Health and the phenomenon of healing have meant different things in different ages. The concept of health, like the concept of life, cannot be defined precisely, and in fact, the two are closely related. What is meant by health depends on one's view of the living organism and its relation to its environment. As this view changes from one culture to another, and from one era to another, the notions of health also change. The broad concept of health that will be needed for our cultural transformation—a concept that includes individual, social, and ecological dimensions—will require a systems view of living organisms and, correspondingly, a systems view of health.[2] To begin with, the definition of health given by the World Health Organization in the preamble of its charter may be useful: "Health is a state of complete physical, mental and social well-being and not merely the absence of disease or infirmity."

Although the WHO definition is somewhat unrealistic, picturing health as a static state of perfect well-being, rather than a continually changing and evolving process, it conveys the holistic nature of health, which has to be grasped if we are to understand the phenomenon of healing. Through the ages healing has been practiced by folk healers who are guided by traditional wisdom that sees illness as a disorder of the whole person, involving not only the patient's body but his mind; his self-image, his dependence on the physical and social environment, as well as his relation to the cosmos and the deities. These healers, who still treat the majority of patients throughout the world, follow many different approaches, which are holistic to different degrees, and they use a wide variety of therapeutic techniques. What they have in common is that they never restrict themselves to purely physical phenomena, as the biomedical model does. Through rituals and ceremonies they attempt to influence the patient's mind, relieving the apprehension that is always a significant component of illness and helping the patient to stimulate the natural healing powers that all living organisms possess. These healing ceremonies usually involve an intense relationship between healer and patient and are often interpreted in terms of supernatural forces channeled through the healer.

In modern scientific terms we could say that the healing process represents the coordinated response of the integrated organism to stressful environmental influences. This view of healing implies a number of concepts that transcend the Cartesian division and cannot be formulated adequately within the framework of current medical science. Because of this, biomedical researchers tend to disregard the practices of folk healers and are reluctant to admit their effectiveness. Such "medical scientism" makes them forget that the art of healing is an essential aspect of all medicine, and that even our scientific medicine had to rely on it almost exclusively until a few decades ago, having little else to offer in terms of specific methods of treatment before that time.[3]

Western medicine emerged from a large reservoir of folk healing and subsequently spread to the rest of the world, where it became transformed to various degrees but still retained its basic biomedical approach. With the global extension of the biomedical system, several writers have abandoned the terms "Western," "scientific," or "modern," and are now referring to it as "cosmopolitan medicine."[4] But the "cosmopolitan" medical system is only one among many others. Most societies show a pluralism of medical systems and medical beliefs, with no sharp dividing line between one system and another. In addition to cosmopolitan medicine and folk medicine, or folk healing, many cultures have developed their own high-tradition medicine. Like cosmopolitan medicine, these systems—Indian, Chinese, Persian, and others—are based on a written tradition, using empirical knowledge and are practiced by a professional elite. Their approach is holistic, if not always in actual practice, then at least in theory. In addition to these systems, all societies have developed a system of popular medicine—beliefs and practices used within a family, or a community, which are passed on by word of mouth and do not require professional healers.

The practice of popular medicine has traditionally been the prerogative of women, since the art of healing in the family is usually associated with the tasks and the spirit of motherhood. Folk healers, typically, are both female and male, with proportions varying from culture to culture. They do not practice within an organized profession but derive their authority from their healing powers—often interpreted as their access to the spirit world—rather than from professional licensing. With the appearance of organized, high-tradition medicine, however, patriarchal patterns assert themselves and medicine becomes male-dominated. This is as true for classical Chinese or Greek medicine as for medieval European medicine, or modern cosmopolitan medicine.

In the history of Western medicine, the grasp of power by a male professional elite involved a long struggle that accompanied the emergence of the rational and scientific approach to health and healing. The outcome of this struggle was not only the establishment of an almost exclusively male medical elite, but also the intrusion of medicine into domains such as childbirth, which had traditionally been the province of women. This trend is now being reversed by the women's movement, which has recognized the patriarchal aspects of medicine as one more manifestation of the control of women's bodies by men, and has come to see the full participation of women in their own health care as one of its central goals.[5]

The greatest change in the history of Western medicine came with the Cartesian revolution. Before Descartes, most healers had addressed themselves to the interplay of body and soul, and had treated their patients within the context of their social and spiritual environment. As their world views changed over the ages, so did their views of illness and their methods of treatment, but their approaches were usually concerned with the whole patient. Descartes' philosophy changed this situation profoundly. His strict division between mind and body led physicians to concentrate on the body machine and to neglect the psychological, social, and environmental aspects of illness. From the seventeenth century on, progress in medicine closely followed the developments in biology and the other natural sciences. As the perspective of biomedical science shifted from the study of bodily organs and their functions to that of cells and, finally, to the study of molecules, study of the phenomenon of healing was progressively neglected, and physicians found it more and more difficult to deal with the interdependence of body and mind.

Descartes himself, although he introduced the separation of mind and body, nevertheless considered the interplay between the two an essential aspect of human nature, and was well aware of its implications for medicine. The union of body and soul was the principal subject of his correspondence with one of his most brilliant disciples, Princess Elizabeth of Bohemia. Descartes considered himself not only the teacher and close friend of the princess, but also her physician, and when Elizabeth suffered from ill health and described her physical symptoms to Descartes, he did not hesitate to diagnose her affliction as being largely due to emotional stress, as we would say today, and to prescribe relaxation and meditation in addition to physical remedies.[6] Thus Descartes showed himself to be far less "Cartesian" than most of today's medical profession.

In the seventeenth century William Harvey explained the phenomenon of blood circulation in purely mechanistic terms, but other attempts to build mechanistic models of physiological functions were far less successful. By the end of the century it was apparent that a straightforward application of the Cartesian approach would not lead to further medical progress, and several countermovements emerged in the eighteenth century, among which the system of homeopathy became the most widespread and most successful.[7]

The rise of modern scientific medicine began in the nineteenth century with the great advances made in biology. At the beginning of the century the structure of the human body, even down to minute details, was almost fully known. In addition, rapid progress was being made in the understanding of physiological processes, largely because of the careful experiments carried out by Claude Bernard. Thus biologists and physicians, faithful to the reductionist approach, turned their attention to smaller entities. This trend proceeded in two directions. One was instigated by Rudolf Virchow, who postulated that all illness involved structural changes at the cellular level, thus establishing cellular biology as the basis of medical science. The other direction of research was pioneered by Louis Pasteur who began the intensive study of microorganisms that has occupied biomedical researchers ever since.

Pasteur's clear demonstration of a correlation between bacteria and disease had a decisive impact. Throughout medical history physicians had debated the question whether a specific disease was caused by a single factor or was the result of a constellation of factors acting simultaneously. In the nineteenth century these two views were emphasized by Pasteur and Bernard respectively. Bernard concentrated on environmental factors, external and internal, and stressed the view of illness as resulting from a loss of internal balance involving, in general, the concurrence of a variety of factors. Pasteur concentrated his efforts on elucidating the role of bacteria in the outbreak of illness, associating specific types of diseases with specific microbes.

Pasteur and his followers won the debate triumphantly, and as a result the germ theory of disease—the doctrine that specific diseases are caused by specific microbes—was swiftly accepted by the medical profession. The concept of specific etiology* was formulated precisely by the physician Robert Koch, who postulated a set of criteria needed to prove conclusively that a particular microbe caused a specific disease. These criteria, known as "Koch's postulates," have been taught in medical schools ever since.

There were several reasons for such a complete and exclusive acceptance of Pasteur's view. One was the great genius of Louis Pasteur, who was not only an outstanding scientist but a skilled and vigorous debater, with a special flair for dramatic demonstrations. Another reason was the outbreak of several epidemics in Europe at that time, which provided ideal models for demonstrating the concept of specific causation. The most important reason, however, was the fact that the doctrine of specific disease causation fitted perfectly into the framework of nineteenth-century biology.

The Linnean classification of living forms was gaining general acceptance at the beginning of the century, and it seemed natural to extend it to other biological phenomena. The identification of microbes with diseases provided a method for isolating and defining disease entities, and thus a taxonomy of diseases was established not unlike the taxonomy of plants and animals. Furthermore, the idea of a disease being caused by a single factor was in perfect agreement with the Cartesian view of living organisms as machines whose breakdown can be traced back to the malfunctioning of a single mechanism.

* Etiology, from the Greek *aitia* ("cause"), is a medical term meaning "cause (or causes) of disease."

As the reductionist view of disease established itself as a fundamental principle of modern medical science, physicians overlooked the fact that Pasteur's own views on the question of disease causation were much more subtle than the simplistic interpretation given by his followers. René Dubos has shown convincingly, with the help of many quotations, that Pasteur's view of life was fundamentally ecological.[8] He was well aware of the effect of environmental factors on the functioning of living organisms, although he did not have time to investigate them experimentally. The primary aim of his research on disease was to establish the causative role of microbes, but he was also intensely interested in what he called the "terrain," by which he meant the internal and external environment of the organism. In his study of the diseases of silkworms, which led to the germ theory, Pasteur recognized that these diseases resulted from a complex interaction among host, germs, and environment, and he wrote, having completed his research: "If I were to undertake new studies on the silkworm diseases, I would direct my effort to the environmental conditions that increase their vigor and resistance."

Pasteur's view of human diseases showed the same ecological awareness. He took it for granted that the healthy body exhibits a striking resistance to many types of microbes. He knew very well that every human organism acts as host to a multitude of bacteria, and he pointed out that these can cause damage only when the body is weakened. Thus, in Pasteur's view, successful therapy will often depend on the physician's ability to restore the physiological conditions favorable to natural resistance. "This is a principle," wrote Pasteur, "which must always be present in the mind of the physician or of the surgeon, because it can often become one of the foundations of the art of healing." Even more boldly, Pasteur suggested that mental states affect resistance to infection: "How often does it occur that the condition of the patient—his weakness, his mental attitude . . .—form but an insufficient barrier against the invasion of the infinitely small ones." The founder of microbiology had a view of illness broad enough so that he intuitively anticipated mind-body approaches to therapy that have been developed only very recently and are still suspect to the medical establishment.

The doctrine of specific etiology has influenced the development of medicine enormously, from the days of Pasteur and Koch to the present, by shifting the focus of biomedical research from the host and the environment to the study of microorganisms. The resulting narrow view of illness represents a serious flaw of modern medicine which is now becoming increasingly apparent. On the other hand, the knowledge that microorganisms not only affected the development of disease but could also cause the infection of surgical wounds revolutionized the practice of surgery. It led first to the antiseptic system, in which surgical instruments and dressings were sterilized, and subsequently to the aseptic method, in which everything that comes in contact with the wound has to be completely free of bacteria. Together with the technique of general anesthesia, these advances put surgery on an entirely new basis, creating the principal elements of the intricate ritual that has become characteristic of modern surgery.

Advances in biology during the nineteenth century were accompanied by the rise of medical technology. New diagnostic tools, like the stethoscope and instruments for taking blood pressure, were invented and surgical technology became more sophisticated. At the same time the attention of physicians gradually shifted from the patient to the disease. Pathologies were located, diagnosed, and labeled according to a definite system of classification, and were studied in hospitals transformed from medieval "houses of mercy" into centers of diagnosis, therapy, and teaching. Thus began the trend toward specialization that was to reach its height in the twentieth century.

The emphasis on the precise definition and location of pathologies was also applied to the medical study of mental disorders, for which the word psychiatry* was coined. Rather than trying to understand the psychological dimensions of mental illness, psychiatrists concentrated their efforts on finding organic causes—infections, nutritional deficiencies, brain damage—for all mental disturbances. This "organic orientation" in psychiatry was furthered by the fact that in several instances researchers could indeed identify organic origins of mental disorders and were able to develop successful methods of treatment. Although these successes were partial and isolated, they established psychiatry firmly as a branch of medicine, committed to the biomedical model. This turned out to be rather a problematic development in the twentieth century. Indeed, even in the nineteenth century the limited success of the biomedical approach to mental illness inspired an alternative movement—the psychological approach—which led to the

* From the Greek *psyche* ("mind") and *iatreia* ("healing").

founding of dynamic psychiatry and psychotherapy of Sigmund Freud[9] and brought psychiatry much closer to the social sciences and to philosophy.

In the twentieth century the reductionist trend in biomedical science continued. There were outstanding achievements, but some of the triumphs themselves demonstrated the problems inherent in its methods, visible since the turn of the century but now apparent to a great number of people, both within and outside the field of medicine. This has brought the practice of medicine and the organization of health care to the center of public debate and has made it evident to many that its problems are thoroughly intertwined with the other manifestations of our cultural crisis.[10]

Twentieth-century medicine is characterized by the progression of biology to the molecular level, and by the understanding of various biological phenomena at that level. This progress, as we have seen, has established molecular biology as a general way of thinking in the life sciences, and has consequently made it the scientific basis of medicine. The great successes of medical science in our century have all been based on detailed knowledge of cellular and molecular mechanisms.

The first major advance, which was really the result of further applications and elaborations of nineteenth-century concepts, was the development of a host of drugs and vaccines to combat infectious diseases. Vaccines were found first against bacterial diseases—typhoid, tetanus, diphtheria, and many others—and later against diseases involving viruses. In tropical medicine the combined use of immunization and insecticides (to control disease-transmitting mosquitoes) has resulted in the virtual conquest of three major diseases of the tropics, malaria, yellow fever, and leprosy. At the same time many years of experience in these programs have taught scientists that the control of tropical diseases involves far more than vaccinations and the spraying of chemicals. Since all insecticides are toxic to humans, and since they accumulate in plant and animal tissue, they should be used very judiciously. In addition, detailed ecological research is needed to understand the interdependencies of the organisms and life cycles involved in the transmission and development of each disease. The complexities are such that none of these diseases can be completely eradicated, but they can be effectively controlled by skillful handling of the ecological situation.[11]

The discovery of penicillin in 1928 ushered in the era of the antibiotics—one of the most dramatic periods of modern medicine—which culminated in the 1950s with the discovery of a profusion of antibacterial agents capable of coping with a wide variety of microorganisms. The other major pharmaceutical novelty, which also appeared in the 1950s, was a broad range of psychoactive drugs, particularly tranquilizers and antidepressants. With these new drugs psychiatrists were able to control a variety of symptoms and behavior patterns of psychotic patients without causing deep clouding of consciousness. This brought about a major transformation in the care of the mentally ill. Techniques of external coercion were now replaced by the subtle internal chains of modern drugs, which dramatically reduced the time of hospitalization and made it possible to treat many people as outpatients. Enthusiasm for these initial successes obscured for a time the fact that psychoactive drugs, besides having a wide range of dangerous side effects, control symptoms but have no effect on the underlying disorders. Psychiatrists are increasingly aware of this, and critical opinions have begun to gain ground over enthusiastic therapeutic claims.

A major triumph of modern medicine came in endocrinology, the study of the various endocrine glands* and their secretions, known as hormones, which circulate in the bloodstream and regulate a great variety of bodily functions. The outstanding event in this study was the discovery of insulin.† The isolation of this hormone, together with the recognition that diabetes was associated with insulin deficiency, made it possible to save countless diabetics from almost certain death and allow them to lead a normal life, sustained by regular insulin injections. Another major advance in the study of hormones came with the discovery of cortisone, a substance isolated from the cortex of the adrenal gland which constitutes a potent antiinflammatory agent. Finally, endocrinology provided greater knowledge and understanding of sex hormones, which led to the development of contraceptive pills.

These examples all illustrate the successes as well as the shortcomings of the biomedical approach. In all cases medical problems are reduced to molecular phenomena with the aim of finding a mechansim

* Glands included in the endocrine system are the pituitary (in the brain), thyroid (throat), adrenals (kidneys), islets of Langerhans (pancreas), and gonads (genitals).

† Insulin is a hormone secreted by the pancreas glands known as the islets of Langerhans.

that is central to the problem. Once this mechanism is understood, it is counteracted by a drug that is often isolated from another organic process whose "active principle" it is said to represent. By reducing biological functions to molecular mechanisms and active principles in this way, biomedical researchers necessarily limit themselves to partial aspects of the phenomena they study. As a consequence they can achieve only a narrow view of the disorders they investigate and the remedies they develop. All aspects that go beyond this view are considered irrelevant, as far as the disorders are concerned, and are listed as "side effects" in the case of the remedies. Cortisone, for example, has become known for its many dangerous side effects, and the discovery of insulin, although extremely useful, has focused the attention of clinicians and researchers on the symptoms of diabetes, preventing them from looking for the underlying causes. In view of this state of affairs, the discovery of vitamins may be seen as perhaps the greatest success of biomedical science. Once the importance of these "accessory food factors" was recognized and their chemical identity established, many nutritional diseases caused by vitamin deficiency, such as rickets and scurvy, could be cured with the greatest ease by appropriate dietary changes.

Detailed knowledge of biological functions at the cellular and molecular levels not only led to the extensive development of drug therapies but was of tremendous help for surgery, allowing surgeons to advance their art to levels of sophistication beyond all previous expectations. To begin with, the three blood groups were discovered, blood transfusions became possible, and a substance that prevented blood clotting was developed. These developments, together with great advances in anesthesia, gave surgeons much more freedom and made them far more adventurous. With the appearance of antibiotics, protection from infections became much more efficient and made it possible to replace damaged bones and tissues with foreign materials, especially plastics. At the same time, surgeons developed supreme skills and great dexterity in treating tissues and controlling the organism's reactions. The new medical technology allowed them to maintain normal physiological processes even during prolonged surgical interventions. In the 1960s Christiaan Barnard transplanted a human heart, and other transplants of organs followed with varying degrees of success. With these developments medical technology not only reached an unprecedented degree of sophistication but also became all-pervasive in modern medical care. At the same time the increasing dependence of medicine on high technology has raised a number of problems which are not only of a medical or technical nature but involve much broader social, economic, and moral issues.[12]

In the long rise of scientific medicine, physicians have gained fascinating insights into the intimate mechanisms of the human body and have developed their technologies to an impressive degree of complexity and sophistication. Yet in spite of these great advances of medical science we are now witnessing a profound crisis in health care in Europe and North America. Many reasons are given for the widespread dissatisfaction with medical institutions—inaccessibility of services, lack of sympathy and care, malpractice—but the central theme of all criticism is the striking disproportion between the cost and effectiveness of modern medicine. Despite a staggering increase in health costs over the past three decades, and amid continuing claims of scientific and technological excellence by the medical profession, the health of the population does not seem to have improved significantly.

The relation between medicine and health is difficult to assess because most health statistics use the narrow, biomedical concept of health, defined as the absence of disease. A meaningful assessment would deal with both the health of the individual and the health of the society; it would have to include mental illnesses and social pathologies. Such a comprehensive view would show that, although medicine has contributed to the elimination of certain diseases, this has not necessarily restored health. In the holistic view of illness physical disease is only one of several manifestations of a basic imbalance of the organism.[13] Other manifestations may take the form of psychological and social pathologies, and when the symptoms of a physical disease are effectively suppressed by medical intervention, an illness may well express itself through some of the other modes.

Indeed, psychological and social pathologies have now become major problems of public health. According to some surveys, as many as 25 percent of our population are sufficiently troubled psychologically to be seriously handicapped and in need of therapeutic attention[14] At the same time there has been an alarming rise in alcoholism, violent crimes, accidents, and suicides, all symptoms of social ill health. Similarly, the current serious health problems of children have to be seen as indicators of social illness,[15] along with the rise in crime and political terrorism.

On the other hand, there has been a great increase in life expectancy in developed countries over the past two hundred years, and this is often cited as an indication of the beneficial effects of modern medicine. However, this argument is quite misleading. Health has many dimensions, all arising from the complex interplay between the physical, psychological, and social aspects of human nature. In its many facets it mirrors the entire social and cultural system and can never be represented by a single parameter, such as the death rate or the average length of the life span. Life expectancy is a useful statistic but is not sufficient to measure the health of a society. To get a more accurate picture we have to shift our attention from quantity to quality. The increase in life expectancy has resulted primarily from a decline in infant mortality, which in turn is related to the level of poverty, the availability of proper nutrition, and many other social, economic, and cultural factors. Just how these multiple forces combine to affect infant mortality is still poorly understood, but it has become apparent that medical care has played almost no role in its decline.[16]

What, then, *is* the relation between medicine and health? To what extent has modern Western medicine been successful in curing disease and in alleviating pain and suffering? Opinions tend to vary considerably and have led to a number of conflicting affirmations. For example, the following statements can be found in a recent study of health in the United States, sponsored by the Johnson Foundation and the Rockefeller Foundation:

> We have developed the finest biomedical research effort in the world, and our medical technology is second to none.
>
> —John H. Knowles, President,
> Rockefeller Foundation

> In most instances, we are relatively ineffective in preventing disease or preserving health by medical intervention.
>
> —David E. Rogers, President,
> Robert Wood Johnson Foundation

> . . . the remarkable, almost unimaginable progress medicine has in fact made in recent decades . . .
>
> —Daniel Callahan, Director,
> Institute of Society,
> Ethics and the Life Sciences,
> Hastings-on-Hudson, New York

> We are left with approximately the same roster of common major diseases which confronted the country in 1950 and, although we have accumulated a formidable body of information about some of them in the intervening time, the accumulation is not yet sufficient to permit either the prevention or the outright cure of any of them.
>
> —Lewis Thomas, President,
> Memorial Sloan-Kettering Cancer Center

> The best estimates are that the medical system (doctors, drugs, hospitals) affects about 10 percent of the usual indices for measuring health.
>
> —Aaron Wildavsky, Dean,
> Graduate School of
> Public Policy, U. C. Berkeley[17]

These seemingly contradictory statements become intelligible when we realize that different people refer to different phenomena when they speak about progress in medicine. Those who say that there has been progress mean the scientific advances in unraveling biological mechanisms, associating them with specific diseases and developing technologies that will affect them. Indeed, biomedical science has made considerable progress in that sense over the past decades. However, since biological mechanisms are very rarely the exclusive causes of illness, understanding them does not necessarily mean making progress in health care. Hence those who say that medicine has made very little progress over the past twenty years are also right. They are talking about healing rather than scientific knowledge. The two kinds of progress are, of course, not incompatible. Biomedical research will remain an important part of future health care, while being integrated into a broader, holistic approach.

In discussing the relation between medicine and health, one also has to realize that there is a whole spectrum of medicine, from general practice to emergency medicine, surgery to psychiatry. In some of these areas the biomedical approach has been highly successful whereas in others it has proven to be rather ineffective. The great success of emergency medicine in dealing with accidents, acute infections, and premature births is well known. Almost everyone knows somebody whose life has been saved, or whose pain and discomfort have been dramatically reduced, by medical intervention. Indeed, our modern medical technologies are superb in dealing with these emergencies. But although such medical care can be decisive in individual cases, it does not seem to make a significant difference for the health of populations as a whole.[18] The great publicity given to such spectacular medical procedures as open-heart surgery and organ transplants tends to make us forget that many of these patients would not have been hospitalized in the first place if preventive measures had not been severely neglected.

A dramatic development in the history of public health, for which modern medicine is usually given credit, has been the sharp decline in infectious diseases during the late nineteenth and early twentieth centuries. A hundred years ago diseases like tuberculosis, cholera, and typhoid were a constant threat. Anyone could catch them at any time, and every family anticipated losing at least one of its children. Today most of these diseases have almost completely disappeared in developed countries, and the very rare occurrences can easily be controlled with antibiotics. The fact that this dramatic change has taken place more or less simultaneously with the rise of modern scientific medicine has led to the widespread belief that it was brought about by the achievements of medical science. This belief, although shared by most doctors, is quite erroneous. Studies of the history of disease patterns have shown conclusively that the contribution of medical intervention to the decline of the infectious diseases has been much smaller than is generally believed. Thomas McKeown, a leading authority in the fields of public health and social medicine, has made one of the most detailed studies of the history of infections.[19] His work provides ample evidence that the striking decline in mortality since the eighteenth century has been due mainly to three effects. The earliest and, over the whole period, most important influence was a vast improvement in nutrition. From the end of the seventeenth century, food production increased rapidly throughout the Western world; there were great advances in agriculture, and the resulting expansion of food supplies made people more resistant to infections. The critical role of nutrition in strengthening the response of the organism to infectious disease is now well established and is consistent with the experience of Third World countries, where malnutrition is recognized as the predominant cause of ill health.[20] The second major reason for the decline of infectious diseases is the improvement in hygiene and sanitation of the second half of the nineteenth century. The nineteenth century not only brought us the discovery of microorganisms and the germ theory of disease; it was also the era in which the influence of the environment on human life became a focal point of scientific thought and public awareness. Lamarck and Darwin saw the evolution of living organisms as the result of environmental pressure; Bernard emphasized the importance of the *milieu intérieur*, and Pasteur was intrigued by the "terrain" in which microbes were active. In the social domain a similar preoccupation with the environment produced popular health movements and sanitary crusades promoting public health and hygiene.

Most nineteenth-century public health reformers did not believe in the germ theory of disease but assumed that bad health originated from poverty, malnutrition, and filth, and they organized vigorous public health campaigns to combat these conditions. This led to improvements in personal hygiene and nutrition and to the introduction of new sanitary measures—purification of water, efficient disposal of sewage, provision of safe milk, and improved food hygiene—all extremely efficient in controlling the infectious diseases. There was also a significant decline in birth rates, which was itself related to the general improvement of living conditions.[21] This reduced the rate of population growth and thus insured that the improvement in health would not be jeopardized by rising numbers.

McKeown's analysis of the various factors that influenced mortality from infections shows quite clearly that medical intervention was much less important than others. The major infectious diseases had all peaked and declined well before the first effective antibiotics and immunization techniques were introduced. This lack of correlation between the change of disease patterns and medical intervention has also found striking confirmation in several experiments in which modern medical technologies were used unsuccessfully to improve the health of various "underdeveloped" populations in the United States and elsewhere.[22] These experiments seem to indicate that medical technology alone is unable to bring about significant changes in basic disease patterns.

The conclusion to be drawn from these studies of the relation between medicine and health seems to be that biomedical interventions, although extremely helpful in individual emergencies, have very little effect on the health of entire populations. The health of human beings is predominantly determined not by medical intervention but by their behavior, their food, and the nature of their environment. Since these variables differ from culture to culture, each culture has its own characteristic illnesses, and as food, behavior, and environmental situations gradually change, so do the patterns of disease. Thus the acute infectious diseases that plagued Europe and North America in the nineteenth century, and that are still the major killers in the Third World today, have been replaced in the industrialized countries by illnesses no longer associated with poverty and deficient living conditions but, on the contrary, with affluence and technological complexity. These are the chronic and degenerative diseases—heart disease, cancer, diabetes—that have aptly been called "diseases of civilization," since they are closely related to the stressful attitudes, rich diet, drug abuse, sedentary living, and environmental pollution characteristic of modern life.

Because of their difficulties in dealing with degenerative diseases within the biomedical framework, physicians, rather than enlarging this framework, often seem to resign themselves to accepting these diseases as inevitable consequences of general "wear-and-tear" for which there is no cure. By contrast, the public has become increasingly dissatisfied with the present system of medical care, noticing painfully that it has generated exorbitant costs without significantly improving people's health, and complaining that doctors treat diseases but are not interested in the patients.

The causes of our health crisis are manifold; they can be found both within and without medical science, and are inextricably linked to the larger social and cultural crisis. Still, increasing numbers of people, both within and outside the medical field, perceive the shortcomings of the current health care system as being rooted in the conceptual framework that supports medical theory and practice, and have come to believe that the crisis will persist unless this framework is modified.[23] So it is useful to study in some detail the conceptual basis of modern scientific medicine, the biomedical model, to see how it affects the practice of medicine and the organization of health care.[24]

Medicine is practiced in many different ways by men and women with different personalities, attitudes, and beliefs. The following characterization therefore does not apply to all physicians, medical researchers, or institutions. There is great variety within the framework of modern scientific medicine; some family physicians are very caring and others care very little; there are surgeons who are highly spiritual and practice their art with a profound reverence for the human condition, and there are others who are cynical and profit-motivated; there are very human experiences in hospitals, and there are others that are inhuman and degrading. In spite of this wide variety, however, one general belief system underlies current medical education, research, and institutional health care. This belief system is based on the conceptual model we have described historically.

The biomedical model is firmly grounded in Cartesian thought. Descartes introduced the strict separation of mind and body, along with the idea that the body is a machine that can be understood completely in terms of the arrangement and functioning of its parts. A healthy person was like a well-made clock in perfect mechanical condition, a sick person like a clock whose parts were not functioning properly. The principal characteristics of the biomedical model, as well as many aspects of current medical practice, can be traced back to this Cartesian imagery.

Following the Cartesian approach, medical science has limited itself to the attempt of understanding the biological mechanisms involved in an injury to various parts of the body. These mechanisms are studied from the point of view of cellular and molecular biology, leaving out all influences of nonbiological circumstances on biological processes. Out of the large network of phenomena that influence health, the biomedical approach studies only a few physiological aspects. Knowledge of these aspects is, of course, very useful, but they represent only a small part of the story. Medical practice, based on such a limited approach, is not very effective in promoting and maintaining good health. In fact its practices now quite often *cause* suffering and disease, according to some critics even more than they cure.[25] This will not change until medical science relates its study of the biological aspects of illness to the general physical and psychological condition of the human organism and its environment.

Like physicists in their study of matter, medical scientists have tried to understand the human body by reducing it to basic "building blocks" and fundamental functions. As Donald Fredrickson, director of the National Institutes of Health, says, "The reduction of life in all its complicated forms to certain fundamentals that can then be resynthesized for a better understanding of man and his ills is the basic concern of biomedical research."[26] In this reductionist spirit medical problems are analyzed by proceeding to smaller and smaller fragments—from organs and tissues to cells, then to cellular fragments, and finally to single molecules—and all too frequently the original phenomenon itself is lost on the way. The history of modern medical science has shown again and again that the reduction of life to molecular phenomena is not sufficient for understanding the human condition in health and illness.

Confronted with environmental or social problems, medical researchers often argue that these are outside the boundaries of medicine. Medical education, so the argument goes, must by definition be dissociated from social concerns, since those are caused by forces over which physicians have no control.[27] But doctors have played a major part in bringing about this dilemma by insisting that they alone are qualified to determine what constitutes illness and to select the appropriate therapy. As long as they maintain their positions at the top of the hierarchy of power within the health care system, they will have the responsibility of being sensitive to all aspects of health.

Public health interests are generally isolated from medical education and practice, which are severely imbalanced by the overemphasis on biological mechanisms. Many issues that are crucial to health—such as nutrition, employment, population density, and housing—are not sufficiently discussed in medical schools, and thus there is little room for preventive health care in contemporary medicine. When physicians talk about disease prevention they often do so within the mechanistic framework of the biomedical model, but preventive measures within such a limited framework can, of course, not go very far. John Knowles, president of the Rockefeller Foundation, says bluntly, "The basic biological mechanisms of most of the common diseases are still not well enough known to give clear direction to preventive measures."[28]

What is true for the prevention of illness is also true for the art of healing the sick. In both cases physicians have to deal with whole individuals and their relation to the physical and social environment. Although the art of healing is still widely practiced, both within and outside medicine, this is not explicitly acknowledged in our medical institutions. The phenomenon of healing will be excluded from medical science as long as researchers limit themselves to a framework that does not allow them to deal significantly with the interplay of body, mind, and environment.

• • •

The Cartesian division has influenced the practice of health care in several important ways. First, it has split the profession into two separate camps with very little communication between them. Physicians are concerned with the treatment of the body, psychiatrists and psychologists with the healing of the mind. The gap between the two groups has been a severe handicap in the understanding of most major diseases, because it has prevented medical researchers from studying the roles of stress and of emotional states in the development of illness. Stress has only very recently been recognized as a significant source of a wide range of diseases and disorders, and the link between emotional states and illness, although known throughout the ages, still receives little attention from the medical profession.

As a result of the Cartesian split, there are now two distinct bodies of literature in health research. In the psychological literature the relevance of emotional states to illness is widely discussed and well documented. This research is carried out by experimental psychologists and reported in psychology journals that biomedical scientists rarely read. For its part, the medical literature is well grounded in physiology but hardly ever deals with the psychological aspects of illness. Cancer studies are typical. The connection between emotional states and cancer has been well known since the late nineteenth century, and the evidence reported in the psychological literature is substantial. But very few physicians are aware of this work, and medical scientists have not integrated the psychological data into their research.[29]

Another phenomenon that is poorly understood because of the inability of biomedical scientists to integrate physical and psychological elements is the phenomenon of pain.[30] Medical researchers still do not know precisely what causes pain, nor do they fully understand its pathways of communication between body and mind. Just as illness as a whole has physical and psychological aspects, so does the pain which is often associated with it. In practice it is frequently impossible to know which sources of pain are physical and which psychological; of two patients with identical physical symptoms, one may be in excruciating pain while the other experiences none at all. To understand pain, and to be able to alleviate it in the process of healing, we must see its wider context, which includes the patient's mental attitudes and expectations, belief system, emotional support from family and friends, and many other circumstances. Instead of dealing with pain in this comprehensive way, current medical practice, operating within the narrow biomedical framework, tries to reduce pain to an indicator of specific physiological breakdown. Most of the time pain is dealt with by means of denial, and is suppressed with the help of pain killers.

A person's psychological state, of course, is not only relevant in the generation of illness but crucial to the process of healing. The patient's psychological response to the physician is an important part, perhaps the most important part, of every therapy. To induce peace of mind and confidence in the healing process has always been a major purpose of the therapeutic encounter between doctor and patient, and it is well known among physicians that this is usually done intuitively and has nothing to do with technical skills. As Leonard Shlain, himself an outstanding surgeon, observes, "Some doctors seem to make people well, while others, regardless of their expertise, have high rates of complications. The art of healing cannot be quantified."[31]

In modern medicine psychological problems and problems of behavior are studied and treated by psychiatrists. Although they are M.D.s with formal training, there is very little communication between them and physicians outside psychiatry, between mental health professionals and physical health professionals. Many doctors even look down on psychiatrists and consider them second-class physicians. This shows once again the power of the biomedical dogma. Biological mechanisms are seen as the basis of life, mental events as secondary phenomena. Physicians who deal with mental illness are considered somehow less important.

In many cases, psychiatrists have reacted to this attitude by adhering rigorously to the biomedical model and trying to understand mental illness in terms of a derangement of underlying physical mechanisms in the brain. According to this view, mental illness is basically the same as physical illness; the only difference is that it affects the brain rather than some other organ of the body, and thus manifests itself through mental rather than physical symptoms. This conceptual development has led to a rather curious situation. Whereas healers through the ages have tried to treat physical illness by psychological means, modern psychiatrists now treat psychological illness by physical means, having convinced themselves that mental problems are diseases of the body.

The organic orientation in psychiatry has resulted in the transplantation of concepts and methods that have been found useful in the treatment of physical diseases into the field of emotional and behavioral disorders. Since these disorders are believed to be based on specific biological mechanisms, great emphasis is placed on establishing the correct diagnosis using a reductionist system of classification. Although this approach has failed for most mental disorders, it is still widely pursued in the hope of finding, ultimately, the specific mechanisms of disease causation and the corresponding specific methods of treatment for all mental disorders.

As for treatment, the preferred method is to treat mental illness with medication, which controls the symptoms of the disorder but does not cure it. And it is becoming increasingly apparent that this kind of treatment is countertherapeutic. From a holistic perspective of health, mental illness can be seen as resulting from a failure to evaluate and integrate experience. In this view the symptoms of a mental disorder reflect the organism's attempt to heal itself and achieve a new level of integration.[32] Standard psychiatric practice interferes with this spontaneous healing process by suppressing the symptoms. True therapy would consist in facilitating the healing by providing an emotionally supportive atmosphere for the patient. Rather than being suppressed, the process that constitutes a symptom would be allowed to intensify in such an atmosphere, and continuing self-exploration would lead to its full experience and conscious integration, thus completing the healing process.

To practice such a therapy, considerable knowledge of the full spectrum of human consciousness is required. Psychiatrists often lack such knowledge, yet they are legally responsible for the treatment of mental patients. Accordingly, mental patients are treated in medical institutions where clinical psychologists, who often have a much more thorough knowledge of psychological phenomena, act merely as ancillary personnel subordinated to psychiatrists.

The extension of the biomedical model to the treatment of mental disorders has been, on the whole, very unfortunate. Although the biological approach has been useful for the treatment of some disorders with a clear organic origin, it is quite inappropriate for many others to which psychological models are of fundamental significance. A great deal of effort has been wasted in trying to arrive at a precise, organically based diagnostic system of mental disorders, without the realization that the search for accurate, objective diagnosis will ultimately be futile for most psychiatric cases. The practical disadvantage of this approach has been that many individuals with no organic malfunctions are treated in medical facilities where they receive therapies of problematic value at extremely high costs.

The limitations of the biomedical approach to psychiatry are now becoming apparent to an increasing number of health professionals, and these practitioners are engaged in a lively debate about the nature of mental illness. Thomas Szasz, who regards mental illness as pure myth, takes perhaps the most extreme position.[33] Szasz condemns the notion of illness as something that attacks people without any relation to their personalities, life styles, belief systems, or social environment. In this sense all illness, whether mental or physical, is a myth. If the term is used in a holistic sense, taking into account the patient's entire organism and personality, as well as the physical and social environment, mental disorders are as real as physical illnesses. But such an understanding of mental illness transcends the conceptual framework of current medical science.

Avoidance of the philosophical and existential issues that arise in connection with every serious illness is a characteristic aspect of contemporary medicine. It is another consequence of the Cartesian division that has led medical scientists to concentrate exclusively on the physical aspects of health. In fact the question "What is health?" is generally not even addressed in medical schools, nor is there any discussion of healthy attitudes and life styles. These are considered philosophical issues that belong to the spiritual realm, outside the domain of medicine. Furthermore, medicine is supposed to be an objective science, not concerned with moral judgments.

This seventeenth-century view of medical science often prevents physicians from seeing the beneficial aspects and potential meaning of illness. Disease is viewed as an enemy to be conquered, and medical scientists pursue the Utopian ideal of eliminating, eventually, all diseases through the application of biomedical research. Such a narrow point of view fails to comprehend the subtle psychological and spiritual aspects of illness, and prevents medical researchers from realizing, as Dubos has noted, that "complete freedom from disease and struggle is almost totally incompatible with the living process."[34]

The ultimate existential issue is, of course, death—and, like all other philosophical and existential questions, the matter of death is avoided as much as possible. The lack of spirituality that has become characteristic of our modern technological society is reflected in the fact that the medical profession, like society as a whole, is death-denying. Within the mechanistic framework of our medical science, death cannot be qualified. The distinction between a good death and a poor death does not make sense; death becomes simply the total standstill of the body-machine.

The age-old art of dying is no longer practiced in our culture, and the fact that it is possible to die in good health seems to have been forgotten by the medical profession. Whereas in the past one of the most important roles of a good doctor was to provide comfort and support for dying patients and their families, physicians and other health professionals today are no longer trained to deal with dying patients and find it extremely difficult to cope with the phenomenon of death in a meaningful way. They tend to see death as a failure; bodies are carried out of hospitals secretly at night, and doctors seem significantly more afraid of death than other people, whether sick or healthy.[35] Although general attitudes toward death and dying have recently begun to change considerably,[36] following the spiritual renaissance of the 1960s and 1970s, the new attitudes have not yet been incorporated into our health care system. To do so will require a fundamental conceptual shift in the medical view of health and illness.

Having discussed some of the consequences of the Cartesian division for contemporary medicine, let us now take a closer look at the image of the body as a machine and its impact on current medical theory and practice. The mechanistic view of the human organism has encouraged an engineering approach to health in which illness is reduced to mechanical trouble and medical therapy to technical manipulation.[37] In many cases this approach has been successful. Medical science and technology have developed highly sophisticated methods for removing or repairing various parts of the body, and even for replacing them by artificial constructs. This has alleviated the suffering and discomfort of countless victims of illnesses and accidents, but it has also helped to distort the views of health and health care held by the medical profession and the general public.

The public image of the human organism—enforced by the content of television programs, and especially by advertising—is that of a machine which is prone to constant failure unless supervised by doctors and treated with medication. The notion of the organism's inherent healing power and tendency to stay healthy is not communicated, and trust in one's own organism is not promoted. Nor is the relation between health and living habits emphasized; we are encouraged to assume that doctors can fix anything, irrespective of our life styles.

It is intriguing and quite ironic that physicians themselves are the ones who suffer most from the mechanistic view of health by disregarding stressful circumstances in their lives. Whereas traditional healers were expected to be healthy people, keeping their body and soul in harmony and in tune with their environment, the typical attitudes and habits of doctors today are quite unhealthy and produce considerable illness. Physicians' life expectancy today is ten to fifteen years less than that of the average population, and they have not only high rates of physical illness but also high rates of alcoholism, drug abuse, suicide, and other social pathologies.[38]

Most doctors adopt their unhealthy attitudes right at the beginning of medical school, where their training has been designed to be a highly stressful experience. The unhealthy value system that dominates our society has found some of its most extreme expressions in medical education. Medical schools, especially in the United States, are by far the most competitive of all professional schools. Like the business world, they present high competitiveness as a virtue and emphasize an "aggressive approach" to patient care. In fact the aggressive stance of medical care is often so extreme that the metaphors used to describe illness and therapy are taken from the language of warfare. For example, a malignant tumor is said to "invade" the body, radiation therapy "bombards" the tissues to "kill" the cancer cells, and chemotherapy is often likened to chemical warfare. Thus medical education and practice perpetuate the attitudes and behavior patterns of a value system that plays a significant role in causing many of the diseases medicine seeks to cure.

Medical schools not only generate stress but also neglect to teach their students how to cope with it. The essence of current medical training is inculcating the notion that the patient's concerns come first and that the doctor's well-being is secondary. This is thought to be necessary to produce commitment and responsibility, and to foster such an attitude the medical training consists of extremely long hours with very few breaks. Many physicians continue this practice in their professional lives. It is not uncommon for a physician to work for a full year with no vacation. This excessive stress is aggravated by the fact that doctors continually have to deal with people in states of high anxiety or deep depression, which adds further intensity to their daily work. On the other hand, they are trained to use a model of health and illness in which emotional forces play no role, and hence they tend to disregard them in their own lives.

The mechanistic view of the human organism and the resulting engineering approach to health has led to an excessive emphasis on medical technology, which is perceived as the only way to improve health. Lewis Thomas, for example, is quite explicit about this in his paper "On the Science and Technology of Medicine." After his remark that medicine has not been able to prevent or cure any of our common major diseases over the past three decades, he goes on to say, "We are, in a sense, stuck with today's technology, and we will stay stuck until we have more scientific knowledge to work with it."[39]

Hard technology has taken a central role in modern medical care. At the turn of the century the ratio of supporting personnel to doctors was about one to two; now it can be as high as fifteen to one. The diagnostic and therapeutic tools operated by this army of technicians are the result of recent advances in physics, chemistry, electronics, computer science, and other related fields. They include computerized blood analyzers and tomography scanners,* machines for renal dialysis,† cardiac pacemakers, equipment for radiation therapy, and many other machines that are not only highly sophisticated but also extremely expensive, some of them costing close to a million dollars.[40] As in other areas, the use of high technology in medicine is often unwarranted. The increasing dependence of medical care on complex technologies has accelerated the trend toward specialization and has enforced the doctors' tendency to look at particular parts of the body, forgetting to deal with the patient as a whole person.

At the same time the practice of medicine has shifted from the office of the general physician to the hospital where it became progressively depersonalized, if not dehumanized. Hospitals have grown into

* The computerized tomography scanner, or "CAT scanner," is a machine used for X-ray diagnosis of abnormalities within the skull. It consists of an X-ray unit directing beams through the skull from multiple directions, coupled to a computer that analyzes the X-ray information and constructs visual images that could not be obtained by conventional techniques.

† A renal dialysis machine filters or "dialyzes" the blood of patients with kidney failure, replacing the function of the kidneys.

large professional institutions, emphasizing technology and scientific competence rather than contact with the patient. In these modern medical centers, which look more like airports than therapeutic environments, patients tend to feel helpless and frightened, which often keeps them from getting well. Some 30 to 50 percent of present hospitalization is medically unnecessary, but alternative services that could be therapeutically more effective and economically more efficient have almost disappeared.[41]

The costs of medical care have increased at a frightening pace over the past three decades. In the United States, they went up from twelve billion dollars in 1950 to a hundred and sixty billion in 1977, rising almost twice as fast as the cost of living during 1974–77.[42] Similar tendencies have been observed in most other countries, including those with socialized medical systems. The development and widespread use of expensive medical technologies is one of the main reasons for this sharp increase in health costs. For example, renal dialysis for one individual may cost as much as $10,000 a year, and coronary bypass surgery, which has yet to be shown to prolong life, is being performed thousands of times at a cost of $10,000 to $25,000 per operation.[43]

The excessive use of high technology in medical care is not only uneconomic but also causes an unnecessary amount of pain and suffering. Accidents in hospitals now occur more frequently than in any other industries except mining and high-rise construction. It has been estimated that one out of every five patients admitted to a typical research hospital will acquire an iatrogenic illness,* with half of these episodes resulting from complications of drug therapy and a surprising 10 percent from diagnostic procedures.[44]

The high risks of modern medical technology have led to a further significant increase in health costs through the growing number of malpractice suits against physicians and hospitals. There is now an almost paranoid fear of litigation among American doctors, who try to protect themselves from lawsuits by practicing "defensive medicine," ordering even more diagnostic technologies which further increase the costs of health care and expose patients to additional risks.[45] This malpractice crisis is the result of several things: excessive use of high technology within a mechanistic model of illness in which all responsibility is delegated to the doctor; considerable pressure from a large number

* Iatrogenic illnesses—from the Greek *iatros* ("physician") and *genesis* ("origin")—are illnesses generated by the medical care process itself.

of profit-motivated lawyers; and a society that prides itself on being democratic but does not have a socialized medical system.

The conceptual problem at the center of contemporary health care is the biomedical definition of disease, according to which diseases are well-defined entities that involve structural changes at the cellular level and have unique causal roots. The biomedical model allows for several kinds of causative factors, but researchers tend to adhere to the doctrine of "one disease, one cause." The germ theory was the first example of specific disease causation. Bacteria and, later on, viruses have been assumed to be the cause of virtually every disease of unknown origin. Then the rise of molecular biology brought the concept of the single lesion,* which includes genetic anomalies; and more recently environmental causes of disease have come under study. In all these cases medical scientists have tried to achieve three objectives: precise definition of the disease under study; identification of its specific cause; and development of the appropriate treatment—usually some technical manipulation—that will eliminate the causal root of the disease.

The theory of specific disease causation has been successful in a few special cases, such as acute infectious processes and nutritional deficiencies, but the overwhelming majority of illnesses cannot be understood in terms of the reductionist concepts of well-defined disease entities and single causes. The main error of the biomedical approach is the confusion between disease processes and disease origins. Instead of asking why an illness occurs, and trying to remove the conditions that lead to it, medical researchers try to understand the biological mechanisms through which the disease operates, so that they can then interfere with them. Among the leading contemporary researchers Thomas has expressed his belief in such an approach with unusual clarity: "For every disease there is a single key mechanism that dominates all others. If one can find it, and then think one's way around it, one can control the disorder . . . In short, I believe that the major diseases of human beings have become approachable biological puzzles, ultimately solvable."[46]

These mechanisms, rather than the true origins, are seen as the causes of disease in current medical thinking, and this confusion lies at the very center of the conceptual problems of contemporary medicine.

* Lesion is a technical term for injury; it denotes an abnormal change in structure of an organ or other bodily part.

As Thomas McKeown has emphasized, "It should be recognized that the most fundamental question in medicine is why disease occurs rather than how it operates after it has occurred; that is to say, conceptually the origins of disease should take precedence over the nature of disease process."[47]

The origins of disease will generally be found in several causative factors that must concur to result in ill health.[48] Moreover, their effects will differ profoundly from person to person, since they depend on the individual's emotional reactions to stressful situations and on the social environment in which these situations occur. The common cold is a good example. It can develop only if a person is exposed to one of several viruses, but not everybody exposed to these viruses will be afflicted. Exposure will result in illness only when the exposed individual is in a receptive state, and this will depend on weather conditions, fatigue, stress, and a host of other circumstances that influence the person's resistance to infection. To understand why a particular person develops a cold, many of these factors have to be assessed and weighed against one another. Only then will the "puzzle of the common cold" be solved.

This situation has its counterpart in almost all illnesses, most of them far more serious than the common cold. An extreme case, in both complexity and severity, is cancer. Over the past decades huge amounts of money have been poured into cancer research with the aim of identifying a virus that caused the disease. When this line of research remained fruitless, attention shifted to environmental causes, which were also investigated within a reductionist framework. Today many researchers still perpetuate the impression that exposure to a carcinogenic substance alone causes cancer. But if we look at the number of people who are exposed, for example, to asbestos and ask how many of them will develop lung cancer, we find that the incidence is something like one in a thousand. Why does that one person develop the disease? The answer is that any noxious influence from the environment involves the organism as a whole, including the psychological state and the social and cultural conditioning of the person. All these factors are significant in the development of cancer and have to be taken into account to understand the disease.

The concept of disease as a well-defined entity has led to a classification of diseases patterned after the taxonomy of plants and animals. Such a classification system has some justification for illnesses with predominantly physical symptoms, but it is extremely problematic for mental illnesses, to which it has been extended. Psychiatric diagnosis is notorious for its lack of objective criteria. Since the patient's behavior toward the psychiatrist is part of the clinical picture on which the diagnosis is based, and since this behavior is influenced by the doctor's personality, attitudes, and expectations, the diagnosis will necessarily be subjective. Thus the ideal of a precise classification of "mental disease" remains largely illusory. Nevertheless, psychiatrists have spent an enormous amount of effort trying to establish objective diagnostic systems for emotional and behavioral disorders that would allow them to include mental illness in the biomedical definition of disease.

In the process of reducing illness to disease, the attention of physicians has moved away from the patient as a whole person. Whereas illness is a condition of the total human being, disease is a condition of a particular part of the body, and rather than treating patients who are ill, doctors have concentrated on treating their diseases.[49] They have lost sight of the important distinction between the two concepts. According to the biomedical view, there is no illness, and thus no justification for medical attention, without the structural or biochemical alterations characteristic of a specific disease. But clinical experience has shown repeatedly that one can be ill without having a disease. Half of all visits to the doctor are for complaints that cannot be associated with any physiological disorder.[50]

Because of the biomedical definition of disease as the basis of illness, medical treatment is directed exclusively at the biological abnormality. But this does not necessarily restore the patient to health, even if the treatment is successful. For example, medical cancer therapy may result in the complete regression of a tumor without making the patient well. Emotional problems may continue to affect the patient's health and, if not dealt with, may produce a recurrence of the malignancy.[51] On the other hand, it may happen that a patient has no demonstrable disease but nevertheless feels quite sick. Because of the limitations of the biomedical approach, physicians are often unable to help such patients, who have been called "the worried well."

Although the biomedical model distinguishes between symptoms and diseases, each disease itself, in a wider sense, can be seen as merely a symptom of an underlying illness whose origins are rarely investigated. To do so would require seeing ill health within the broad context of the human condition, recognizing that any illness or behavioral disorder of a particular individual can be understood only in relation to the whole network of interactions in which that person is embedded.

Perhaps the most striking example of the emphasis on symptoms rather than underlying causes is the drug approach of contemporary medicine. It has its roots in the erroneous view that bacteria are the primary causes of disease, rather than symptomatic manifestations of underlying physiological disorder. For many decades after Pasteur advanced his germ theory medical research was focused on the bacteria and neglected to study the host organism and its environment. Because of this one-sided emphasis, which began to change only in the second half of our century with the rise of immunology, physicians have tended to concentrate on destroying the bacteria instead of looking for the causal roots of the disorder. They have been very successful in suppressing or alleviating the symptoms but at the same time often cause further damage to the organism.

The overemphasis on bacteria has given rise to the view that disease is the consequence of an attack from outside, rather than of a breakdown within the organism. Lewis Thomas, in his popular *Lives of a Cell*, has given a vivid description of this widespread misconception:

> Watching television, you'd think we lived at bay, in total jeopardy, surrounded on all sides by human-seeking germs, shielded against infection and death only by a chemical technology that enables us to keep killing them off. We are instructed to spray disinfectants everywhere . . . We apply potent antibiotics to minor scratches and seal them with plastic. Plastic is the new protector; we wrap the already plastic tumblers of hotels in more plastic, and seal the toilet seats like state secrets after irradiating them with ultraviolet light. We live in a world where the microbes are always trying to get at us, to tear us cell from cell, and we only stay alive through diligence and fear.[52]

These rather grotesque attitudes, more noticeable in the United States than anywhere else, are of course promoted not only by medical science but even more forcefully by the chemical industry. Whatever their motivation, they are hardly justified on the basis of biological fact. It is well known that many types of bacteria and viruses associated with disease are commonly present in the tissues of healthy individuals without causing any harm. Only under special circumstances that lower the general resistance of the host do they produce pathological symptoms. Our society makes it hard to believe, but the functioning of many essential organs requires the presence of bacteria. Animals raised under totally germ-free conditions have been shown to develop gross anatomical and physiological abnormalities.[53]

Out of the huge population of bacteria on the earth, only a small number is capable of generating diseases in human organisms, and these are usually destroyed in due course by the organism's immune mechanisms. As Thomas says, "The man who catches a meningococcus* is in considerably less danger for his life, even without chemotherapy, than the meningococci with the bad luck to catch a man."[54] On the other hand, bacteria that are relatively harmless for a particular group of people who have built up resistance to them may be extremely virulent for others if they have never been exposed to these microbes before. The catastrophic epidemics that afflicted Polynesians, American Indians, and Eskimos at their first contacts with European explorers provide striking illustrations of this.[55]

The point is that the development of infectious diseases depends as much on the response of the host as on the specific characteristics of the bacteria. This view is further enforced by a careful study of the detailed mechanism of infection. There seem to be very few infectious diseases in which the bacteria cause actual direct damage to the cells or tissues of the host organism. There are some, but in most cases the damage is caused by an overreaction of the organism, a kind of panic in which a number of powerful, unrelated defense mechanisms are all turned on at once.[56] Infectious diseases, then, arise most of the time from a lack of coordination within the organism, rather than from injury caused by invading bacteria.

Given these facts, it would seem extremely useful, as well as intellectually challenging, to study the complex interactions of mind, body, and environment that affect resistance to bacteria. However, very little research of this kind is being done. The major research effort in this century has been directed toward identifying specific microorganisms and developing medicines to kill them. This effort has been extremely successful, providing doctors with an arsenal of drugs that are highly effective in the treatment of acute bacterial infections. But while the

* Meningococcus is the bacterium associated with meningitis, an inflammation of the membranes covering the brain and spinal cord.

proper use of antibiotics in emergency situations will continue to be justified, it will also be essential to study and enhance the natural resistance of human organisms to bacteria.

Antibiotics, of course, are not the only type of drugs used in modern medicine. Drugs have become the key to all medical therapy. They are used to regulate a wide variety of physiological functions through their effects on nerves, muscles, and other tissues, as well as on the blood and other bodily fluids. Drugs can improve the functioning of the heart and correct irregularities in the heartbeat; they can raise or lower blood pressure, prevent blood clotting or control excessive bleeding, induce muscle relaxation, affect the secretion of various glands, and regulate a number of digestive processes. By acting on the central nervous system, they can alleviate or temporarily eliminate pain, relieve tension and anxiety, induce sleep, or increase alertness. Drugs can affect a wide range of regulatory functions, from the visual accommodation of the eye to the destruction of cancer cells. Many of these functions involve subtle biochemical processes that are barely understood, if not completely mysterious.

The extensive development of chemotherapy* in modern medicine has allowed physicians to save innumerable lives and alleviate much suffering and discomfort, but, unfortunately, it has also led to the well-known overuse and misuse of drugs, both by doctors through prescription and by individuals through self-medication. Until recently it was believed that the toxic side effects of medical drugs, although sometimes serious, were so rare that they were generally insignificant. This turned out to be a grave misjudgment. During the past two decades adverse drug reactions have become a public health problem of alarming proportions, producing considerable pain and discomfort for millions each year.[57] Some of these effects are inevitable, and many of them are clearly the fault of patients, but many others are the result of careless and inappropriate prescriptions by doctors who adhere rigidly to the biomedical approach. It has been argued that high-quality medicine can be practiced without the use of any of the twenty most commonly prescribed drugs.[58]

The central role of drugs in contemporary health care is often justified with the observation that today's most effective drugs—including digitalis, penicillin, and morphine—all come from plants, many of

* Chemotherapy is the treatment of disease with chemicals, that is, with drugs.

them used as medicines throughout the ages. The medical use of drugs, according to this argument, is merely the continuation of a custom that is probably as old as humanity itself. Although this is certainly true, there is a crucial difference between the use of herbal medicines and chemical drugs. The drugs prepared in modern pharmaceutical laboratories are purified and highly concentrated samples of substances that occur naturally in plants. These purified products turn out to be less efficient and more hazardous than the original unpurified remedies. Recent experiments with herbal medicine indicate that the purified active principle is less effective as a medicine than the crude extract from the plant, because the latter contains trace elements and molecules that were considered unimportant but turn out to play a vital role by limiting the effect of the main active ingredient. They ensure that the body's reaction does not go too far and cause unwanted side effects. Crude extracts of herbal mixtures also have very special antibacterial properties. They do not destroy the bacteria but prevent them from multiplying, so that mutations cannot occur and strains of bacteria resistant to the medication are unlikely to develop.[59] Furthermore, the dosage of herbal medicines is much less problematic than that of chemical drugs. Herbal mixtures that have been tried out empirically for thousands of years need not be quantified precisely because of their in-built moderating effects. Approximate dosages, according to age, body weight, and size of the patient, are sufficient. Thus modern science is now validating empirical knowledge that has been passed on from generation to generation by folk healers in all cultures and traditions.

An important aspect of the mechanistic view of living organisms and the resulting engineering approach to health is the belief that the cure of illness requires some outside intervention by the physician, which can be either physical, through surgery or radiation, or chemical, through drugs. Current medical therapy is based on this principle of medical intervention, relying on outside forces for cure, or at least for the alleviation of suffering and discomfort, without taking into account the healing potential within the patient. This attitude derives directly from the Cartesian view of the body as a machine that requires somebody to repair it when it breaks down. Accordingly, medical intervention is carried out with the aim of correcting a specific biological mechanism in a particular part of the body, with different parts treated by different specialists.

To associate a particular illness with a definite part of the body is, of course, very useful in many cases. But modern scientific medicine has overemphasized the reductionist approach and has developed its specialized disciplines to a point where doctors are often no longer able to view illness as a disturbance of the whole organism, nor to treat it as such. What they tend to do is to treat a particular organ or tissue, and this is generally done without taking the rest of the body into account, let alone considering the psychological and social aspects of the patient's illness.

Even though such a fragmentary medical intervention can be very successful in alleviating pain and suffering, this alone is not always enough to justify it. From a broader point of view, not everything that alleviates suffering temporarily is necessarily good. If the intervention is carried out without taking other aspects of the illness into account, the result will generally be unhealthy for the patient in the long run. For example, somebody may develop arteriosclerosis, a narrowing and hardening of the arteries, as the result of an unhealthy way of life—heavy diet, lack of exercise, excessive smoking. Surgical treatment of a blocked artery may temporarily alleviate pain but will not make the person well. The surgical intervention merely treats a local effect of a systemic disorder that will continue until the underlying problems are identified and resolved.

Medical therapy, of course, will always be based on some form of intervention. It need not, however, take the excessive and fragmentary form we see so often in contemporary health care. It could be the kind of therapy practiced by wise physicians and healers for millennia, a subtle interference with the organism to stimulate it in a specific way so that it will, by itself, complete the process of healing. Therapies of that kind are based on a profound respect for self-healing; on the view of the patient as a responsible individual who can herself initiate the process of getting well. Such an attitude is contrary to the biomedical approach, which delegates all authority and responsibility to the doctor.

According to the biomedical model, only the doctor knows what is important for an individual's health, and only he can do anything about it, because all knowledge about health is rational, scientific knowledge, based on objective observation of clinical data. Thus laboratory tests and measurement of physical parameters in the examining room are generally considered more relevant to the diagnosis than the assessment of the patient's emotional state, family history, or social situation.

The physician's authority and his responsibility for the patient's health make him assume a paternal role. He can be a benevolent parent or a dictatorial parent, but his position is clearly superior to that of the patient. Moreover, since most doctors are men, the paternal role of the physician encourages and perpetuates sexist attitudes in medicine, with respect to both women patients and women doctors.[60] These attitudes include some of the most dangerous manifestations of sexism, not provoked by medicine as such but reflecting the patriarchal bias in society as a whole, and especially in science.

In today's health care system physicians play a unique and decisive role in the health teams that share the tasks of patient care.[61] It is the physician who sends patients to the hospital and sends them home, who orders tests and X-rays, recommends surgery and prescribes drugs. Nurses, although often highly trained as therapists and health educators, are considered merely assistants of doctors and can rarely use their full potential. Because of the narrow biomedical view of illness and the patriarchal patterns of power in the health care system, the important role that nurses play in the healing process through their human contacts with the patients is not fully recognized. From these contacts nurses often acquire much more extensive knowledge of the patient's physical and psychological condition than doctors, but this knowledge is considered less relevant than the M.D.'s "scientific" assessment based on laboratory tests. Spellbound by the mystique that surrounds the medical profession, our society has conferred on physicians the exclusive right to determine what constitutes illness, who is ill and who is well, and what should be done to the sick. Numerous other healers, such as homeopaths, chiropractors, and herbalists, whose therapeutic techniques are based on different, but equally coherent, conceptual models have been legally excluded from the mainstream of health care.

Although physicians have considerable power to influence the health care system, they are also very conditioned by it. Since their training is heavily oriented toward hospital care, they feel more comfortable in doubtful cases when their patients are in the hospital, and since they receive very little reliable information about drugs from noncommercial sources, they tend to be unduly influenced by the pharmaceutical industry. However, the essential aspects of contemporary health care are determined by the nature of medical education. The emphasis on hard technology, the overuse of drugs, and the practice of centralized, highly specialized medical care all originate in the medical schools and academic medical centers. Any attempt to change the current system of health care will therefore have to begin by changing medical education.

American medical education was cast into its present form at the beginning of the century, when the American Medical Association commissioned a national survey of medical schools with the aim of putting medical education on a sound scientific basis. A related purpose of the survey was to channel the huge funds of newly formed foundations—especially those of the Carnegie and Rockefeller foundations—into a few carefully chosen medical institutions.[62] This established the link between medicine and big business that has dominated the entire health care system ever since.

The result of the survey was the Flexner Report, published in 1910, which decisively shaped American medical education by setting up strict guidelines that are still followed today.[63] The modern medical school was to be part of a university, with a permanent faculty committed to teaching and research. Its primary purpose was the education of students and the study of disease, not the care of the sick. Accordingly, the M.D. degree it granted was to certify the successful mastery of medical science, not the ability to care for patients. The science to be taught, and the research to be pursued, were firmly embedded in the reductionist biomedical framework; in particular, they were to be dissociated from social concerns, which were considered outside the boundaries of medicine.

The Flexner Report found that only about 20 percent of all American medical schools met its "scientific" standards. The others were declared "second-rate" and were forced to close through legal and financial pressures. Although many of the schools had indeed been inadequate, they were also the institutions that had admitted female, black, and poor students, all now effectively barred from access to medical training. In particular, the medical establishment vehemently opposed the admission of women into medicine and erected a number of barricades against the training and practice of female physicians.

Under the impact of the Flexner Report, scientific medicine became more and more biologically oriented, specialized, and hospital-based.[64] Specialists increasingly replaced generalists as teachers and became the models for aspiring physicians. By the late 1940s medical students in the academic medical centers had almost no contact with physicians practicing general medicine, and since their training took place more and more within hospitals, they were effectively removed from contact with most of the illnesses that confront people in their daily lives. This situation has persisted to the present day. Whereas two-thirds of the complaints encountered in everyday medical practice are for minor illnesses of brief duration, which usually cure themselves, and less than 5 percent for major illnesses carrying a threat to life, this proportion is reversed in a university hospital.[65] Thus medical students are given a thoroughly distorted view of illness. Their major experience involves only a tiny portion of common health problems, and these problems are not studied out in the community, where their broader context could be assessed, but in the hospital, where students concentrate exclusively on the biological aspects of illness. As a consequence, interns and residents develop a disdain for the ambulatory patient—the walking, living person with complaints that usually involve emotional as well as physical problems—and come to see the hospital as an ideal place to practice specialized and technology-oriented medicine.

A generation ago more than half of all physicians were general practitioners; now over 75 percent are specialists, confining their attention to a particular age group, disease, or part of the body. According to David Rogers,[66] this has resulted in "the apparent inability of American medicine to deal with the simple day-to-day medical needs of our population." On the other hand, there is a "surplus" of surgeons in the United States which, according to some critics, results in considerable overuse of surgical procedures.[67] These are some of the reasons why many people see the need for primary health care—the broad range of general care traditionally rendered by physicians in community practice—as the central problem facing American medicine.

The problem with primary care is not only the small number of general practitioners but also their approach to patient care, which is often restricted by the heavily biased training they received in medical school. The task of the general practitioner requires not only scientific knowledge and technical skills but wisdom, compassion, and patience, an ability to provide human comfort and reassurance, sensitivity to the patient's emotional problems, and therapeutic skills in the management of psychological aspects of illness. These attitudes and skills are generally not emphasized in the present programs of medical training, in which the identification and treatment of a specific disease is presented as the essence of medical care. Moreover, medical schools vigorously promote an imbalanced, "macho" value system and actively suppress the qualities of intuition, sensitivity, and nurturance in favor of a rational, aggressive, and competitive approach. As Scott May, a student at the University of California School of Medicine in San Francisco, said in his graduation speech, "Medical school felt like a family where the mother was gone and only the hard father remained at home."[68] Because of this imbalance, physicians often regard an empathic discussion of personal issues as quite unnecessary, and in turn patients tend to perceive them as cold and unfriendly and complain that the doctor fails to understand their worries.

The purpose of our academic medical centers is not only training but research. As in medical education, the biological orientation is heavily favored in the support and funding of research projects. Although epidemiological, social, and environmental research would often be much more useful and efficient for improving human health than the strict biomedical approach,[69] projects of this kind are little encouraged and poorly financed. The reason for this resistance is not merely the strong conceptual appeal of the biomedical model to most researchers but also its vigorous promotion by the various interest groups in the health industry.[70]

Although there is widespread dissatisfaction with medicine and with doctors among the general public, most people are not aware that one of the main reasons for the current state of affairs is the narrow conceptual basis of medicine. On the contrary, the biomedical model is generally accepted, and its basic principles are so thoroughly ingrained in our culture that it has even become the dominant folk model of illness. Most patients do not understand its intricacies very well, but they have been conditioned to believe that the doctor alone knows what made them sick and that technological intervention is the only thing that will get them well.

This public attitude makes it very difficult for progressive physicians to change the patterns of current health care. I know several who try to explain their patients' symptoms to them, relating the illness to the patients' living habits, and who find again and again that patients are not satisfied with that approach. They want something else, and often they will not be content until they can leave the doctor's office with a prescription in their hands. Many physicians make great efforts to change people's attitudes about health, so that they will not insist on having an antibiotic prescribed for a cold, but the power of the patients' belief system often makes these efforts ineffective. As one general practitioner tells me, "You have a mother with a child who is running a fever, and who says, 'Give him a penicillin shot'; and then you say, 'You don't understand, penicillin won't help in that case,' and then she says, 'What kind of a doctor are you? If you don't want to do it I'll go somewhere else.' "

The biomedical model today is much more than a model. Among the medical profession it has acquired the status of a dogma, and for the general public it is inextricably linked to the common cultural belief system. To go beyond it will require nothing less than a profound cultural revolution. And such a revolution is necessary if we want to improve, or even maintain, our health. The shortcomings of our current health care system—in terms of health costs, effectiveness, and fulfillment of human needs—are becoming more and more conspicuous and are increasingly recognized as stemming from the restrictive nature of the conceptual model on which it is based. The biomedical approach to health will still be extremely useful, just as the Cartesian-Newtonian framework remains useful in many areas of classical science, as long as its limitations are recognized. Medical scientists will need to realize that the reductionist analysis of the body-machine cannot provide them with a complete understanding of human problems. Biomedical research will have to be integrated into a broader system of health care in which the manifestations of all human illness are seen as resulting from the interplay of mind, body, and environment, and are studied and treated accordingly.

To adopt such a holistic and ecological concept of health, in theory and in practice, will require not only a radical conceptual shift in medical science but also a major public reeducation. Many people obstinately adhere to the biomedical model because they are afraid to have their life styles examined and to be confronted with their unhealthy behavior. Rather than face such an embarrassing and often painful situation, they insist on delegating all responsibility for their health to the doctor and the drugs. Furthermore, as a society we tend to use medical diagnosis as a cover-up of social problems. We prefer to talk about our children's "hyperactivity" or "learning disability," rather than examine the inadequacy of our schools; we prefer to be told that we suffer from "hypertension" rather than change our overcompetitive business world; we accept ever increasing rates of cancer rather than investigate how the chemical industry poisons our food to increase its profits. These health problems go far beyond the concerns of the medical profession, but they are brought into focus, inevitably, as soon as we seriously try to go beyond current medical care. Transcending the biomedical model will be possible only if we are willing to change other things as well; it will be linked, ultimately, to the entire social and cultural transformation.

6 · Newtonian Psychology

Like biology and medicine, the science of psychology has been shaped by the Cartesian paradigm. Psychologists, following Descartes, adopted the strict division between the *res cogitans* and the *res extensa,* which made it extremely difficult for them to understand how mind and body interacted with each other. The current confusion about the role and nature of the mind, as distinct from that of the brain, is a manifest consequence of the Cartesian division.

Descartes not only made a sharp distinction between the impermanent human body and the indestructible soul, but also suggested different methods for studying them. The soul, or mind, should be studied by introspection, the body by the methods of natural science. However, psychologists in the subsequent centuries did not follow Descartes' suggestion but adopted both methods for the study of the human psyche, thus creating two major schools of psychology. The structuralists studied the mind through introspection and tried to analyze consciousness into its basic elements, while behaviorists concentrated exclusively on the study of behavior and so were led to ignore or deny the existence of mind altogether. Both these schools emerged at a time when scientific thought was dominated by the Newtonian model of reality. Accordingly, they both modeled themselves after classical physics, incorporating the basic concepts of Newtonian mechanics into their theoretical frameworks.

Meanwhile, working in the clinic and the consulting room rather than the laboratory, Sigmund Freud used the method of free association to develop psychoanalysis. Although this was a very different, even revolutionary, theory of the human mind, its basic concepts were again Newtonian in nature. Thus the three main currents of psychological thinking in the first decades of the twentieth century—two in the academy and one in the clinic—were based not only on the Cartesian paradigm but also on specifically Newtonian concepts of reality.

Psychology as a science is commonly believed to date from the nineteenth century, and its historical roots are usually traced back to the philosophies of Greek antiquity.[1] The Western belief that this tradition has produced the only serious psychological theories is now being recognized as a rather narrow and culturally conditioned view. Recent developments in consciousness research, psychotherapy, and transpersonal psychology have stimulated interest in Eastern systems of thought, and particularly those of India, which exhibit a variety of profound and sophisticated approaches to psychology. The rich tradition of Indian philosophy has generated a spectrum of philosophical schools, from extreme materialism to extreme idealism, from absolute monism through dualism to complete pluralism. Accordingly, these schools have developed numerous and often conflicting theories about human behavior, the nature of consciousness, and the relation between mind and matter.

In addition to this wide range of philosophical schools, Indian and other Eastern cultures have also developed spiritual traditions that are based on empirical knowledge and thus more akin to the approach of modern science.[2] These traditions are grounded in mystical experiences that have led to elaborate and extremely refined models of consciousness that cannot be understood within the Cartesian framework but are in surprising agreement with recent scientific developments.[3] Eastern mystical traditions are not, however, primarily concerned with theoretical concepts. They are, above all, ways of liberation, concerned with the transformation of consciousness. During their long history they have developed subtle techniques to change their followers' awareness of their own existence and of their relation to human society and the natural world. Thus traditions like Vedanta, Yoga, Buddhism, and Taoism resemble psychotherapies much more than religions or philosophies, and it is therefore not surprising that some Western psychotherapists have recently shown a keen interest in Eastern mysticism.[4]

The psychological speculations of ancient Greek philosophers also show strong influences of Eastern ideas, which the Greeks assimilated, according to history and legend, during extended studies in Egypt. This early Western philosophical psychology fluctuates between idealistic and materialistic views of the soul. Among the pre-Socratics, Empedocles taught a materialistic theory of the psyche, according to which all thought and perception were dependent on bodily change. Pythagoras, on the other hand, expounded strongly mystical views that included the belief in the transmigration of souls. Socrates introduced a new concept of the soul into Greek philosophy. Whereas it had been described before either as a vital force—the "breath of life"—or as a transcendental principle in the mystical sense, Socrates used the word "psyche" in the sense that modern psychology does, as the seat of intelligence and character.

Plato was the first to deal explicitly with the problem of consciousness, and Aristotle wrote the first systematic treatise on it, *On the Soul*, in which he developed a biological and materialistic approach to psychology. This materialistic approach, which was further elaborated by the Stoics, found its most eloquent opponent in Plotinus, the founder of Neoplatonism and last of the great philosophers of antiquity, whose teachings resembled the Indian Vedanta philosophy in many aspects and had a powerful influence on early Christian doctrine. According to Plotinus, the soul is immaterial and immortal; consciousness is the image of the Divine One and as such is present at all levels of reality.

One of the most powerful and influential images of the psyche is found in Plato's philosophy. In the *Phaedrus*, the soul is pictured as a charioteer driving two horses, one representing the bodily passions and the other the higher emotions. This metaphor encapsulates the two approaches to consciousness—the biological and the spiritual—which have been pursued, without being reconciled, throughout Western philosophy and science. This conflict generated the "mind-body problem" that is reflected in many schools of psychology, most notably in the conflict between the psychologies of Freud and Jung.

In the seventeenth century, the mind-body problem was cast into the form that shaped the subsequent development of Western scientific psychology. According to Descartes, mind and body belonged to two parallel but fundamentally different realms, each of which could be studied without reference to the other. The body was governed by mechanical laws, but the mind—or soul—was free and immortal. The soul was clearly and specifically identified with consciousness and could affect the body by interacting with it through the brain's pineal gland. Human emotions were seen as combinations of six elementary "passions" and described in a semimechanical way. As far as knowledge and perception were concerned, Descartes believed that knowing was a primary function of human reason, that is, of the soul, which could take place independently of the brain. Clarity of concepts, which played such an important role in Descartes' philosophy and science,[5] could not be derived from the confused performance of the senses but was the result of an innate cognitive disposition. Learning and experience merely provided the occasions for the manifestation of innate ideas.

The Cartesian paradigm provided inspiration, as well as challenge, for two great philosophers of the seventeenth century, Baruch Spinoza and Gottfried Wilhelm Leibniz. Spinoza could not accept Descartes' dualism and replaced it with a rather mystical monism;* Leibniz introduced the idea of an infinite number of substances which he called "monads," meaning organismic units of an essentially psychic nature, with the human soul occupying a special position among them. According to Leibniz, monads "have no windows"; they merely mirror one another.[6] There is no interaction between mind and body, but both act in "preestablished harmony."

The subsequent development of psychology followed neither the spiritual views of Spinoza nor the organismic ideas of Leibniz. Instead, philosophers and scientists turned to Newton's precise mathematical formulation of Descartes' mechanistic paradigm, and attempted to use its principles to understand human nature. While La Mettrie in France applied Descartes' mechanical model of animals in a straightforward way to the human organism, including its mind, British empiricist philosophers used Newtonian ideas to develop more sophisticated psychological theories. Hobbes and Locke refuted Descartes' concept of innate ideas and maintained that there was nothing in the mind that was not first in the senses. At birth, the human mind was, in Locke's famous phrase, a *tabula rasa*, a blank tablet upon which ideas were imprinted through sensory perceptions. This notion served as the starting

* Monism, from the Greek *monos* ("single"), is a philosophical view which holds that there is only one kind of ultimate substance or reality.

point for a mechanistic theory of knowledge, in which sensations were the basic elements of the mental realm and were combined into more complex structures by the process of association.

The concept of association represented a significant step in the development of the Newtonian approach to psychology, since it allowed philosophers to reduce the complexity of mental functioning to certain elementary rules. David Hume in particular elevated association to the central principle in the analysis of the human mind, seeing it as an "attraction in the mental world" that played a role comparable to the force of gravity in the material Newtonian universe. Hume was also deeply influenced by Newton's method of inductive reasoning, based on experience and observation, and he used it to construct an atomistic psychology in which the self was reduced to a "bundle of perceptions."

David Hartley took a further step, combining the concept of the association of ideas with that of the neurological reflex, to develop a detailed and ingenious mechanistic model of the mind in which all mental activity was reduced to neurophysiological processes. This model was further elaborated by several empiricists and in the 1870s was incorporated into the work of Wilhelm Wundt, generally regarded as the founder of scientific psychology.

The modern science of psychology was a result of nineteenth-century developments in anatomy and physiology. Intensive studies of the brain and the nervous system established specific relations between mental functions and brain structures, clarified various functions of the nervous system, and brought detailed knowledge of the anatomy and physiology of the sensory organs. As a result of these advances, the ingenious but naïve mechanistic models outlined by Descartes, La Mettrie, and Hartley were reformulated in modern terms, and the Newtonian orientation of psychology became firmly established.

The discovery of correlations between mental activity and brain structure created great enthusiasm among neuroanatomists and led some of them to postulate that human behavior could be reduced to a set of independent mental faculties, or traits, that were localized in specific regions of the brain. Although this hypothesis could not be sustained, its basic aim of associating various functions of the mind with precise locations in the brain is still very popular among neuroscientists. At first researchers were able to demonstrate a high degree of localization for the primary motor and sensory functions, but when the approach was extended to higher cognitive processes, such as learning and memory, it did not lead to any consistent picture of these phenomena. Nevertheless, most neuroscientists continued their research along the established reductionist lines.

Nineteenth-century studies of the nervous system produced another field of research, reflexology, which had a profound influence on subsequent psychological theories. The neurological reflex, with its clear causal relation between stimulus and response and its machinelike reliability, became the prime candidate for the elementary physiological building block that formed the basis of more complex patterns of behavior. The discovery of new forms of reflexive responses gave hope to many psychologists that, ultimately, all human behavior would be understood in terms of complex combinations of basic reflex mechanisms. This view was put forth by Ivan Sechenov, founder of the influential Russian school of reflexology whose most prominent member was Ivan Pavlov. Pavlov's discovery of the principle of conditioned reflexes had a decisive impact on subsequent learning theories.

Detailed investigation of the central nervous system was complemented by increased understanding of the structure and function of the sensory organs, which helped establish systematic relations between the quality of sensory experiences and the physical characteristics of their stimuli. Pioneering experiments by Ernst Weber and Gustav Fechner resulted in formulation of the celebrated Weber-Fechner law, which postulates a mathematical relation between the intensities of sensations and their stimuli. Physicists made major contributions to this field of sense physiology; Hermann von Helmholtz, for example, developed comprehensive theories of hearing and color vision.

These experimental approaches to the study of perception and behavior culminated in Wundt's research. Founder of the first psychology laboratory, he remained the most influential figure in scientific psychology for almost four decades. During that time he was the chief representative of the so-called elementist orientation, which maintained that all mental functioning could be analyzed into specific elements. The object of psychology, according to Wundt, was to study how these elements could be combined to form perceptions, ideas, and various associative processes.

The orthodox experimental psychologists of the nineteenth century were dualists who tried to draw a clear distinction between mind and matter. They believed that introspection was a necessary source of information about the mind, but saw it as an analytical method that would allow them to reduce consciousness to well-defined elements associated with specific nerve currents in the brain. These reductionist and materialistic theories of psychological phenomena evoked strong opposition among psychologists who emphasized the unitary nature of consciousness and perception. The holistic approach gave rise to two influential schools, gestalt psychology and functionalism. Neither was able to change the Newtonian orientation of the majority of psychologists during the nineteenth and early twentieth centuries, but both strongly influenced the new trends in psychology and psychotherapy that emerged in the second half of our century.

Gestalt psychology, founded by Max Wertheimer and his associates, was based on the assumption that living organisms do not perceive things in terms of isolated elements but in terms of *Gestalten*, that is, as meaningful wholes which exhibit qualities that are absent in their individual parts. Kurt Goldstein then applied the gestalt view to the treatment of brain disorders in what he called an organismic approach, with the aim of helping people to come to terms with themselves and their environment.

The development of functionalism was a consequence of nineteenth-century evolutionary thought, which established an important connection between structure and function. For Darwin each anatomical structure was a functioning component of an integrated living organism engaged in the evolutionary struggle for survival. This dynamic emphasis inspired many psychologists to turn from the study of mental structure to that of mental processes, to view consciousness as a dynamic phenomenon, and to investigate its modes of functioning, especially in relation to the life of the whole organism. These psychologists, known as functionalists, were highly critical of the tendencies of their contemporaries to analyze the mind into atomistic elements; instead, they emphasized the unity and dynamic nature of the "stream of consciousness."

The foremost exponent of functionalism was William James, whom many people consider the greatest American psychologist. Certainly his work contains a unique mixture of ideas that have stimulated psychologists of many different schools. James taught physiology before he moved into psychology to become a pioneer of the scientific experimental approach. He was the founder of the first American psychology laboratory and played a major role in changing the status of his discipline from a branch of philosophy to a laboratory science.

In spite of his thoroughly scientific orientation, William James was a fervent critic of the atomistic and mechanistic tendencies in psychology, and an enthusiastic advocate of the interaction and interdependence of mind and body. He reinterpreted the findings of contemporary experimenters with a determined emphasis on consciousness as a personal, integral, and continuous phenomenon. It was not sufficient to study the elements of mental functioning and the rules for the association of ideas. These elements were merely arbitrary cross sections of a continuous "stream of thought" that had to be understood in relation to the conscious actions of human beings in their daily confrontations with a variety of environmental challenges.

In 1890 James published his innovative views on the human psyche in the monumental *Principles of Psychology*, which soon became a classic. After its completion, his interest shifted to more philosophical and esoteric pursuits, such as the study of unusual states of consciousness, psychic phenomena, and religious experiences. The aim of these investigations was to probe the full range of human consciousness, as he stated eloquently in his *Varieties of Religious Experience:*

> Our normal waking consciousness, rational consciousness as we call it, is but one special type of consciousness, whilst all about it, parted from it by the filmiest of screens, there lie potential forms of consciousness entirely different. We may go through life without suspecting their existence; but apply the requisite stimulus, and at a touch they are there in all their completeness. . . .
>
> No account of the universe in its totality can be final which leaves these other forms of consciousness quite disregarded. How to regard them is the question . . . At any rate, they forbid our premature closing of accounts with reality.[7]

This broad view of psychology is probably the strongest aspect of James's influence on recent psychological research.

In the twentieth century psychology made great progress and gained increasing reputation. It benefited considerably from cooperation with other disciplines—from biology and medicine to statistics, cybernetics, and communication theory—and found important applications in health care, education, industry, and many other areas of practical human activity. During the early decades of the century, psychological thinking was dominated by two powerful schools—behaviorism and psychoanalysis—which differed markedly in their methods and their views of consciousness but nevertheless adhered, basically, to the same Newtonian model of reality.

Behaviorism represents the culmination of the mechanistic approach in psychology. Based on detailed knowledge of human physiology, behaviorists created a "psychology without a soul," a sophisticated version of La Mettrie's human machine.[8] Mental phenomena were reduced to patterns of behavior, and behavior to physiological processes governed by the laws of physics and chemistry. John Watson, who founded behaviorism, was strongly influenced by several trends in the life sciences around the turn of the century.

Wundt's experimental approach had been brought to the United States from Germany by Edward Titchener, the acknowledged leader of the "structuralist" school of psychology. He attempted a rigorous reduction of the contents of consciousness to "simple" elements and emphasized that the "meaning" of mental states was nothing but the context in which mental structures occurred and had no further significance for psychology. At the same time, the reductionist and materialist view of mental phenomena was decisively influenced by Loeb's mechanistic biology, and in particular by his theory of tropism—the tendency of plants and animals to turn certain parts in certain directions. Loeb explained this phenomenon in terms of "forced movements" imposed on living organisms by the environment in strictly mechanistic fashion. This new theory, which made tropism one of the key mechanisms of life, had a tremendous appeal for many psychologists, who applied the notion of forced movements to a wider range of animal behaviors and, eventually, to those of human beings.

In the description of mental phenomena in terms of patterns of behavior, the study of the learning process played a central role. Quantitative experiments on animal learning opened up the new field of experimental animal psychology, and theories of learning were developed by most schools of psychology, with the notable exception of psychoanalysis. Among these learning theories behaviorism was most influenced by Pavlov's work on conditioned reflexes. When Pavlov studied salivation in response to stimuli coinciding with the provision of food, he took great care to avoid all psychological concepts and to describe the dogs' behavior exclusively in terms of their reflex systems. This approach suggested to psychologists that a more general theory of behavior could be formulated in purely physiological terms. Vladimir Bekhterev, founder of the first Russian laboratory of experimental psychology, outlined such a theory, describing the learning process in strictly physiological language by reducing complex behavior patterns to compounds of conditioned responses.

The general trend of moving away from concern with consciousness and toward strictly mechanistic views, the new methods of animal psychology, the principle of the conditioned reflex, and the concept of learning as a modification of behavior were all assimilated into Watson's new theory, which identified psychology with the study of behavior. To him behaviorism represented an attempt to apply to the experimental study of human behavior the same procedures and the same language of description that had been found useful in the study of animals. Indeed, Watson, like La Mettrie two centuries before him, saw no essential difference between humans and animals. "Man," he wrote,[9] "is an animal different from other animals only in the types of behaviors he displays."

It was Watson's ambition to raise the status of psychology to that of an objective natural science, and to do so he adhered as closely as possible to the methodology and principles of Newtonian mechanics, the eminent example of scientific rigor and objectivity. To subject psychological experiments to the criteria used in physics required that psychologists focus exclusively on phenomena that can be registered and described objectively by independent observers. Thus Watson became a vigorous critic of the introspective method used by James and Freud as well as Wundt and Titchener. The whole concept of consciousness, which resulted from introspection, was to be excluded from psychology, and all related terms—like "mind," "thinking," and "feeling"—were to be eliminated from psychological terminology. "Psychology, as the behaviorist views it," wrote Watson,[10] "is a purely objective, experimental branch of natural science which needs consciousness as little as do the sciences of chemistry and physics." It would certainly have been a great shock to him had he known that only a few decades later a leading physicist, Eugene Wigner, would state, "It was not possible to formulate the laws of [quantum theory] in a fully consistent way without reference to consciousness."[11]

In the behaviorist view, according to Watson, living organisms were complex machines reacting to external stimuli, and this stimulus-response mechanism was of course modeled after Newtonian physics. It implied a rigorous causal relation that would allow psychologists to predict the response for a given stimulus and, conversely, to specify the stimulus for a given response. In actual fact, behaviorists seldom dealt with simple stimuli and responses but studied entire constellations of stimuli and complex responses, which were referred to respectively as "situations" and "adjustments." The basic behaviorist assumption was that these complex phenomena could always, at least in principle, be reduced to combinations of simple stimuli and responses. Thus the laws derived from simple experimental situations were expected to apply to more complex phenomena, and conditioned responses of ever increasing complexity were seen as adequate explanations of all human expressions, including science, art, and religion.

A logical consequence of the stimulus-response model was a tendency to look for the determinants of psychological phenomena in the external world rather than within the organism. Watson applied this approach not only to perception but also to imagery, thinking, and emotions. All these phenomena were seen not as subjective experiences but as implicit modes of behavior in response to external stimuli.

Since the process of learning is especially suitable for objective experimental research, behaviorism became primarily a psychology of learning. Its original formulation did not contain the concept of conditioning, but after Watson had studied Bekhterev's work, conditioning became the main method and explanatory principle of behaviorism. Accordingly, there was a strong emphasis on control that was in keeping with the Baconian ideal that has become characteristic of Western science.[12] The aim of dominance and control of nature was applied to animals, and later on, with the notion of "behavioral engineering," to human beings.

One consequence of this approach was the development of behavior therapy that attempted to apply conditioning techniques to the treatment of psychological disorders through the modification of behavior. Although these efforts can be traced back to the pioneering work of Pavlov and Bekhterev, they were not developed in a systematic way until the middle of this century. Today "pure" behavior therapy is totally symptom- or problem-oriented. Psychiatric symptoms are not regarded as manifestations of underlying disorders but as isolated instances of learned maladjustive behavior, to be corrected by appropriate conditioning techniques.

The first three decades of the twentieth century are usually regarded as the period of "classical behaviorism," dominated by John Watson and characterized by fierce polemics against introspective psychologists. This classical phase of behaviorist psychology gave rise to an enormous amount of experimentation but failed to produce a comprehensive theory of human behavior. In the 1930s and 1940s Clark Hull tried to construct such a comprehensive theory, based on highly refined experiments and formulated in terms of a system of definitions and postulates, not unlike Newton's *Principia.* The cornerstone of Hull's theory was the principle of reinforcement, meaning that the response to a specific stimulus is strengthened by the satisfaction of a basic need or drive. Hull's approach dominated learning theories and his system was applied to the investigation of practically all known problems of learning.[13] In the 1950s, however, Hull's influence declined and his theory was gradually replaced by the Skinnerian approach, which revitalized behaviorism in the second half of the century.

B. F. Skinner has been the main exponent of the behaviorist view for the past three decades. His special talent for designing simple and clean experimental situations led him to develop a much more rigorous but also more subtle theory, which has been highly popular, especially in the United States, and has helped behaviorism maintain its dominant role in academic psychology. The main innovations in Skinner's behaviorism were a strictly operational definition of reinforcement—anything that increases the probability of a preceding response—and a strong emphasis on precise "schedules of reinforcement." To test his theoretical concepts Skinner developed a new method of conditioning, called "operant conditioning," which differs from the classical, Pavlovian conditioning process in that reinforcement occurs only after the animal executes a predesignated operation, such as pressing a lever or pecking at an illuminated disk. This method was greatly refined by extreme simplification of the animal's environment. For example, rats were confined in boxes, called "Skinner boxes," which contained simply a horizontal bar that the animal could depress to release a pellet of food. Other experiments involved the pecking response of pigeons, which can be very precisely controlled.

While the notion of operant behavior—behavior controlled by its entire past history rather than by direct stimuli—was a great advance in behaviorist theory, the whole framework remained strictly Newtonian. In his well-known text, *Science and Human Behavior*, Skinner makes it clear from the very outset that he regards all phenomena associated with human consciousness, such as mind or ideas, as nonexisting entities, "invented to provide spurious explanations." The only serious explanations, according to Skinner, are those based on the mechanistic view of living organisms and satisfying the criteria of Newtonian physics. "Since mental or psychic events are asserted to lack the dimensions of physical science," he writes, "we have an additional reason for rejecting them."[14]

Although the title of Skinner's book refers explicitly to human behavior, the concepts discussed in it are based almost exclusively on conditioning experiments with rats and pigeons. These animals are reduced, as Paul Weiss has put it, to "puppets operated by environmental strings."[15] Behaviorists largely ignore the mutual interplay and interdependence between a living organism and its natural environment, which is itself an organism. From their narrow perspective on animal behavior, they then make a huge conceptual leap to human behavior, asserting that human beings, like animals, are machines whose activity is limited to conditioned responses to environmental stimuli. Skinner has firmly rejected the image of human beings acting in accordance with the decisions of their inner selves, and instead has proposed an engineering approach to create a new type of "man," a human being who will be conditioned to behave in the way that is best for himself and for society. According to Skinner, this will be the only way to overcome our current crisis: not through an evolution of consciousness, because there is no such thing; not through a change of values because values are nothing but positive or negative reinforcements—but through scientific control of human behavior. "What we need," he writes, "is a technology of behavior . . . comparable in power and precision to physical and biological technology."[16]

This, then, is Newtonian psychology par excellence, a psychology without consciousness that reduces all behavior to mechanistic sequences of conditioned responses and asserts that the only scientific understanding of human nature is one that remains within the framework of classical physics and biology; a psychology, furthermore, that reflects our culture's preoccupation with manipulative technology, designed for domination and control. In recent years behaviorism has begun to change by assimilating elements of many other disciplines and thus losing much of its former rigid stance. But behaviorists still adhere to the mechanistic paradigm and often defend it as the only scientific approach to psychology, thus clearly limiting science to the classical Newtonian framework.

Psychoanalysis, the other dominant school of twentieth-century psychology, did not originate in psychology but came from psychiatry, which in the nineteenth century was firmly established as a branch of medicine. At that time psychiatrists were thoroughly committed to the biomedical model and were bent on finding organic causes for all mental disturbances. This organic orientation had a promising beginning, but it failed to reveal a specific organic basis for neuroses* and other mental disorders, and some psychiatrists began to look for psychological approaches to mental illness.

A decisive stage in this development was reached during the last quarter of the nineteenth century, when Jean-Martin Charcot successfully used hypnosis for the treatment of hysteria.† In dramatic demonstrations Charcot showed that patients could be freed from the symptoms of hysteria purely by hypnotic suggestion, and that these symptoms could also be brought back again by the same method. This called in question the whole organic approach to psychiatry, and it made a deep impression on Sigmund Freud, who went to Paris in 1885 to hear Charcot's lectures and witness his demonstrations. When he returned to Vienna, Freud, in collaboration with Joseph Breuer, began to use the hypnotic technique to treat neurotic patients.

The publication of *Studies in Hysteria* by Breuer and Freud in 1895 is often regarded as the birth of psychoanalysis because it described the new method of free association which Freud and Breuer had discovered and found far more useful than hypnosis. It consisted of putting patients into a drowsy, dreamlike state and then letting them talk freely about their problems, with special emphasis on traumatic emo-

* Psychoneuroses, also called simply neuroses, are functional nervous disorders without demonstrable physical lesions; psychoses are more severe mental disturbances characterized by a loss of contact with popularly accepted views of reality.

† Hysteria is a psychoneurosis marked by emotional excitability and disturbances of various psychological and physiological functions.

tional experiences. This use of free association was to become the cornerstone of the "psychoanalytic" method.

Trained as a neurologist, Freud believed that in principle one should be able to understand all mental problems in terms of neurochemistry. In the same year that saw the publication of his work on hysteria he also wrote a remarkable document, *Project for a Scientific Psychology,* in which he outlined a detailed scheme for a neurological explanation of mental illness.[17] Freud never published this work, but two decades later he again expressed his belief that "all our provisional ideas in psychology will some day be based on an organic substructure."[18] For the time being, however, neurological science was not advanced enough, and thus Freud pursued a different avenue to study the "intrapsychic apparatus." His collaboration with Breuer ended after their joint research on hysteria, and Freud embarked alone on a unique exploration of the human mind which resulted in the first systematic psychological approach to mental illness.

Freud's contribution was truly extraordinary, considering the state of psychiatry in his time. For over thirty years he sustained a continuous flow of creativity that culminated in several momentous discoveries, any one of which would be admirable as a product of an entire lifetime. To begin with, Freud almost single-handedly discovered the unconscious and its dynamics. Whereas behaviorists, later on, refused to recognize the existence of the human unconscious, Freud saw it as an essential source of behavior. He pointed out that our conscious awareness represents only a thin layer resting on a vast unconscious realm—the tip of the iceberg, as it were, whose hidden regions are governed by powerful instinctual forces. Through the process of psychoanalysis these deeply submerged tendencies of human nature could be revealed, and thus Freud's system also became known as depth psychology.

Freud's theory was a dynamic approach to psychiatry that studied the forces leading to psychological disorders and emphasized the importance of childhood experiences for the future development of the individual. He identified the libido, or sexual drive, as one of the principal psychological forces and considerably extended the concept of human sexuality, introducing the notion of infantile sexuality and outlining the principal stages of early psychosexual development. Another major discovery was Freud's interpretation of dreams, which he called "the royal road to the unconscious."

In 1909, at Clark University in Massachusetts, Freud delivered an epoch-making lecture, "Origin and Development of Psychoanalysis," which brought him world-wide fame and established the psychoanalytic school in the United States. The publication of the lecture was followed by an autobiographical essay, "On the History of the Psychoanalytic Movement," published in 1914, which marked the end of the first great phase of psychoanalysis.[19] This phase had produced a coherent theory of unconscious dynamics based on instinctual drives of an essentially sexual nature, whose complex interplay with various inhibiting tendencies generated the rich variety of psychological patterns.

During the second phase of his scientific life Freud formulated a new theory of personality based on three distinct structures of the intrapsychic apparatus, which he called Id, Ego, and Superego. This period was also marked by significant changes in Freud's understanding of the psychotherapeutic process, especially his discovery of transference,* which would become of central importance in the practice of psychoanalysis. These systematic steps in the development of Freud's theory and practice were followed by the psychoanalytic movement in Europe and the United States and established psychoanalysis as a major school of psychology, dominating psychotherapy for many decades. Moreover, Freud's deep insights into the functioning of the mind and the development of human personality had far-reaching consequences for the interpretation of a variety of cultural phenomena—art, religion, history, and many others—and significantly shaped the world view of the modern era.

From the early years of his psychoanalytic explorations to the end of his life, Freud was deeply concerned with the aim of establishing psychoanalysis as a scientific discipline. He firmly believed that the same organizing principles which had shaped nature in all her forms were also responsible for the structure and functioning of the human mind. Although the science of his time was quite far from demonstrating such a unity of nature, Freud assumed that this goal would be reached sometime in the future, and he repeatedly emphasized the descent of psychoanalysis from the natural sciences, particularly from physics and

* Transference denotes the tendency of patients to transfer onto the analyst, during the analytic procedure, an entire gamut of feelings and attitudes that are characteristic of their early relationships with relevant figures of their childhood, particularly parents.

medicine. Although he was the originator of the psychological approach to psychiatry, he remained under the influence of the biomedical model, both in theory and in practice.

To formulate a scientific theory of the psyche and of human behavior, Freud tried as much as possible to use the basic concepts of classical physics in his description of psychological phenomena and thus to establish a conceptual relationship between psychoanalysis and Newtonian mechanics.[20] He made this quite clear when he said, in an address to a group of psychoanalysts: "Analysts . . . cannot repudiate their descent from exact science and their community with its representatives . . . Analysts are at bottom incorrigible mechanists and materialists." At the same time, Freud—unlike many of his followers—was well aware of the limited nature of scientific models and expected that psychoanalysis would have to be continually modified in the light of new developments in the other sciences. Thus he continued his exhortative description of psychoanalysts:

> They are content with fragmentary pieces of knowledge and with basic hypotheses lacking in preciseness and ever open to revision. Instead of waiting for the moment when they will be able to escape from the constraint of the familiar laws of physics and chemistry, they hope for the emergence of more extensive and deeper-reaching natural laws, to which they are ready to submit.[21]

The close relation between psychoanalysis and classical physics becomes strikingly apparent when we consider the four sets of concepts that form the basis of Newtonian mechanics:

(1) The concepts of absolute space and time, and of separate material objects moving in this space and interacting mechanically with one another;

(2) The concept of fundamental forces, essentially different from matter;

(3) The concept of fundamental laws describing the motion and mutual interactions of the material objects in terms of quantitative relations;

(4) The concept of rigorous determinism, and the notion of an objective description of nature based on the Cartesian division between mind and matter.[22]

These concepts correspond to the four basic perspectives from which psychoanalysts have traditionally approached and analyzed mental life. They are known respectively as the topographic, dynamic, economic, and genetic* points of view.[23]

As Newton established absolute Euclidean space as the frame of reference in which material objects are extended and located, so Freud established psychological space as a frame of reference for the structures of the mental "apparatus." The psychological structures on which Freud based his theory of human personality—Id, Ego, and Superego—are seen as some kind of internal "objects," located and extended in psychological space. Thus spatial metaphors, such as "depth psychology," "deep unconscious," and "subconscious," are prominent throughout the Freudian system. The psychoanalyst is seen as delving into the psyche almost like a surgeon. In fact, Freud advised his followers to be "cold as a surgeon," which reflects the classical ideal of scientific objectivity as well as the spatial and mechanistic conception of the mind.

In Freud's topographic description the unconscious contains "matter" that has been forgotten or repressed, or has never reached conscious awareness. In its deeper realms lies the Id, an entity that is the source of powerful instinctual drives which are in conflict with a well-developed system of inhibiting mechanisms residing in the Superego. The Ego is a frail entity located between these two powers and engaged in a continual existential struggle.

Although Freud sometimes described these psychological structures as abstractions and resisted all attempts to associate them with specific structures and functions of the brain, they had all the properties of material objects. No two of them could occupy the same place, and thus any portion of the psychological apparatus could expand only by displacing other parts. As in Newtonian mechanics, the psychological objects were characterized by their extension, position, and motion.

* "Genetic," as used by psychoanalysts, refers to the origin, or genesis, of mental phenomena and should not be confused with the sense in which it is used in biology.

The dynamic aspect of psychoanalysis, like the dynamic aspect of Newtonian physics, consists in describing how the "material objects" interact with one another through forces that are essentially different from "matter." These forces have definite directions and can reinforce or inhibit one another. The most fundamental among them are instinctual drives, in particular the sexual drive. Freudian psychology is basically a conflict psychology. In his emphasis on existential struggle, Freud was undoubtedly influenced by Darwin and the Social Darwinists, but for the detailed dynamics of psychological "collisions" he turned to Newton. In the Freudian system the mechanisms and machineries of the mind are all driven by forces modeled after classical mechanics.

A characteristic aspect of Newtonian dynamics is the principle that forces always come in pairs; for every "active" force there is an equal "reactive" force of opposite direction. Freud adopted this principle, calling the active and reactive forces "drives" and "defenses." Other pairs of forces, developed at different stages of Freud's theory, were Libido and Destrudo, or Eros and Thanatos, in both of which one force was life-oriented, the other death-oriented. As in Newtonian mechanics, these forces were defined in terms of their effects, which were studied in great detail, but the intrinsic nature of the forces was not investigated. The nature of the force of gravity had always been a problematic and controversial issue in Newton's theory, and so was the nature of the libido in Freud's.[24]

In psychoanalytic theory, understanding of the dynamics of the unconscious is essential for understanding of the therapeutic process. The basic picture is one of instinctual drives striving for discharge, and of various counterforces that inhibit and thereby distort them. Thus the skillful analyst will concentrate on eliminating the obstacles that prevent the direct expression of the primary forces. Freud's conception of the detailed mechanisms through which this goal would be achieved underwent considerable changes during his lifetime, but in all his speculations one can clearly recognize the influence of the Newtonian system of thought.

Freud's earliest theory of the origin and treatment of neuroses, and especially of hysteria, was formulated in terms of a hydraulic model. The primary causes of hysteria were identified as traumatic situations in the patient's childhood which occurred under circumstances that prevented an adequate expression of the emotional energy generated by the incidents. This pent-up, or jammed energy remained stored in the organism and would continue to seek discharge until it found a modified expression through various neurotic "channels." Therapy, according to this model, consisted in remembering the original trauma under conditions that would allow a belated emotional release of the trapped energies.

Freud abandoned the hydraulic model as being too simplistic when he found evidence that the patients' symptoms did not stem from isolated pathological processes but were consequences of the overall mosaic of their life histories. This new view saw the roots of neuroses in instinctual, predominantly sexual tendencies that were unacceptable and were therefore repressed by psychic forces, which converted them into neurotic symptoms. Thus the basic conception had shifted from the hydraulic image of an explosive release of pent-up energies to the more subtle but still Newtonian image of a constellation of mutually inhibiting dynamic forces.

The latter concept implies the notion of entities separated in psychological space but unable to move or expand without displacing one another. Thus there is no room for qualitative development and improvement of the Ego in the framework of classical psychoanalysis; its expansion can occur only at the expense of the Superego or the Id. As Freud saw it,[25] "Where Id was, there shall Ego be." In classical physics the interactions between material objects and the effects of the various forces on them are described in terms of certain measurable quantities—mass, velocity, energy, and so on—which are interrelated through mathematical equations. Although Freud could not go that far in his theory of the mind, he did attach great importance to the quantitative or "economic" aspect of psychoanalysis, endowing the mental images representing instinctual drives with definite quantities of emotional energy that could not be directly measured but could be inferred from the intensity of the manifest symptoms. The "mental energy exchange" was seen as a crucial aspect of all psychological conflicts. "The final outcome of the struggle," wrote Freud, "depends on *quantitative* relations."[26]

As in Newtonian physics so also in psychoanalysis, the mechanistic view of reality implies a rigorous determinism. Every psychological event has a definite cause and gives rise to a definite effect, and the whole psychological state of an individual is uniquely determined by "initial conditions" in early childhood. The "genetic" approach of psychoanalysis consists of tracing the symptoms and behavior of a patient back to previous developmental stages along a linear chain of cause-and-effect relations.

A closely related notion is that of the objective scientific observer. Classical Freudian theory is based on the assumption that the observation of the patient during analysis can take place without any interference or appreciable interaction. This belief is reflected in the basic arrangement of psychoanalytic practice, with the patient lying on the couch and the invisible therapist sitting behind her head maintaining a cold and uninvolved attitude, objectively observing the data. The Cartesian division between mind and matter, which is the philosophical origin of the concept of scientific objectivity, is reflected in psychoanalytic practice in the exclusive focus on mental processes. Physical consequences of psychological events are discussed during the psychoanalytic process, but the therapeutic technique itself does not involve any direct physical interventions. Freudian psychotherapy neglects the body just as medical therapy neglects the mind. The taboo against physical contact is so strong that some analysts do not even shake hands with their patients.

Freud himself was actually far less rigid in his psychoanalytic practice than in his theory. The theory had to adhere to the principle of scientific objectivity if it was to be accepted as a science, but in practice Freud was often able to transcend the limitations of the Newtonian framework. Being an excellent clinical observer, he recognized that his analytic observation represented a powerful intervention that induced significant changes in the patient's psychological condition. Prolonged analysis would even produce an entirely new clinical picture—the transference neurosis—which was not uniquely determined by the individual's early history but depended on the interaction between therapist and patient. This observation led Freud to abandon the ideal of the cool and uninvolved observer in his clinical work and to emphasize serious interest and sympathetic understanding. "Personal influence is our most powerful dynamic weapon," he wrote in 1926. "It is the new element we introduce into the situation and by means of which we make it fluid."[27]

The classical theory of psychoanalysis was the brilliant result of Freud's attempts to integrate his many revolutionary discoveries and ideas into a coherent conceptual framework that satisfied the criteria of the science of his time. Given the scope and depth of his work, it is not surprising that we can now recognize shortcomings in his approach which are due partly to the limitations inherent in the Cartesian-Newtonian framework and in part to Freud's own cultural conditioning. Recognizing these limitations of the psychoanalytic approach in no way diminishes the genius of its founder, but is crucial for the future of psychotherapy.

Recent developments in psychology and psychotherapy have begun to produce a new view of the human psyche, one in which the Freudian model is recognized as extremely useful for dealing with certain aspects, or levels, of the unconscious but as severely limiting when applied to the totality of mental life in health and illness. The situation is not unlike that in physics, where the Newtonian model is extremely useful for the description of a certain range of phenomena but has to be extended, and often radically changed, when we go beyond that range.

In psychiatry, some of the necessary extensions and modifications of the Freudian approach were pointed out even during Freud's lifetime by his immediate followers. The psychoanalytic movement had attracted many extraordinary individuals, some of whom formed an inner circle around Freud in Vienna. There was a rich intellectual exchange and much cross-fertilization of ideas in the inner circle, but also a considerable amount of conflict, tension, and dissent. Several of Freud's prominent disciples left the movement because of basic theoretical disagreements and started their own schools, emphasizing various modifications of the Freudian model. The most famous of these psychoanalytic renegades were Jung, Adler, Reich, and Rank.

The first to leave mainstream psychoanalysis was Alfred Adler, who developed what he called Individual Psychology. He rejected the dominant role of sexuality in the Freudian theory and put crucial emphasis on the will to power and the tendency to compensate for real or imaginary inferiority. Adler's study of the individual's role in the family led him to emphasize the social roots of mental disorders, which are generally neglected in classical psychoanalysis. Moreover, he was one of the first to formulate a feminist critique of Freud's views on female psychology.[28] He pointed out that what Freud called masculine and feminine psychologies were not so much rooted in biological differences between men and women but were essentially consequences of the social order prevailing under patriarchy.

The feminist critique of Freud's ideas on women was elaborated later on by Karen Horney and has since been discussed by many authors, both within and outside the field of psychoanalysis.[29] According to these critics, Freud took the masculine as the cultural and sexual norm and thus failed to reach a genuine understanding of the female psyche. Female sexuality in particular remained for him—in his own expressive metaphor—"the 'dark continent' for psychology."[30]

Wilhelm Reich broke with Freud because of conceptual differences which led him to formulate several unorthodox ideas that have had considerable influence on recent developments in psychotherapy. During his pioneering research in character analysis Reich discovered that mental attitudes and emotional experiences provoke resistances in the physical organism that are expressed in muscular patterns, resulting in what he called the "character armor." He also extended Freud's concept of libido by associating it with concrete energy flowing through the physical organism. Accordingly he emphasized the direct release of sexual energy in his therapy by breaking the Freudian taboo against touching the patient and by developing techniques of bodywork that many therapists are now elaborating upon.[31]

Otto Rank left the Freudian school after formulating a theory of psychopathology that put primary emphasis on the trauma of birth and regarded many of the neurotic patterns Freud had discovered as derivatives of the anxiety experienced during the birth process. In his analytic practice Rank moved directly to the anxiety-producing issue of birth and focused his therapeutic efforts on helping the patient to relive the traumatic event, rather than remembering and analyzing it. Rank's insights into the significance of the birth trauma were truly remarkable. It was not until several decades later that they were taken up again and further elaborated by psychiatrists and psychotherapists.

Among all Freud's disciples Carl Gustav Jung is probably the one who went furthest in expanding the psychoanalytic system. He was originally Freud's favorite student and was considered the crown prince of psychoanalysis, but he parted with his teacher because of irreconcilable theoretical difficulties that challenged the Freudian theory at its very core. Jung's approach to psychology has had a profound impact on subsequent developments in the field and will be discussed in great detail later on.[32] His basic concepts clearly transcended the mechanistic models of classical psychology and brought his science much closer to the conceptual framework of modern physics than any other psychological school. More than that, Jung was well aware that the rational approach of Freudian psychoanalysis would have to be transcended if psychologists wanted to explore the subtler aspects of the human psyche that lie far beyond our everyday experience.

The strictly rational and mechanistic approach made it especially difficult for Freud to deal with religious, or mystical, experiences. Although he showed a deep interest in religion and spirituality throughout his life, he never acknowledged mystical experience as their source. Instead he equated religion with ritual, seeing it as "an obsessive-compulsive neurosis of mankind" that reflected unresolved conflicts from infantile stages of psychosexual development. This limitation of Freudian thought has had a strong influence on subsequent psychoanalytic practice. In the Freudian model there is no room for experiences of altered states of consciousness that challenge all the basic concepts of classical science. Consequently, experiences of this nature, which occur spontaneously much more frequently than is commonly believed, have often been labeled as psychotic symptoms by psychiatrists who could not incorporate them into their conceptual framework.[33]

In this area especially an awareness of modern physics could have a very salutary effect on psychotherapy. The extension of their research to atomic and subatomic phenomena has led physicists to adopt concepts that contradict all our common-sense views, as well as the basic principles of Newtonian science, but are nevertheless scientifically sound. Knowledge of these concepts, and of their similarities to those of mystical traditions, may well make it easier for psychiatrists to go beyond the traditional Freudian framework when dealing with the full range of human consciousness.

7 · The Impasse of Economics

The triumph of Newtonian mechanics in the eighteenth and nineteenth centuries established physics as the prototype of a "hard" science against which all other sciences were measured. The closer scientists could come to emulating the methods of physics, and the more of its concepts they were able to use, the higher the standing of their discipline in the scientific community. In our century this tendency to model scientific concepts and theories after those of Newtonian physics has become a severe handicap in many fields, but more than anywhere else, perhaps, in the social sciences.* These have been traditionally regarded as the "softest" among the sciences, and social scientists have tried very hard to gain respectability by adopting the Cartesian paradigm and the methods of Newtonian physics. However, the Cartesian framework is often quite inappropriate for the phenomena they are describing, and consequently their models have become increasingly unrealistic. This is now especially apparent in economics.

Present-day economics is characterized by the fragmentary and reductionist approach that typifies most social sciences. Economists generally fail to recognize that the economy is merely one aspect of a whole ecological and social fabric; a living system composed of human beings in continual interaction with one another and with their natural resources, most of which are, in turn, living organisms. The basic error

* The social sciences deal with the social and cultural aspects of human behavior. They include the disciplines of economics, political science, sociology, social anthropology, and—in the view of many of its practitioners—history.

of the social sciences is to divide this fabric into fragments, assumed to be independent and to be dealt with in separate academic departments. Thus political scientists tend to neglect basic economic forces, while economists fail to incorporate social and political realities into their models. These fragmentary approaches are also reflected in government, in the split between social and economic policies and, especially in the United States, in the maze of congressional committees and subcommittees where these policies are discussed.

The fragmentation and compartmentalization in economics has been noted and criticized throughout its modern history. But at the same time those critical economists who wished to study economic phenomena as they actually existed, embedded within society and the ecosystem, and who therefore dissented from the narrow economic viewpoint, were virtually forced to place themselves outside of economic "science," thus saving the economics fraternity from dealing with the issues their critics raised. For example, Max Weber, the nineteenth-century critic of capitalism, is generally regarded as an economic historian; John Kenneth Galbraith and Robert Heilbroner are often thought of as sociologists; and Kenneth Boulding is referred to as a philosopher. Karl Marx, by contrast, refused to be called an economist and saw himself as a social critic, asserting that economists were merely apologists for the existing capitalist order. In fact, the term "socialist" originally merely described those who did not accept the economists' view of the world. More recently Hazel Henderson has continued this tradition by calling herself a futurist and subtitling one of her books "The End of Economics."[1]

Another aspect of economic phenomena, crucially important but severely neglected by economists, is the economy's dynamic evolution. In their dynamic nature the phenomena described by economics differ profoundly from those covered by the natural sciences. Classical physics applies to a well-defined and unchanging range of natural phenomena. Although it has to be replaced by quantum and relativistic physics beyond this range, the Newtonian model remains valid within the classical domain and continues to be an efficient theoretical basis for a large part of contemporary technology. Similarly, the concepts of biology apply to a reality that has changed very little over the centuries, although knowledge of biological phenomena has progressed considerably and much of the old Cartesian framework is now recognized as too restrictive. But biological evolution tends to proceed over very long time spans and, in general, does not produce entirely new phenomena but rather advances through a continual reshuffling and recombination of a limited number of structures and functions.[2]

The evolution of economic patterns, by contrast, takes place at a much faster pace. An economy is a continually changing and evolving system, dependent on the changing ecological and social systems in which it is embedded. To understand it we need a conceptual framework that is also capable of change and continual adaptation to new situations. Such a framework is sadly lacking in the work of most contemporary economists, who are still fascinated by the absolute rigor of the Cartesian paradigm and the elegance of Newtonian models, and so are increasingly out of touch with current economic realities.

The evolution of a society, including the evolution of its economic system, is closely linked to changes in the value system that underlies all its manifestations. The values a society lives by will determine its world view and religious institutions, its scientific enterprise and technology, and its political and economic arrangements. Once the collective set of values and goals has been expressed and codified, it will constitute the framework of the society's perceptions, insights, and choices for innovation and social adaptation. As the cultural value system changes—often in response to environmental challenges—new patterns of cultural evolution will emerge.

The study of values is thus of paramount importance for all social sciences; there can be no such thing as a "value-free" social science. Social scientists who consider the question of values "nonscientific" and think they are avoiding it are attempting the impossible. Any "value-free" analysis of social phenomena is based on the tacit assumption of an existing value system that is implicit in the selection and interpretation of data. By avoiding the issue of values, then, social scientists are not more scientific but, on the contrary, less scientific, because they neglect to state explicitly the assumptions underlying their theories. They are open to the Marxist critique that "all social sciences are ideologies in disguise."[3]

Economics is defined as the discipline dealing with the production, distribution, and consumption of wealth. It attempts to determine what is valuable at a given time by studying the relative exchange values of goods and services. Economics is therefore the most clearly value-dependent and normative among the social sciences. Its models and theories will always be based on a certain value system and on a certain view of human nature; on a body of assumptions that E. F. Schumacher calls "meta-economics" because it is rarely included explicitly in contemporary economic thought.[4] Schumacher has illustrated the value dependence of economics very eloquently by comparing two economic systems embodying entirely different values and goals.[5] One is our present materialist system, in which the "standard of living" is measured by the amount of annual consumption, and which therefore tries to achieve the maximum consumption along with an optimal pattern of production. The other is a system of Buddhist economics, based on the notions of "right livelihood" and the "Middle Way," in which the aim is to achieve a maximum of human well-being with an optimal pattern of consumption.

Contemporary economists, in a misguided attempt to provide their discipline with scientific rigor, have consistently avoided the issue of unstated values. Kenneth Boulding, speaking as president of the American Economic Association, has called this concerted attempt "a monumentally unsuccessful exercise . . . which has preoccupied a whole generation of economists (indeed, several generations) with a dead end, to the almost total neglect of the major problems of our age."[6] The evasion of value-related issues has led economists to retreat to easier but less relevant problems, and to disguise value conflicts by using elaborate technical language. This trend is particularly strong in the United States, where there is now a widespread belief that all problems—economic, political, or social—have technical solutions. Thus industry and business hire armies of economists to prepare cost/benefit analyses that convert social and moral choices into pseudotechnical ones and thereby conceal value conflicts that can only be resolved politically.[7]

The only values appearing in current economic models are those that can be quantified by being assigned monetary weightings. This emphasis on quantification gives economics the appearance of an exact science. At the same time, however, it severely restricts the scope of economic theories by excluding qualitative distinctions that are crucial to understanding the ecological, social, and psychological dimensions of economic activity. For example, energy is measured only in kilowatts, regardless of its origins; no distinction is made between renewable and nonrenewable goods; and the social costs of production are added, incomprehensibly, as positive contributions to the gross national product. Furthermore, economists have completely disregarded psychological research on people's behavior as income earners, consumers, and investors because the results of such research cannot be integrated into the current quantitative analyses.[8]

The fragmentary approach of contemporary economists, their preference for abstract quantitative models, and their neglect of the economy's structural evolution, have resulted in a tremendous gap between theory and economic reality. In the *Washington Post*'s opinion, "Ambitious economists elaborate elegant mathematical solutions to theoretical problems with little if any relevance to public issues."[9] Economics today is in a profound conceptual crisis. The social and economic anomalies it can no longer address—global inflation and unemployment, maldistribution of wealth, and energy shortages, among others—are now painfully visible to everyone. The failure of the economic profession to come to terms with these problems is recognized by an increasingly skeptical public, by scientists from other disciplines, and by economists themselves.

Opinion polls taken during the 1970s have consistently shown a drastic decline in the American public's confidence in its business institutions. Thus the percentage of people who believe that major companies have become too powerful rose to 75 in 1973; in 1974, 53 percent thought many major companies should be dismantled, and over half of the American citizenry wanted more federal regulation of utilities, insurance companies, and the oil, drug, and automobile industries.[10]

Attitudes are also shifting within the corporations. According to a study published in 1975 in the *Harvard Business Review*, 70 percent of the corporate executives questioned preferred the old ideologies of individualism, private property, and free enterprise, but 73 percent believed that these values would be superseded by collective models of problem-solving during the next ten years, and 60 percent thought such a collective orientation would be more effective in finding solutions.[11]

And economists themselves are beginning to acknowledge that their discipline has reached an impasse. In 1971 Arthur Burns, then in the chair of the Federal Reserve Board, remarked that "the rules of economics are not working quite the way they used to,"[12] and Milton Friedman, addressing the American Economic Association in 1972, was even more frank: "I believe that we economists in recent years have done vast harm—to society at large and to our profession in particular—by claiming more than we can deliver."[13] By 1978 the tone had changed from caution to despair when Treasury Secretary Michael Blumenthal declared: "I really think the economics profession is close to bankruptcy in understanding the present situation, before or after the fact."[14] Juanita Kreps, outgoing Secretary of Commerce in 1979, said flatly that she found it impossible to go back to her old job as professor of economics at Duke University, because "I would not know what to teach."[15]

The current mismanagement of our economy calls into question the basic concepts of contemporary economic thought. Most economists, although acutely aware of the current state of crisis, still believe that solutions to our problems can be found within the existing theoretical framework. This framework, however, is based on concepts and variables that originated several hundred years ago and have been hopelessly outdated by social and technological changes. What economists need to do most urgently is reevaluate their entire conceptual foundation and redesign their basic models and theories accordingly. The current economic crisis will be overcome only if economists are willing to participate in the paradigm shift that is now occurring in all fields. As in psychology and medicine, the shift from the Cartesian paradigm to a holistic and ecological vision will not make the new approaches any less scientific, but on the contrary will make them more consistent with recent developments in the natural sciences.

At the deepest level, reexamination of economic concepts and models needs to deal with the underlying value system and to recognize its relation to the cultural context. From such a perspective many of the current social and economic problems are seen to have their roots in the painful adjustments of individuals and institutions to the shifting values of our time.[16] The emergence of economics as a separate discipline from philosophy and politics coincided with the emergence of Western Europe's sensate culture at the end of the Middle Ages. As this culture unfolded, it embodied in its social institutions the masculine, and "yang-oriented" values that now dominate our society and form the basis of our economic system. Economics, with its basic focus on material wealth, is today the quintessential expression of sensate values.[17]

Attitudes and activities that are highly valued in this system include material acquisition, expansion, competition, and an obsession with "hard technology" and "hard science." In overemphasizing these values, our society has encouraged the pursuit of goals that are both dangerous and unethical, and has institutionalized several of the sins known in Christianity as deadly—gluttony, pride, selfishness, and greed.

The value system that developed during the seventeenth and eighteenth centuries gradually replaced a coherent set of medieval values and attitudes—belief in the sacredness of the natural world; moral strictures against money-lending for interest; the requirement that prices should be "just"; convictions that personal gain and hoarding should be discouraged, that work was for the use value of the group and the well-being of the soul, that trade was justified only to restore the group's sufficiency, and that all true rewards were in the next world. Until the sixteenth century there was no isolation of purely economic phenomena from the fabric of life. Throughout most of history food, clothing, shelter, and other basic resources were produced for use value and were distributed within tribes and groups on a reciprocal basis.[18] A national system of markets is a relatively recent phenomenon that arose in seventeenth-century England and spread from there over the entire world, resulting in today's interlinked "global marketplace." Markets, of course, had existed since the Stone Age, but they were based on barter, not cash, and thus were bound to be local. Even early trading had little economic motivation but was more often a sacred and ceremonial activity related to kinship and family customs. For example, the Trobriand Islanders of the Southwest Pacific undertook circular voyages along sea-trading routes that stretched for thousands of miles, without significant profit, barter, or exchange motives. Their incentive was the etiquette and magic symbolism of carrying white sea-shell jewelry in one direction and red sea-shell ornaments in the other, so as to encircle their entire archipelago every ten years.[19]

Many archaic societies used money, including metal currencies, but these were for payment of taxes and salaries, not for general circulation. The motive of individual gain from economic activities was generally absent; the very idea of profit, let alone interest, was either inconceivable or banned. Economic organizations of great complexity, involving an elaborate division of labor, were operated entirely by the mechanism of storing and redistribution of common commodities, such as grain, as indeed were all systems of feudalism. Of course this did not preclude the age-old motives of power, domination, and exploitation, but the idea that human needs were boundless was generally not held before the Enlightenment.

An important principle in all early societies was that of "householding," the Greek *oikonomia*, which is the root of our modern term "economics." Private property was justified only to the extent that it served the welfare of all. In fact, the word "private" comes from the Latin *privare* ("to deprive"), which shows the widespread ancient view that property was first and foremost communal. As societies moved from this communal, participatory viewpoint to more individualistic and self-assertive views, people no longer thought of private property as those goods that individuals deprived the group from using, but actually inverted the meaning of the term, holding that property should be private in the first place and that society should not deprive the individual without due process of law.

With the Scientific Revolution and the Enlightenment, critical reasoning, empiricism, and individualism became the dominant values, together with a secular and materialistic orientation that led to the production of worldly goods and luxuries, and to the manipulative mentality of the Industrial Age. The new customs and activities resulted in the creation of new social and political institutions and gave rise to a new academic pursuit: the theorizing about a set of specific *economic* activities—production, exchange, distribution, money-lending—which suddenly stood out in sharp relief and required not only description and explanation but also rationalization.

One of the most important consequences of the shift of values at the end of the Middle Ages was the rise of capitalism in the sixteenth and seventeenth centuries. The development of the capitalist mentality, according to an ingenious thesis by Max Weber, was closely related to the religious idea of a "calling," which emerged with Martin Luther and the Reformation, together with the notion of a moral obligation to fulfill one's duty in worldly endeavors. This idea of a worldly calling projected religious behavior into the secular world. It was emphasized even more forcefully by the Puritan sects, which saw worldly activity and the material rewards resulting from industrious behavior as a sign of divine predestination. Thus arose the famous Protestant Work Ethic, in which hard, self-denying work and worldly success were equated with virtue. On the other hand, the Puritans abhorred all but the most frugal consumption, and consequently the accumulation of wealth was sanctioned, as long as it was combined with an industrious career. In Weber's theory these religious values and motives provided the essential emotional drive and energy for the rise and rapid development of capitalism.[20]

The Weberian tradition of criticizing economic activities on the basis of an analysis of their underlying values provided avenues for many later critics, among them Kenneth Boulding, Erich Fromm, and Barbara Ward.[21] Continuing this tradition, but going to an even deeper level, the recent feminist critique of economic systems—both capitalist and Marxist—has focused on the patriarchal value system that underlies virtually all of today's economies.[22] The connection between patriarchal values and capitalism was pointed out in the nineteenth century by Friedrich Engels, and has been emphasized by succeeding generations of Marxists. For Engels, however, the oppression of women had its roots in the capitalist economic system and would come to an end with the overthrow of capitalism. What feminist critics are pointing out forcefully today is that patriarchal attitudes are far older than capitalist economies and much more deeply ingrained in most societies. Indeed, the majority of socialist and revolutionary movements exhibit an overwhelming male bias, promoting social revolutions that leave male leadership and control essentially untouched.[23]

During the sixteenth and seventeenth centuries, while the new values of individualism, property rights, and representative government led to the decline of the traditional feudal system and eroded the power of the aristocracy, the old economic order was still defended by theorists who believed that a nation's path to riches was in the accumulation of money through foreign trade. This theory was given the name "mercantilism" later on. Its practitioners did not call themselves economists but were politicians, administrators, and merchants. They applied the ancient notion of economy—in the sense of managing a household—to the state as the household of the ruler, and thus their policies became known as "political economy." This term remained in use until the twentieth century, when it was replaced by the modern term economics.

The mercantilist idea of the balance of trade—the belief that a nation will grow rich when its exports exceed its imports—became a central concept of subsequent economic thought. It was undoubtedly influenced by the concept of equilibrium in Newtonian mechanics and was quite consistent with the limited world views of the insular, sparsely populated monarchies of its time. But today, in our overpopulated and tightly interdependent world, it is evident that not all nations can win simultaneously at the mercantilist game. The fact that many nations—most recently, notably, Japan—still attempt to maintain trade balances in their favor is bound to lead to trade wars, depressions, and international conflict.

Modern economics, strictly speaking, is a little over three hundred years old. It was founded in the seventeenth century by Sir William Petty, professor of anatomy at Oxford and of music at London, as well as physician to the army of Oliver Cromwell. Among his circle of friends were Christopher Wren, the architect of many London landmarks, and Isaac Newton. Petty's *Political Arithmetick* seemed to owe much to Newton and to Descartes, its method consisting of replacing words and arguments by numbers, weights, and measures, and "to use only Arguments of Sense and to consider only such Causes as have visible Foundations in Nature."[24]

In this and other works, Petty put forth a set of ideas that became indispensable ingredients of the theories of Adam Smith and other later economists. Among these ideas were the labor theory of value—adopted by Smith, Ricardo, and Marx—according to which the value of a product is derived only from the human labor required to produce it; and the distinction between price and value, which, in various formulations, has preoccupied economists ever since. Petty also expounded the notion of "just wages," described the advantages of the division of labor, and defined the concept of monopoly. He discussed the "Newtonian" notions of the quantity of money and its velocity in circulation, which are still debated by the monetarist school today, and suggested public works as a remedy for unemployment, thus anticipating Keynes by more than two centuries. Today's economic policies, as they are debated in Washington, Bonn, or London, would not be any surprise to Petty, except for the fact that they have changed so little.

Along with Petty and the mercantilists, John Locke helped lay the foundation stones of modern economics. He was the outstanding philosopher of the Enlightenment, and his ideas about psychological, social, and economic phenomena—strongly influenced by Descartes and Newton—became the core of eighteenth-century thought. Locke's atomistic theory of human society[25] led him to the idea of a representative government whose function it was to safeguard individuals' rights to property and to the fruits of their labor. Locke held that once individuals created a government as the trustee of their rights, liberties, and property, its legitimacy depended on protecting these rights. If the government failed to do so, the people should have the power to dissolve it. A number of economic and political theories were influenced by these radical moral concepts of the Enlightenment. In economics, however, one of Locke's most innovative theories had to do with prices. Whereas Petty had held that prices and commodities should reflect justly the amount of labor they embodied, Locke came up with the idea that prices were also determined objectively, by demand and supply. This not only liberated the merchants of the day from the moral law of "just" prices; it also became another cornerstone of economics and was elevated to equal status with the laws of mechanics, where it stands even today in most economic analyses.

The law of supply and demand also fit perfectly with the new mathematics of Newton and Leibniz—the differential calculus—since economics was perceived as dealing with continuous variations of very small quantities which could be described most efficiently by this mathematical technique. This notion became the basis of subsequent efforts to turn economics into an exact mathematical science. However, the problem was—and is—that the variables used in these mathematical models cannot be rigorously quantified but are defined on the basis of assumptions that often make the models quite unrealistic.

A distinct school of eighteenth-century thought that had a significant influence on classical economic theory, and notably on Adam Smith, was that of the French Physiocrats. These thinkers were the first to call themselves economists, to regard their theories as "objectively" scientific, and to develop a complete view of the French economy as it existed just before the Revolution. Physiocracy meant "the rule of nature" and the Physiocrats bitterly criticized mercantilism and the growth of cities. They claimed that only agriculture and the land were truly productive of real wealth, thus promoting an early "ecological" view. Their leader, like William Petty and John Locke, was a physician, François Quesnay, who was surgeon to the Royal Court. Quesnay expounded on the idea that Natural Law, if left unimpeded, would govern economic affairs for the greatest benefit of all. Thus the doctrine of laissez faire was introduced as another keystone of economics.

The period of "classical political economy" was inaugurated in 1776, when Adam Smith published *An Inquiry into the Nature and*

Causes of the Wealth of Nations. Smith, a Scottish philosopher and friend of David Hume, was by far the most influential of all economists. His *Wealth of Nations* was the first full-scale treatise on economics and has been called "in its ultimate results, probably the most important book that has ever been written."[26] Smith was not only influenced by the Physiocrats and the philosophers of the Enlightenment, but was also friendly with James Watt, the inventor of the steam engine, met Benjamin Franklin and probably Thomas Jefferson, and lived at a time when the Industrial Revolution had begun to change the face of Britain. When Smith wrote *Wealth of Nations* the transition from an agrarian, handicraft economy to one dominated by steam power and by machines operated in large factories and mills was well under way. The spinning jenny had been invented and machine looms were used in cotton factories employing up to three hundred workers. The new private enterprise, factories, and power-driven machinery shaped Smith's ideas, so that he enthusiastically advocated the social transformation of his time and criticized the remnants of the land-based feudal system.

Like most of the great classical economists, Adam Smith was not a specialist but a broad, imaginative thinker with many fresh insights. He set out to investigate how the wealth of a nation is increased and distributed—the basic theme of modern economics. In countering the mercantilist view that wealth is increased by foreign trade and by hoarding gold and silver bullion, Smith held that its true base is production resulting from human labor and natural resources. A nation's wealth would depend on the percentage of its people engaged in such production and on their efficiency and skill. The basic means of increasing production was the division of labor, Smith contended, as Petty had before him. From the prevailing Newtonian idea of natural law Smith deduced that it was "human nature to barter and exchange," and he also thought it "natural" that workers would gradually have to facilitate their work and improve their productivity with the help of labor-saving machinery. At the same time the early manufacturers had a much darker view of the role of machines; they well understood that machines could replace workers and thus could be used to keep them afraid and docile.[27]

From the Physiocrats Smith adopted the theme of laissez faire, which he immortalized in the metaphor of the Invisible Hand. According to Smith, the Invisible Hand of the market would guide the individual self-interest of all entrepreneurs, producers, and consumers for the harmonious betterment of all; "betterment" being equated with the production of material wealth. In this way a social result would be achieved that was independent of individual intentions, and thus an objective science of economic activity was made possible.

Smith believed in the labor theory of value, but he also accepted the idea that prices would be determined in "free" markets by the balancing effects of supply and demand. He based his economic theory on the Newtonian notions of equilibrium, laws of motion, and scientific objectivity. One of the difficulties in applying these mechanistic concepts to social phenomena was the lack of appreciation for the problem of friction. Because the phenomenon of friction is generally neglected in Newtonian mechanics, Smith imagined that the balancing mechanisms of the market would be almost instantaneous. He described their adjustments as "prompt," "occurring soon," and "continual," while prices were "gravitating" in the proper direction. Small producers and small consumers would meet in the marketplace with equal power and information.

This idealistic picture underlies the "competitive model" widely used by economists today. Its basic assumptions include perfect and free information for all participants in a market transaction; the belief that each buyer and seller in a market is small and has no influence on price; and the complete and instant mobility of displaced workers, natural resources, and machinery. All these conditions are violated in the vast majority of today's markets, yet most economists continue to use them as the basis of their theories. As Lucia Dunn, professor of economics at Northwestern University, describes the situation, "They use these assumptions in their work almost unconsciously. In fact, in many economists' minds, they have ceased to be assumptions and have become a picture of how the world really is."[28]

For international trade Smith developed the doctrine of comparative advantage, according to which each nation should excel in some types of production, the result being an international division of labor and free trade. This model of international free trade still underlies much of today's thinking on the global economy and is now producing its own set of social and environmental costs.[29] Within a nation Smith thought that the self-balancing market system was one of slow and steady growth, with continually increasing demands for goods and labor. The idea of continual growth was adopted by succeeding generations of economists, who, paradoxically, continued to use mechanistic equilibrium assumptions while at the same time postulating continuing economic growth. Smith himself predicted that economic progress would eventually come to an end when the wealth of nations had been pushed to the natural limits of soil and climate, but unfortunately he thought this point was so far in the future that it was irrelevant to his theories.

Smith alluded to the idea of the growth of social and economic structures such as monopolies when he denounced people in the same trade who conspired to raise prices artificially, but he did not see the broad implications of such practices. The growth of these structures, and in particular of the class structure, was to become a central theme in Marx's economic analysis. Adam Smith justified capitalists' profits by arguing that they were needed to invest in more machines and factories for the common good. He noted the struggle between workers and employers and the efforts of both "to interfere with the market," but he never referred to the unequal power of workers and capitalists—a point that Marx would drive home with force.

When Smith wrote that workers and "other inferior ranks of people" produced too many children which would cause wages to fall to the level of mere subsistence, he showed that his views of society were similar to those of other Enlightenment philosophers. Their educated middle-class status allowed them to conceive of radical ideas of equality, justice, and liberty, but did not allow them to extend these concepts to include the "inferior classes"; nor did they ever include women.

At the beginning of the nineteenth century, economists began to systematize their discipline in an attempt to cast it into the form of a science. The first and most influential among these systematic economic thinkers was David Ricardo, a stockbroker who became a multimillionaire at the age of thirty-five and then, after reading *Wealth of Nations,* devoted himself to the study of political economy. Ricardo built on the work of Adam Smith but defined a narrower scope for economics and thus began a process that would become characteristic of most subsequent non-Marxist economic thought. Ricardo's work contained very little social philosophy and instead introduced the concept of an "economic model," a logical system of postulates and laws, involving a limited number of variables, that could be used to describe and predict economic phenomena.

Central to the Ricardian system was the idea that progress would sooner or later come to an end because of the rising cost of growing food on a limited area of land. Underlying this ecological perspective was the gloomy view, evoked earlier by Thomas Malthus, that the population would increase faster than the food supply. Ricardo accepted the Malthusian principle but analyzed the situation in greater detail. He wrote that as the population increased, poorer marginal land would have to be cultivated. At the same time the relative value of the superior land would increase, and the higher rent charged for it would be a surplus received by the landlords for merely owning the land. This concept of "marginal" land became the basis for today's economic schools of marginal analysis. Ricardo, like Smith, accepted the labor theory of value but, significantly, included in his definition of prices the cost of the labor required to build machines and factories. In his view the owner of a factory, in receiving profit, was taking something that labor had produced, a point upon which Marx built his theory of surplus value.

The systematic efforts of Ricardo and other classical economists consolidated economics into a set of dogmas that supported the existing class structure and countered all attempts at social improvement with the "scientific" argument that "laws of nature" were operating and the poor were responsible for their own misfortune. At the same time, workers' uprisings were becoming frequent and the new body of economic thought engendered its own horrified critics long before Marx.

One well-meaning but unrealistic approach led to a long series of unworkable formulations known later as welfare economics. The proponents of this school shifted their focus from the earlier view of welfare as material production to the subjective criteria of individual pleasure and pain, constructing elaborate charts and curves based on "units of pleasure" and "units of pain." Vilfredo Pareto improved these rather crude schemes with his theory of optimality, based on the assumption that social welfare would be increased if the satisfaction of some individuals could be increased without decreasing that of others. In other words, any economic change that made someone "better off" without making anyone "worse off" was desirable for social welfare. However,

Pareto's theory still neglected the facts of unequal power, information, and income. Welfare economics has persisted to the present day, although it has been shown conclusively that individual preferences cannot be made to add up to social choice.[30] Many contemporary critics see it as a thinly disguised excuse for selfish behavior that undermines any cohesive set of social goals and is now playing havoc with environmental policies.[31]

While welfare economists were constructing elaborate mathematical schemes, another school of reformers tried to counter the deficiencies of capitalism in frankly idealistic experiments. These Utopians set up factories and mills according to humanitarian principles—with reduced working hours, increased wages, recreation, insurance, and sometimes housing—founded workers' cooperatives, and promoted ethical, esthetic, and spiritual values. Many of these experiments were very successful for a while, but all of them ultimately failed, unable to survive in a hostile economic environment. Karl Marx, who owed much to the imagination of the Utopians, believed that their communities could not last, since they had not emerged "organically" from the existing stage of material economic development. From the perspective of the 1980s, it seems that Marx may well have been right. Perhaps we had to wait until today's "post-industrial" weariness with mass consumption and awareness of the mounting social and environmental costs, not to mention the diminishing resource base, to reach conditions in which the Utopians' dream of a cooperative-based, ecologically harmonious social order can become reality.

The greatest among the classical economic reformers was John Stuart Mill, who joined in the social criticism, having absorbed most of the work of the philosophers and economists of his time by the time he was thirteen. In 1848 he published his own *Principles of Political Economy*, a herculean reassessment that came to a radical conclusion. Economics, he wrote, had only one province—production and the scarcity of means. Distribution was not an economic but a political process. This narrowed the scope of political economy to a "pure economics," later to be called "neoclassical," and allowed a more detailed focus on the "economic core process," while excluding social and environmental variables in analogy to the controlled experiments of the physical sciences. After Mill, economics became split between the neoclassical, "scientific," and mathematical approach, on the one hand, and the "art" of broader social philosophy on the other. Eventually this split led to today's disastrous confusion between the two approaches, resulting in policy tools that are derived from abstract, unrealistic mathematical models.

John Stuart Mill meant well with his emphasis on the political nature of all economic distribution. His pointing out that the distribution of a society's wealth depended on the laws and customs of that society, which were very different in different cultures and ages, should have forced the issue of values back onto the agenda of political economy. Mill not only saw the ethical choices at the heart of economics, but was also keenly aware of their psychological and philosophical implications.

Anyone who seriously tries to understand the social condition of humankind has to deal with the thought of Karl Marx and will experience its continuing intellectual fascination. According to Robert Heilbroner, this fascination is rooted in the fact that Marx was "the first to discover a whole mode of inquiring that would forever after belong to him. This was done previously only once, when Plato 'discovered' the mode of philosophical inquiry."[32] Marx's mode of inquiry was that of social critique, and that was why he referred to himself not as philosopher, historian, or economist—though he was all of those—but as a social critic. It is also why his social philosophy and science continue to exert a strong influence on social thought.

As a philosopher Marx taught a philosophy of action. "The philosophers," he wrote, "have only *interpreted* the world, in various ways; the point, however, is to *change* it."[33] As an economist Marx criticized classical economics more expertly and efficiently than any of its practitioners. His main influence, however, has been not intellectual but political. As a revolutionary, if judged by the number of worshiping followers, "Marx must be considered a religious leader to rank with Christ or Mohammed."[34]

While Marx the revolutionary has been canonized by millions all over the world, economists have had to deal with—but more often ignore or misquote—his embarrassingly accurate predictions, among them the occurrence of "boom" and "bust" business cycles and the tendency of market-oriented economies to develop "reserve armies" of unemployed, which today usually consist of ethnic minorities and women. Marx's main body of work, set forth in his three-volume *Das*

Kapital, represents a thorough critique of capitalism. He viewed society and economics from the explicitly stated perspective of the struggle between workers and capitalists, but his broad ideas about social evolution allowed him to see economic processes in much larger patterns.

Marx recognized that capitalist forms of social organization would speed the process of technological innovation and increase material productivity, and he predicted that this, dialectically, would change social relationships. Thus he was able to foresee phenomena like monopolies and depressions and to predict that capitalism would foster socialism—as it has—and that it would, eventually, disappear—as it may. In the first volume of *Kapital,* Marx stated his indictment of capitalism in the following words:

> Hand in hand with [the] centralization [of capital] . . . develop, on an ever-extending scale . . . the entanglement of all peoples in the net of the world-market, and with this, the international character of the capitalistic régime. Along with the constantly diminishing number of the magnates of capital, who usurp and monopolize all advantages of this process of transformation, grows the mass of misery, oppression, slavery, degradation, exploitation . . .[35]

Today, in the context of our crisis-ridden, corporate-dominated global economy with its megarisk technologies and its enormous social and ecological costs, this statement has lost none of its power.

It is generally pointed out by Marx's critics that the labor force in the United States, which one would have expected to be the first to organize politically and rise up to create a socialist society, failed to do so because workers received high enough wages to begin identifying with the upward mobility of the middle class. But, there are several other explanations for the failure of socialism to take hold in the United States.[36] American workers were extremely transient, moving with their jobs along a changing frontier; they were divided by their languages and other ethnic differences which factory owners did not fail to exploit; and enormous numbers of them went back to the old country as soon as they acquired the means to provide a better life for their waiting families. Thus the opportunities for organizing a European-type socialist party were very limited. On the other hand, it is true that

American workers have not been continually immiserated but have ridden the escalator of material wealth, although at relatively low levels and with much struggle.

Another important point is that in the late twentieth century the Third World has taken on the role of the proletariat because of the development of multinational corporations, which Marx did not foresee. Today these multinationals play off workers in one country against those in another, exploiting racism, sexism, and nationalism. Thus advantages won by American workers are generally to the detriment of those in Third World countries; the Marxist slogan "Workers of the world unite!" has become even more difficult to achieve.

In his "Critique of Political Economy," as he subtitled *Das Kapital*, Marx used the labor theory of value to raise issues of justice, and developed powerful new concepts to counter the reductionist logic of the neoclassical economists of his time. He knew that to a large extent wages and prices are politically determined. Starting from the premise that human labor creates all values, Marx observed that continuing and reproducing labor must, at the very least, produce subsistence for the worker plus enough to replace the materials used up. But in general, there will be a surplus over and above that minimum. The form this "surplus value" takes will be a key to the structure of society, to both its economy and its technology.[37]

In capitalist societies, Marx pointed out, the surplus value is appropriated by capitalists, who own the means of production and determine the conditions of labor. This transaction between people of unequal power allows the capitalists to make more money from the labor of the workers, and thus money is turned into capital. In this analysis Marx emphasized that the precondition for capital to arise was a specific social class relationship, itself the product of a long history.[38] Marx's basic critique of neoclassical economics, as valid today as it was then, is that economists, by narrowing their field of inquiry to the "economic core process," evade the ethical issue of distribution. As the non-Marxist economist Joan Robinson has put it, they shifted "from a measure of value . . . to the much less burning question of relative prices."[39] Value and prices, however, are very different concepts. Another non-Marxist, Oscar Wilde, said it best: "It's possible to know the price of everything and the value of nothing."

Marx was not rigid in his labor theory of value but always seemed to allow for change. He predicted that labor would become more "mental," as science and knowledge were increasingly applied to the production process, and he also acknowledged the important role of natural resources. Thus he wrote in his early *Economic and Philosophic Manuscripts*, "The worker can create nothing without nature, without the sensuous, external world. It is the material on which his labor is manifested, in which it is active, from which and by means of which it produces."[40]

In Marx's time, when resources were plentiful and the population was small, human labor was indeed the most important contribution to production. But as the twentieth century unfolded, the labor theory of value made less sense, and today the production process has become so complex that it is no longer possible to neatly separate the contributions of land, labor, capital, and other factors.

Marx's view of the role of nature in the process of production was part of his organic perception of reality, as Michael Harrington has emphasized in his persuasive reassessment of Marxian thought.[41] This organic, or systems view is often overlooked by Marx's critics, who claim that his theories are exclusively deterministic and materialistic. In dealing with the reductionist economic arguments of his contemporaries, Marx fell into the trap of expressing his ideas in "scientific" mathematical formulas that undermined his larger sociopolitical theory. But that larger theory consistently reflected a keen awareness of society and nature as an organic whole, as in this beautiful passage from the *Economic and Philosophic Manuscripts*:

> Nature is man's inorganic body—nature, that is, in so far as it is not itself the human body. "Man lives on nature" means that nature is his body, with which he must remain in continuous intercourse if he is not to die. That man's physical and spiritual life is linked to nature means simply that nature is linked to itself, for man is part of nature.[42]

Marx emphasized the importance of nature in the social and economic fabric throughout his writings, but it was not the central issue for an activist of the day. Nor was ecology a burning problem then, and Marx could not have been expected to emphasize it strongly. But he was aware of the ecological impact of capitalist economics, as we can see in many of his statements, however incidental they may be. To quote just one example, "All progress in capitalist agriculture is progress in the art, not only of robbing the laborer but of robbing the soil."[43]

It seems, then, that although Marx did not strongly emphasize ecological concerns, his approach *could* have been used to predict the ecological exploitation that capitalism produced and socialism perpetuated. One can certainly fault his followers for not grasping the ecological issue earlier, since it provided yet another devastating critique of capitalism and confirmed the vigor of the Marxian method. Of course, if Marxists had faced the ecological evidence honestly, they would have been forced to conclude that socialist societies had not done much better, their ecological impact being diminished only by their lower consumption (which in any case they were trying to increase).

Ecological knowledge is subtle and difficult to use as a basis for social activism, since other species—whether whales, redwoods, or insects—do not provide revolutionary energies to change human institutions. This is probably why Marxists have ignored the "ecological Marx" for so long. Recent scholarship has brought to light some of the subtleties in Marx's organic thinking, but these are inconvenient for most social activists, who prefer to organize around simpler issues. Perhaps this is why Marx finally stated at the end of his life, "I am not a Marxist."[44]

Marx, like Freud, had a long and rich intellectual life, with many creative insights that have decisively shaped our age. His social critique has inspired millions of revolutionaries around the world, and the Marxian economic analysis is respected academically not only in the socialist world but also in most European countries, as well as Canada, Japan, and Africa—in fact, in virtually the entire world except the United States. Marxian thought is capable of a wide range of interpretations and thus continues to fascinate scholars. Of particular interest for our analysis is the relation of the Marxian critique to the reductionist framework of the science of its time.

Like most nineteenth-century thinkers, Marx was very concerned about being scientific, using the term "scientific" constantly in the description of his critical approach. Accordingly he often attempted to formulate his theories in Cartesian and Newtonian language. Still, his broad view of social phenomena allowed him to transcend the Cartesian framework in significant ways. He did not adopt the classical stance of the objective observer, but fervently emphasized his role as participator by asserting that his social analysis was inseparable from social critique. In his critique he went beyond social issues and often revealed deep humanistic insights, for example in his discussion of the concept of alienation.[45] Finally, although Marx often argued for technological determinism, which made his theory more acceptable as a science, he also had profound insights into the interrelatedness of all phenomena, seeing society as an organic whole in which ideology and technology were equally important.

By the middle of the nineteenth century, classical political economy had branched into two broad streams. On one side were the reformers: Utopians, Marxists, and the minority of classical economists who followed John Stuart Mill. On the other side were the neoclassical economists, who concentrated on the economic core process and developed the school of mathematical economics. Some of them tried to establish objective formulas for the maximization of welfare, others retreated into ever more abstruse mathematics to escape the devastating critiques of the Utopians and the Marxists.

Much of mathematical economics was—and is—devoted to studying the "market mechanism" with the help of curves for demand and supply, always expressed as functions of prices and based on various assumptions about economic behavior, many of them highly unrealistic in today's world. For example, perfect competition in free markets, as postulated by Adam Smith, is assumed in most models. The essence of the approach can be illustrated by the basic supply-demand graph featured in all introductory economics textbooks (p. 210).

The interpretation of this graph is based on the Newtonian assumption that the participants in a market will "gravitate" automatically (and of course without any "friction") to the "equilibrium" price given by the point of intersection between the two curves.

As mathematical economists refined their models during the late nineteenth and early twentieth centuries, the world economy headed for the worst depression in its history, which shook the foundations of capitalism and seemed to verify all the Marxian predictions. However, after the Great Depression there was one more turn of the wheel of capitalism's fortunes, stimulated by the social and economic interventions of governments. These policies were based on the theory of John Maynard Keynes, who had a decisive influence on modern economic thought.

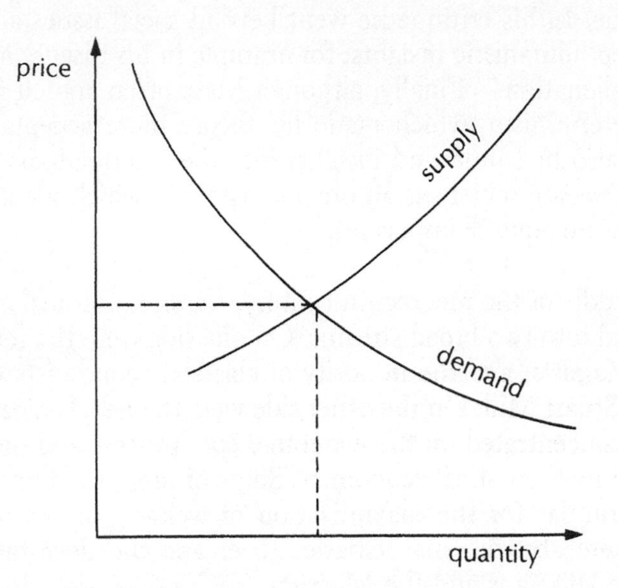

Supply-demand graph: supply curve gives the number of units of a product put on the market as a function of the product's price—the higher the price, the more producers will be attracted to producing this particular product; demand curve shows the demand for the product as a function of its price—the higher the price, the smaller the demand.

Keynes was keenly interested in the entire social and political scene and viewed economic theory as an instrument of policy. He bent the so-called value-free methods of neoclassical economics to serve instrumental purposes and goals, and in doing so made economics once again political, but this time in a new way. This, of course, involved giving up the ideal of the objective scientific observer, which neoclassical economists were very reluctant to do. But Keynes calmed their fears of interfering with the balancing operations of the market system by showing them that he could *derive* his policy interventions from the neoclassical model. To do this he demonstrated that economic equilibrium states were "special cases," exceptions rather than the rule in the real world.

To determine the nature of government interventions, Keynes shifted his focus from the microlevel to the macrolevel—to economic variables like the national income, total consumption and total investment, the total volume of employment, and so on. By establishing simplified relations between these variables, he was able to show that they were susceptible to short-term changes that could be influenced by appropriate policies. According to Keynes, these fluctuating business cycles were an intrinsic property of national economies. This theory was in opposition to orthodox economic thought, which postulated full employment, but Keynes defended his heresy by appealing to experience, pointing out that it was "an outstanding characteristic of the economic system in which we live that . . . it is subject to severe fluctuations with respect to output and employment."[46]

In the Keynesian model additional investment will always increase employment, and thus increase the total level of income, which will in turn lead to higher demand for consumer goods. In this way investment stimulates economic growth and increases national wealth, which will, eventually, "trickle down" to the poor. However, Keynes never said that this process would culminate in full employment; it will only move the system in that direction—*or* peter out at some level of underemployment, *or* even go into reverse, depending on many assumptions that are not part of the Keynesian model.

This explains the crucial role of advertising as a means for big companies to "manage" the demand in the marketplace. Consumers must not only keep increasing their spending; they must do so *predictably* for the system to work. Today the classical economic theory has almost been turned on its head. Economists of whatever persuasion, in their different ways, all create business cycles, consumers are forced to become involuntary investors, and the market is managed by business and government actions, while neoclassical theorists still invoke the Invisible Hand.

In the twentieth century the Keynesian model became thoroughly assimilated into the mainstream of economic thought. Most economists have remained uninterested in the political problem of unemployment, and instead have continued their attempts to "fine tune" the economy by applying the Keynesian remedies of printing money, raising or lowering interest rates, cutting or increasing taxes, and so on. However, these methods ignore the detailed structure of the economy and the qualitative nature of its problems and hence are generally unsuccessful. By the 1970s, the flaws of Keynesian economics had become apparent.

The Keynesian model has now become inappropriate because it ignores so many factors that are crucial to understanding the economic situation. It concentrates on the domestic economy, dissociating it from the global economic network and disregarding international economic agreements; it neglects the overwhelming political power of multinational corporations, pays no attention to political conditions, and ignores the social and environmental costs of economic activities. At best the Keynesian approach can provide a set of possible scenarios but cannot make specific predictions. Like most of Cartesian economic thought, it has outlived its usefulness.

Contemporary economics is a mixed bag of concepts, theories, and models stemming from various epochs of economic history. The main schools of thought that have emerged are the Marxist school and "mixed" economics, a modern version of neoclassical economics using more sophisticated mathematical techniques but still based on classical notions. In the late 1930s and the 1940s a new "neoclassical-Keynesian synthesis" was proclaimed, but such a synthesis actually never took place. The neoclassical economists simply took the Keynesian tools and grafted them onto their own models, trying to manipulate the so-called market forces while at the same time, schizophrenically, retaining the old equilibrium concepts.

More recently a very heterogeneous group of economists has been called collectively the "post-Keynesian" school. The more conservative proponents of post-Keynesian thought are advertising a new brand of so-called supply-side economics that has found powerful adherents in Washington. Their basic argument is that after the failure of Keynesians to stimulate demand without increasing inflation, one should now stimulate supply, for example by investing more in factories and automation and by removing "unproductive" environmental controls. This approach is manifestly antiecological, likely to result in rapid exploitation of natural resources and thus bound to aggravate our problems. Other post-Keynesians have begun to analyze the structure of the economy in a more realistic way. They reject the free-market model and the concept of the Invisible Hand, recognizing that the economy is now dominated by massive corporate institutions and by government agencies which often cater to them. But most post-Keynesians still use too highly aggregated data, inappropriately derived from microanalyses; neglect to qualify the concept of growth; and do not seem to have a clear view of the ecological dimensions of our economic problems. Their elaborate quantitative models describe fragmented segments of economic activity; these are supposed to have an empirical basis and to represent nothing but "facts," but are actually based on tacitly assumed neoclassical concepts.

All these models and theories—Marxist as well as non-Marxist—are still deeply rooted in the Cartesian paradigm, and thus inappropriate to describe today's closely interrelated and continually changing global economic system. It is not too easy for the uninitiated to understand the rather abstract and technical language of modern economics, but once this has been mastered the major flaws of contemporary economic thought become readily apparent.

One of the outstanding characteristics of today's economies, both capitalist and communist, is an obsession with growth. Economic and technological growth are seen as essential by virtually all economists and politicians, although it should by now be abundantly clear that unlimited expansion in a finite environment can only lead to disaster. The belief in the necessity of continuing growth is a consequence of overemphasis on yang values—expansion, self-assertion, competition—and can also be related to the Newtonian notions of absolute, infinite space and time. It is a reflection of linear thinking; of the erroneous belief that if something is good for an individual or group, then more of the same will necessarily be better.

The competitive, self-assertive approach to business is part of the legacy of John Locke's atomistic individualism, which in America was vital to the small band of early settlers and explorers but has now become unhealthy, unable to deal with the intricate web of social and ecological relations characteristic of mature industrial economies. The prevailing creed in government and business is still that the common good will be maximized if all individuals, groups, and institutions maximize their own material wealth—what is good for General Motors is good for the United States. The whole is identified with the sum of its parts, and the fact that it can be either more or less than this sum, depending on the mutual interference between the parts, is ignored. The consequences of this reductionist fallacy are now becoming painfully visible, as economic forces increasingly collide with each other, tear the social fabric, and ruin the natural environment.

The global obsession with growth has resulted in a remarkable similarity between capitalist and communist economies. The two dominant representatives of these so-called opposing value systems, the

United States and the Soviet Union, no longer seem to be all that different. Both are dedicated to industrial growth and hard technology, with increasingly centralized and bureaucratic control, whether by the state or by so-called "private" multinational corporations. The universal addiction to growth and expansion is becoming stronger than all other ideologies; to borrow Marx's phrase, it has become the opium of the people.

In a sense, the common belief in growth is justified, because growth is an essential feature of life. Women and men have known this since ancient times, as we can see from the terms used in antiquity to describe reality. The Greek word *phusis*—the root of our modern terms physics, physiology, and physician—as well as the Sanskrit *brahman*, both of which denoted the essential nature of all things, derive from the same Indo-European root *bheu*, "to grow." Indeed, evolution, change, and growth seem to be essential aspects of reality. What is wrong with current notions of economic and technological growth, however, is the lack of any qualification. It is commonly believed that all growth is good without recognizing that, in a finite environment, there has to be a dynamic balance between growth and decline. While some things have to grow, others have to diminish, so that their constituent elements can be released and recycled.

Most of today's economic thought is based on the notion of *undifferentiated* growth. The idea that growth can be obstructive, unhealthy, or pathological is not entertained. What we urgently need, therefore, is differentiation and qualification of the concept of growth. From excessive production and consumption in the private sector, growth will have to be channeled into public service areas such as transportation, education, and health care. And this change will have to be accompanied by a fundamental shift of emphasis from material acquisition to inner growth and development.

There are three closely interrelated dimensions of growth in most industrial societies—economic, technological, and institutional. Continuing economic growth is accepted as a dogma by virtually all economists, who assume, with Keynes, that it is the only way to insure that material wealth will trickle down to the poor. This "trickle-down" model of growth has long been shown to be unrealistic. High rates of growth not only do little to ease urgent social and human problems but in many countries have been accompanied by increasing unemployment and a general deterioration of social conditions.[47] Yet economists and politicians still insist on the importance of economic growth. Thus Nelson Rockefeller asserted as late as 1976, at a meeting of the Club of Rome: "More growth is essential if all the millions of Americans are to have the opportunity to improve their quality of life."[48]

What Rockefeller was referring to, of course, was not the quality of life but the so-called standard of living, which is equated with material consumption. Manufacturers spend enormous amounts of money on advertising to keep up a pattern of competitive consumption; many of the goods thus consumed are unnecessary, wasteful, and often outright harmful. The price we pay for this excessive cultural habit is the continual degradation of the real quality of life—the air we breathe, the food we eat, the environment we live in, and the social relations that constitute the fabric of our lives. These costs of wasteful overconsumption were well documented several decades ago and have continued to increase.[49]

The most severe consequence of continuing economic growth is the depletion of the planet's natural resources. The rate of this depletion was predicted with mathematical precision in the early 1950s by the geologist M. King Hubbert, who tried to present his hypothesis to President John Kennedy and to subsequent American presidents but was generally considered a crank. In the meantime history has confirmed Hubbert's predictions to the finest details, and he has lately received many awards.

Hubbert's estimates and calculations show that the production/depletion curves for all nonrenewable natural resources are bell-shaped, not unlike the curves picturing the rise and fall of civilizations.[50] They first increase gradually, then rise dramatically, peak, decline sharply, and eventually peter out. Thus Hubbert predicted that petroleum and natural gas production in the United States would peak in the 1970s, as it did, and then begin the descent that continues today. The same model predicts that world petroleum production will reach its highest point in the 1990s, and world coal production during the twenty-first century. The important aspect of these curves is that they describe the depletion of every single natural resource, from coal, petroleum, and natural gas to metals, forest, and fish reserves, and even oxygen and ozone. We may find alternatives to energy production from fossil fuels, but this will not stop the depletion of our other resources. If we continue the current patterns of undifferentiated growth, we will soon exhaust the reserves of metals, food, oxygen, and ozone that are crucial to our survival.

To slow down the rapid depletion of our natural resources we need not only to abandon the idea of continuing economic growth, but to control the worldwide increase in population. The dangers of this "population explosion" are now generally recognized, but views on how to achieve "zero population growth" differ widely, with proposed methods ranging from education and voluntary family planning to coercion by legal means and by brute force. Most of these proposals are based on the view of the problem as a purely biological phenomenon, related only to fertility and contraception. But there is now conclusive evidence, collected by demographers around the world, that population growth is affected as much, if not more, by powerful social factors. The view that this research demands sees the rate of growth as affected by a complex interplay of biological, social, and psychological forces.[51]

Demographers have discovered that the significant pattern is a transition between two levels of stable populations that has been characteristic of all Western countries. In premodern societies birth rates were high, but so were death rates, and thus the population size was stable. As living conditions improved during the time of the Industrial Revolution, death rates began to fall, and, with birth rates remaining high, populations increased rapidly. However, with continuing improvement of living standards, and with the decline in death rates continuing, birth rates began to decline as well, thus reducing the rate of population growth. The reason for this decline has now been observed worldwide. Through the interplay of social and psychological forces, the quality of life—the fulfillment of material needs, a sense of well-being, and confidence in the future—becomes a powerful and effective motivation for controlling population growth. There is, in fact, a critical level of well-being which has been shown to lead to a rapid reduction in birth rate and an approach to a balanced population. Human societies, then, have developed a self-regulating process, based on social conditions, which results in a demographic transition from a balanced population, with high birth rates and high death rates and a low standard of living, to a population with a higher standard of living which is larger but again in balance, and in which both birth and death rates are low.

The present global population crisis is due to the rapid increase of population in the Third World, and the considerations outlined above show clearly that this increase continues because the conditions for the second phase of the demographic transition have not been met. During their colonial past the Third World countries experienced an improvement in living conditions that was sufficient to reduce death rates and thus initiated population growth. But the rise of living standards did not continue, because the wealth generated in the colonies was diverted to the developed countries, where it helped *their* populations to become balanced. This process continues today, as many Third World countries remain colonized in the economic sense. This exploitation continues to increase the affluence of the colonizers and prevents Third World populations from reaching the standard of living conducive to a reduction of their rate of growth.

The world population crisis, then, is an unanticipated effect of international exploitation, a consequence of the fundamental interrelatedness of the global ecosystem, in which every exploitation eventually comes back to haunt the exploiters. From this point of view it becomes quite apparent that ecological balance also requires social justice. The most effective way to control population growth will be to help the peoples of the Third World achieve a level of well-being that will induce them to limit their fertility voluntarily. This will require a global redistribution of wealth in which some of the world's wealth is returned to the countries that have played a major role in producing it.

An important aspect of the population problem, which is not generally known, is that the cost of bringing the standard of living of poor countries to a level that appears to convince people that they should not have excessive numbers of children is very small compared to the wealth of developed countries. That is to say, there is enough wealth to support the entire world at a level that leads to a balanced population.[52] The problem is that this wealth is unevenly distributed, and much of it is wasted. In the United States, where excessive consumption and waste have become a way of life, 5 percent of the world's population now consumes a third of its resources, with energy consumption per capita about twice as high as in most European countries. Simultaneously the frustrations created and sustained by massive doses of advertising, combined with social injustice within the nation, contribute to ever increasing crime, violence, and other social pathologies. This sad state of affairs is well illustrated by the schizophrenic content of our weekly magazines. Half their pages are filled with grim stories about violent crimes, economic disaster, international political tension, and the race toward global destruction, while the other half portray carefree, happy people behind packs of cigarettes, bottles of alcohol, and shiny new cars. Advertising on television influences the content and form of all programs, including the "news shows," and uses the tremendous suggestive power of this medium—switched on for six and a half hours a day by the average American family—to shape people's imagery, distort their sense of reality, and determine their views, tastes, and behavior.[53] The exclusive aim of this dangerous practice is to condition the audience into buying products advertised before, after, and during each program.

Economic growth, in our culture, is inextricably linked with technological growth. Individuals and institutions are mesmerized by the wonders of modern technology and have come to believe that every problem has a technological solution. Whether the nature of the problem is political, psychological, or ecological, the first reaction, almost automatically, is to deal with it by applying or developing some new technology. Wasteful energy consumption is countered by nuclear power, lack of political insight is compensated for by building more missiles and bombs, and the poisoning of the natural environment is remedied by developing special technologies that, in turn, affect the environment in still unknown ways. By looking for technological solutions to all problems, we usually just shift these around in the global ecosystem, and very often the side effects of the "solution" are more harmful than the original problem.

The ultimate manifestation of our obsession with high technology is the widely entertained fantasy that our current problems can be solved by creating artificial habitats in outer space. I do not exclude the possibility that such space colonies may be built some day, although from what I have seen of existing plans and of the mentality underlying them I would certainly not want to live there. However, the basic fallacy of the whole idea is not technological; it is the naïve belief that space technology can solve the social and cultural crisis on this planet.

Technological growth is not only regarded as the ultimate problem solver but is also seen as determining our life styles, our social organizations, and our value system. Such "technological determinism" seems to be a consequence of the high status of science in our public life—as compared to philosophy, art, or religion—and of the fact that scientists have generally failed to deal with human values in significant ways. This has led most people to believe that technology determines the nature of our value system and our social relations, rather than recognizing that it is the other way round; that our values and social relations determine the nature of our technology.

The yang, masculine consciousness that dominates our culture has found its fulfillment not only in "hard" science but also in the "hard" technology derived from it. This technology is fragmented rather than holistic, bent on manipulation and control rather than cooperation, self-assertive rather than integrative, and suitable for centralized management rather than regional application by individuals and small groups. As a result, this technology has become profoundly antiecological, antisocial, unhealthy, and inhuman.

The most dangerous manifestation of our hard, "macho" technology is the expansion of nuclear weapons, which amounts to the most expensive military boom in history.[54] By brainwashing the American public and effectively controlling its representatives, the military-industrial complex has succeeded in extracting regularly increasing defense budgets that are used to design weapons to be employed in a "science-intensive" war ten or twenty years from now. A third to a half of America's scientists and engineers work for the military, using all their imagination and creativity to invent ever more sophisticated means for total destruction—laser communication systems, particle beams, and other complex technologies for computerized warfare in outer space.[55]

It is striking that all these efforts focus exclusively on hardware. America's defense problems, like all its others, are perceived as simply problems of hard technology. The idea that psychological, social, and behavioral research—let alone philosophy or poetry—could also be relevant is not mentioned. Moreover, the question of national security is analyzed predominantly in terms of "power blocks," "action and reaction," the "political vacuum," and similar Newtonian notions.

The effects of the military's excessive use of hard technology are similar to those encountered in the civilian economy. The complexity of our industrial and technological systems has now reached a point where many of those systems can no longer be either modeled or managed. Breakdowns and accidents occur with increasing frequency, unanticipated social and environmental costs are continually generated, and more time is spent on maintaining and regulating the system than on providing useful goods and services. Such enterprises, therefore, are highly inflationary, in addition to their severe effects on our physical and mental health. Thus it is becoming increasingly apparent, as Henderson has pointed out, that we may reach the social, psychological, and conceptual limits to growth even before the physical limits are reached.[56]

What we need, then, is a redefinition of the nature of technology, a change of its direction, and a reevaluation of its underlying value system. If technology is understood in the broadest sense of the term, as the application of human knowledge to the solution of practical problems, it becomes clear that we have concentrated too much on "hard," highly complex, and resource-intensive technologies and must now shift our attention to the "soft" technologies of conflict resolution, social agreements, cooperation, recycling and redistribution, and so on. As Schumacher says in his book *Small Is Beautiful,* we need a "technology with a human face."[57]

The third aspect of undifferentiated growth, which is inseparable from economic and technological growth, is the growth of institutions—from companies and corporations to colleges and universities, churches, cities, governments, and nations. Whatever the original purpose of the institution, its growth beyond a certain size invariably distorts that purpose by making the self-preservation and further expansion of the institution its overriding aim. At the same time the people belonging to the institution and those who have to deal with it feel increasingly alienated and depersonalized, while families, neighborhoods, and other small-scale social organizations are threatened and often destroyed by institutional domination and exploitation.[58]

One of the most dangerous manifestations of institutional growth today is that of corporations. The largest of them have now transcended national boundaries and have become major actors on the global stage. The assets of these multinational giants exceed the gross national products of most nations; their economic and political power surpasses that of many national governments, threatening national sovereignty and world monetary stability. In most countries of the Western world, but especially in the United States, corporate power permeates virtually every facet of public life. Corporations largely control the legislative process, distort the information received by the public through the media, and determine, to a significant extent, the functioning of our educational system and the direction of academic research. Corporate and business leaders are prominent on the boards of trustees of academic institutions and foundations, where they inevitably use their influence to perpetuate a value system consistent with corporate interests.[59]

The nature of large corporations is profoundly inhuman. Competition, coercion, and exploitation are essential aspects of their activities, all motivated by the desire for indefinite expansion. Continuing growth is built into the corporate structure. For example, corporate executives who knowingly bypass an opportunity for increasing the corporation's profits, for whatever reason, are liable to lawsuit. Thus the maximizing of profits becomes the ultimate goal, to the exclusion of all other considerations. Corporate executives have to leave their humaneness behind when they attend their board meetings. They are not expected to show any feelings, nor to express any regrets; they can never say "I am sorry" or "we made a mistake." What they talk about instead is coercion, control, and manipulation.

Large corporations, then, work like machines rather than human institutions once they have grown beyond a certain size. However, there are no laws, national or international, to deal effectively with these giant institutions. The growth of corporate power has outstripped the development of an appropriate legal framework. Laws made for humans are applied to corporations that have lost all resemblance to human beings. The concepts of private property and enterprise have become confused with corporate property and state capitalism, and "commercial speech" is now protected by the First Amendment. On the other hand, these corporations do not assume the responsibilities of individuals, being designed so that none of their executives can be held fully accountable for corporate activities. Many corporate leaders, in fact, believe that corporations are value-free and should be allowed to function outside the moral and ethical order. This dangerous notion was expressed quite candidly by Walter Wriston, who chairs Citibank, the second largest bank in the world. In a recent interview, Wriston made this chilling comment: "Values are topsy-turvy. . . Now college students have a mixed dormitory, men live on one floor and women on the next, and they all sit around worrying about whether or not General Motors is being honest . . . I believe that there are no institutional values, only personal ones."[60]

As multinational corporations intensify their global search for natural resources, cheap labor, and new markets, the environmental disasters and social tensions created by their obsession with indefinite growth become ever more apparent. Thousands of small businesses are driven from the marketplace by the power of large companies that are able to win federal subsidies for their complex, capital-intensive, and resource-consuming technologies. At the same time there is a tremendous need for simple skills like carpentry, plumbing, tailoring, and all kinds of repair and maintenance jobs which have been socially devalued and severely neglected although they are as vital as ever. Instead of regaining self-sufficiency by changing occupations and practicing these skills, most workers remain totally dependent on large corporate institutions and, in times of economic hardship, see no other alternative to collecting unemployment checks and accepting passively that the situation is beyond their control.

If the consequences of corporate power are harmful in the industrialized countries, they are altogether disastrous in the Third World. In those countries, where legal restrictions are often nonexistent or impossible to enforce, the exploitation of people and of their land has reached extreme proportions. With the help of skillful manipulation of the media, emphasizing the "scientific" nature of their enterprises, and often with full support from the U.S. government, multinational corporations ruthlessly extract the Third World's natural resources. To do so, they often use polluting and socially disruptive technologies, thus causing environmental disaster and political chaos. They abuse the soil and wilderness resources of Third World countries to produce profitable cash crops for export instead of food for the local population, and promote unhealthy patterns of consumption, including the sale of highly dangerous products that have been outlawed in the United States. The numerous horror stories of corporate behavior in the Third World which have emerged in recent years show convincingly that respect for people, for nature, and for life are not part of the corporate mentality. On the contrary, large-scale corporate crime is today the most widespread and least prosecuted criminal activity.[61]

Many of the large corporations are now obsolete institutions that lock up capital, management, and resources but are unable to adapt their functioning to changing needs. A well-known example is the automobile industry, which is unable to adjust to the fact that the global limitations of energy and resources will force us to drastically restructure our transportation system, shifting to mass transit and to smaller, more efficient, and more durable cars. Similarly, the utility companies require ever expanding demands for electricity to justify their corporate growth, and have thus embarked on a vigorous campaign for nuclear power, rather than promoting the small-scale, decentralized solar technology which alone can provide the environment conducive to our survival.

Although these corporate giants are often close to bankruptcy, they still have the political power to persuade government to bail them out with taxpayers' money. Their argument is, invariably, that their efforts are motivated by the preservation of jobs, although it has been clearly shown that small-scale, labor-intensive enterprises create more jobs and generate much lower social and environmental costs.[62] We will always need some large-scale operations, but many of the giant corporations, dependent on energy- and resource-intensive means of production to provide marginally useful goods, must either be fundamentally transformed or pass from the scene. They will then release the capital, resources, and human ingenuity that can build a sustainable economy and develop alternative technologies.

The question of scale—which Schumacher pioneered with the slogan "Small is beautiful"—will play a crucial role in the reassessment of our economic system and our technology. The universal obsession with growth has been accompanied by an idolatry of gigantism, of "the bigness of things," as Theodore Roszak has called it.[63] Size, of course, is relative, and small structures are not always better than big ones. In our modern world we need both, and our task will be to achieve a balance between the two. Growth will have to be qualified and the notion of scale will play a crucial role in the restructuring of our society. The qualification of growth and the integration of the notion of scale into economic thought will cause a profound revision of the basic conceptual framework of economics. Many economic patterns that are now tacitly assumed to be inevitable will have to be changed; all economic activity will have to be studied within the context of the global ecosystem; most concepts used in current economic theory will have to be expanded, modified, or abandoned.

Economists tend to freeze the economy arbitrarily in its current institutional structure instead of seeing it as an evolving system that generates continually changing patterns. To grasp this dynamic evolution of the economy is extremely important, because it shows that strategies which are acceptable at one stage may become totally inappropriate at another. Many of our present problems come from the fact that we have often overshot the mark in our technological enterprises and economic planning. As Hazel Henderson likes to say, we have reached a point where "nothing fails like success." Our economic and institutional structures are dinosaurs, unable to adapt to environmental changes and therefore bound to die out.

Today's world economy is based on past configurations of power, perpetuating class structures and unequal distribution of wealth within national economies, as well as exploitation of Third World countries by rich industrialized nations. These social realities are largely ignored by economists, who tend to avoid moral issues and accept the current distribution of wealth as given and unchangeable. In most Western countries economic wealth is highly concentrated and tightly controlled by a small number of people who belong to the "corporate class" and derive their income largely from ownership.[64] In the United States, 76 percent of all corporate securities are owned by 1 percent of the stockholders while, at the bottom, 50 percent of the people own only 8 percent of the nation's wealth.[65] Paul Samuelson illustrates this skewed distribution of wealth in his well-known textbook *Economics* with a graphic analogy: "If we made today an income pyramid out of a child's blocks, with each layer portraying $1,000 of income, the peak would be far higher than the Eiffel Tower, but almost all of us would be within a yard of the ground."[66] This social inequality is not an accident but is built into the very structure of our economic system and is perpetuated by our emphasis on capital-intensive technologies. The necessity of continuing exploitation for the growth of the American economy was pointed out quite bluntly by the *Wall Street Journal* in an editorial on "Growth and Ethics," which insisted that the United States would have to choose between growth and greater equality, since the maintenance of inequality was necessary to create capital.[67]

The grossly unequal distribution of wealth and income within industrialized countries is paralleled by similar patterns of maldistribution between developed countries and the Third World. Programs of economic and technological aid to Third World countries are often used by multinational corporations to exploit those countries' labor and natural resources and to fill the pockets of a small and corrupt elite. As the cynical saying goes, "Economic aid is taking money from the poor people in rich countries and giving it to the rich people in poor countries." The result of these practices is the perpetuation of an "equilibrium of poverty" in the Third World, with life near the bare level of subsistence.[68]

The avoidance of social issues in current economic theory is closely related to the striking inability of economists to adopt an ecological perspective. The debate between ecologists and economists has now been going on for two decades and has shown very clearly that the main body of contemporary economic thought is inherently antiecological.[69] Economists neglect social and ecological interdependence, treating all goods equally without considering the many ways in which these goods are related to the rest of the world—whether they are human-made or naturally occurring, renewable or nonrenewable, and so on. Ten dollars' worth of coal equals ten dollars' worth of bread, transportation, shoes, or education. The only criterion for determining the relative value of these goods and services is their monetary market value: all values are reduced to the single criterion of private profit-making.

Since the conceptual framework of economics is ill suited to account for the social and environmental costs generated by all economic activity, economists have tended to ignore these costs, labeling them "external" variables that do not fit into their theoretical models. And since most economists are employed by private interest groups to prepare cost/benefit analyses that are usually heavily biased in favor of their employers' projects, there are very few data on even the easily quantifiable "externalities." Corporate economists not only treat the air, water, and various reservoirs of the ecosystem as free commodities, but also the delicate web of social relations, which is severely affected by continuing economic expansion. Private profits are being made increasingly at public costs in the deterioration of the environment and the general quality of life. As Henderson writes, "They tell us about the sparkling dishes and clothes, but forget to mention the loss of those sparkling rivers and lakes."[70]

Economists' inability to see economic activities within their ecological context prevents them from understanding some of the most significant economic problems of our time, foremost among them the tenacious persistence of inflation and unemployment. There is no single cause of inflation, but several major sources can be identified, and most economists fail to understand inflation because all these sources involve variables that have been excluded from current economic models. Economists often do not take into account the fact that wealth is based on natural resources and energy, though this is increasingly hard to ignore. As the resource base declines, raw materials and energy must be extracted from ever more degraded and inaccessible reservoirs, and thus more and more capital is needed for the extraction process. As a consequence, the inevitable decline of natural resources, following the well-known bell-shaped curves, is accompanied by an unremitting exponential climb of the price of resources and energy, and this becomes one of the main driving forces of inflation.

The excessive energy- and resource-dependence of our economy is reflected in the fact that it is capital-intensive rather than labor-intensive. Capital represents a potential for work, extracted from past exploitation of natural resources. As these resources diminish, capital itself is becoming a very scarce resource. In spite of this, and because of a narrow notion of productivity, there is a strong tendency to substitute capital for labor, in both capitalist and Marxist economies. The business community lobbies incessantly for tax credits for capital investments, many of which reduce employment through automation by means of such highly complex technologies as automated check-out lines in supermarkets, and electronic funds transfer systems in banks. Both capital and labor produce wealth, but a capital-intensive economy is also resource- and energy-intensive, and thus highly inflationary.

A striking example of such a capital-intensive enterprise is the American system of agriculture, which exerts its inflationary impact on the economy at many levels. Production is achieved with the help of energy-intensive machines and irrigation systems, and by means of heavy doses of oil-based pesticides and fertilizers. These methods not only destroy the organic balance in the soil and produce poisonous chemical substances in our food, but are also yielding ever diminishing returns and thus making farmers prime victims of inflation. The food industry then turns the agricultural products into overprocessed, overpackaged, and overadvertised food, transported across the country to be sold in large supermarkets, all of which requires excessive consumption of energy and thus fuels inflation.

The same is true for animal farming, which is heavily promoted by the petrochemical industry, since it takes about ten times more fossil-fuel energy to produce a unit of animal protein than a unit of plant protein. Most of the grain grown in the United States is not consumed by humans but is fed to livestock which is then eaten by people. As a result, most Americans suffer from an unbalanced diet that often leads to obesity and illness, thus contributing to inflation in health care. Similar patterns can be observed throughout our economic system. Excessive investment in capital, energy, and natural resources taxes the environment, affects our health, and is a major cause of inflation.

Conventional economic wisdom holds that there is a free market which naturally stays in balance. Inflation and unemployment are seen as interdependent temporary aberrations of the equilibrium state, one being the trade-off of the other. In today's reality, however, with economies dominated by huge institutions and interest groups, equilibrium models of this kind are no longer valid. The presumed trade-off between inflation and unemployment—expressed mathematically by the so-called Phillips curve—is an abstract and utterly unrealistic concept. Inflation and unemployment combined, known as "stagflation," have become a structural feature of all industrial societies committed to undifferentiated growth. Excessive dependence on energy and natural resources, and excessive investment in capital rather than labor, are not only highly inflationary but also bring massive unemployment. In fact, unemployment has become such an intrinsic feature of our economy that government economists now speak of "full employment" when more than 5 percent of the labor force is out of work.

The second major cause of inflation is the ever increasing social costs engendered by undifferentiated growth. In their attempts to maximize their profits, individuals, companies, and institutions try to "externalize" all social and environmental costs; they try to exclude them from their own balance sheets and to push them onto each other, passing them on around the system, to the environment and to future generations. Gradually these costs accumulate and manifest themselves as the costs of litigation, crime control, bureaucratic coordination, federal regulation, consumer protection, health care, and so on. None of these activities adds anything to real production; all of them contribute significantly to inflation.

Instead of incorporating such crucial social and environmental variables into their theories, economists tend to work with elegant but unrealistic equilibrium models, most of them based on the classical idea of free markets, where buyers and sellers meet each other with equal power and information. In most industrial societies large corporations control the supply of goods, create artificial demands through advertising, and have a decisive influence on national policies. An extreme example is the oil companies, which shape the energy policy of the United States to such an extent that the crucial decisions are not made in the national interest but rather in the interest of the dominant corporations. This corporate interest, of course, has nothing to do with the welfare of American citizens but is exclusively concerned with corporate profits. John Sweringen, chief executive officer of Standard Oil of Indiana, made this quite clear in a recent interview. "We're not in the energy business," he said. "We're in the business of trying to use the assets entrusted to us by our shareholders to give them the best return on the money they've invested in the company."[71] Corporate giants like Standard Oil now have the power to determine, to a large extent, not only the national energy policy, but also our systems of transportation, agriculture, health care, and many other aspects of our social and economic life. Free markets, balanced by supply and demand, have long disappeared; they exist only in our economics textbooks. Equally outdated in our global economy is the Keynesian idea that fluctuating business cycles can be ironed out by using appropriate national policies. Yet today's economists still apply the traditional Keynesian tools to inflate or deflate the economy and thereby create short-term oscillations that obscure ecological and social realities.

To deal with economic phenomena from an ecological perspective, economists will need to revise their basic concepts in a drastic way. Since most of these concepts were narrowly defined and have been used without their social and ecological context, they are no longer appropriate to map economic activities in our fundamentally interdependent world. The gross national product, for example, is supposed to measure a nation's wealth, but all economic activities associated with monetary values are added up indiscriminately for the GNP while all nonmonetary aspects of the economy are ignored. Social costs, like those of accidents, litigation, and health care, are added as positive contributions to the GNP; education is still often treated as an expenditure rather than an investment, whereas work done in households and goods produced by such work are not counted. Although the inadequacy of such a method of accounting is now widely recognized, there has been no serious effort to redefine the GNP as an effective measure of production and wealth.

Similarly, the concepts of "efficiency," "productivity," and "profit" are used in such a narrow context that they have become quite arbitrary. Corporate efficiency is measured in terms of corporate profits, but with these profits being made increasingly at public cost, we have to ask "efficient for whom?" When economists talk about efficiency, do they mean efficiency at the individual level, at the level of the corporation, at the social level, or at the level of the ecosystem? A striking example of a highly biased use of the notion of efficiency is that of utility companies, which have been trying to persuade us that nuclear power is our most efficient source of energy, completely disregarding the tremendous social and environmental costs that arise from the handling of radioactive material. Such a skewed use of "efficiency" is typical of the energy industry, which has misinformed us not only about social and environmental costs but also about the political realities behind the cost of energy. Having obtained heavy subsidies for conventional energy technology through their political power, the utility companies then turned around and declared that solar energy was inefficient because it could not compete with other energy sources in the "free" market.

Examples of this kind abound. The American system of farming, highly mechanized and petroleum-subsidized, is now the most inefficient in the world when measured in terms of the amount of energy used for a given output of calories; yet agribusiness, which is largely owned by the petrochemical industry, makes huge profits. In fact the whole American industrial system, with its huge use of the planet's resources for a tiny percentage of its population, must be seen as highly inefficient from a global ecological point of view.

Closely related to "efficiency" is the concept of "productivity," which has been similarly distorted. Productivity is usually defined as the output per employee per working hour. To increase this quantity, manufacturers tend to automate and mechanize the production process as much as possible. However, in doing so they increase the number of unemployed workers and lower *their* productivity to zero by adding them to the welfare rolls.

Together with the redefinition of "efficiency" and "productivity" we shall need a thorough revision of the concept of "profit." Private profits are now too often reaped at the expense of social or environmental exploitation. These costs must be fully taken into account, so that the notion of profit becomes associated with the creation of real wealth. Many of the goods produced and sold "profitably" today will then be recognized as wasteful and will be priced out of the market.

One of the reasons why the concept of "profit" has become so distorted is the artificial division of the economy into private and public sectors, which has led economists to ignore the link between private profits and public costs. The relative roles of the private and public sectors in providing goods and services are now being increasingly questioned, as more and more people ask themselves why we should accept the "need" for multimillion-dollar industries devoted to pet foods, cosmetics, drugs, and all kinds of energy-wasting gadgets, when we are told at the same time that we cannot "afford" adequate sanitary services, fire protection, or public transport systems in our cities.

The remapping of the economy is not merely an intellectual task but will involve profound changes in our value system. The idea of wealth itself, which is central to economics, is inextricably linked to human expectations, values, and life styles. To define wealth within an ecological framework will mean to transcend its present connotations of material accumulation and give it the broader sense of human enrichment. Such a notion of wealth, together with "profit" and other related concepts, will not be amenable to rigorous quantification, and thus economists will no longer be able to deal with values exclusively in monetary terms. In fact, our current economic problems make it quite evident that money alone no longer provides an adequate tracking system.[72]

An important aspect of the necessary revision of our value system will be the redefinition of "work."[73] In our society work is identified with a job; it is done for an employer and for money; unpaid activities do not count as work. For example, the work performed by women and men in households is not assigned any economic value; yet this work equals, in monetary terms, two-thirds of the total amount of wages and salaries paid by all the corporations in the United States.[74] On the other hand, work in paid jobs is no longer available for many who want it. Being unemployed carries a social stigma; people lose status and respect in their own and others' eyes because they are unable to get work.

At the same time, those who do have jobs very often have to perform work in which they cannot take any pride, work that leaves them profoundly alienated and dissatisfied. As Marx clearly recognized, this alienation comes from the fact that workers do not own the means of production, have no say about the use to which their work is put, and cannot identify in any meaningful way with the production process. The modern industrial worker no longer feels responsible for his work nor takes pride in it. The result is products that show less and less craft, artistic quality, or taste. Thus work has become profoundly degraded; from the worker's point of view, its only purpose is to earn a living, while the employer's exclusive aim is to increase profits.

Lack of responsibility and pride, together with the overriding profit motive, have resulted in a situation where most of the work carried out today is wasteful and unjustified. As Theodore Roszak has forcefully stated:

> Work that produces unnecessary consumer junk or weapons of war is wrong and wasteful. Work that is built upon false needs or unbecoming appetites is wrong and wasteful. Work that deceives or manipulates, that exploits or degrades is wrong and wasteful. Work that wounds the environment or makes the world ugly is wrong and wasteful. There is no way to redeem such work by enriching it or restructuring it, by socializing it or nationalizing it, by making it "small" or decentralized or democratic.[75]

This state of affairs is in sharp contrast to traditional societies in which ordinary women and men were engaged in a wide variety of activities—farming, fishing, hunting, weaving, making clothes, building, making pottery and tools, cooking, healing—all of them useful, skilled and dignified work. In our society most people are unsatisfied by their work and see recreation as the main focus of their lives. Thus work has become opposed to leisure, and the latter is served by a huge industry featuring resource- and energy-intensive gadgets—computer games, speedboats, and snowmobiles—and exhorting people to ever more wasteful consumption.

As far as the status of different kinds of work is concerned, there is an interesting hierarchy in our culture. Work with the lowest status tends to be that work which is most "entropic,"* i.e., where the tangible evidence of the effort is most easily destroyed. This is work that has to be done over and over again without leaving a lasting impact—

* Entropy is a measure of disorder; see Chapter 2, p. 73.

cooking meals which are immediately eaten, sweeping factory floors which will soon be dirty again, cutting hedges and lawns which keep growing. In our society, as in all industrial cultures, jobs that involve highly entropic work—housework, services, agriculture—are given the lowest value and receive the lowest pay, although they are essential to our daily existence.[76] These jobs are generally delegated to minority groups and to women. High-status jobs involve work that creates something lasting—skyscrapers, supersonic planes, space rockets, nuclear warheads, and all the other products of high technology. High status is also granted to all administrative work connected with high technology, however dull it may be.

This hierarchy of work is exactly opposite in spiritual traditions. There high-entropy work is highly valued and plays a significant role in the daily ritual of spiritual practice. Buddhist monks consider cooking, gardening, or housecleaning part of their meditative activities, and Christian monks and nuns have a long tradition of agriculture, nursing, and other services. It seems that the high spiritual value accorded to entropic work in those traditions comes from a profound ecological awareness. Doing work that has to be done over and over again helps us recognize the natural cycles of growth and decay, of birth and death, and thus become aware of the dynamic order of the universe. "Ordinary" work, as the root meaning of the term indicates, is work that is in harmony with the order we perceive in the natural environment.

Such ecological awareness has been lost in our culture, where the highest value has been associated with work that creates something "extraordinary," something out of the natural order. Not surprisingly, most of this highly valued work is now generating technologies and institutions which are extremely harmful to the natural and social environment. What we need, therefore, is to revise the concept and practice of work in such a way that it becomes meaningful and fulfilling for the individual worker, useful for society, and part of the harmonious order of the ecosystem. To reorganize and practice our work in this way will allow us to recapture its spiritual essence.

The inevitable revision of our basic economic concepts and theories will be so radical that the question arises whether economics itself, as a social science, will survive it. Indeed, several critics have predicted the end of economics. I believe that the most useful approach would be not to abandon economics as such, but to regard the framework of current economic thought, so deeply rooted in the Cartesian paradigm, as a scientific model that has become outdated. It may well continue to be useful for limited microeconomic analyses, but will certainly need to be modified and expanded. The new theory, or set of models, is likely to involve a systems approach that will integrate biology, psychology, political philosophy, and several other branches of human knowledge, together with economics, into a broad ecological framework. The outlines of such a framework are already being shaped by numerous men and women who refuse to be labeled economists or be associated with any single conventional, narrowly defined academic discipline.[77] Their approach is still scientific, but goes far beyond the Cartesian-Newtonian image of science. Its empirical basis includes not only ecological data, social and political facts, and psychological phenomena, but also a clear reference to cultural values. From this basis such scientists will be able to build realistic and reliable models of economic phenomena.

Explicit reference to human attitudes, values, and life styles in future economic thought will make this new science profoundly humanistic. It will deal with human aspirations and potentialities, and will integrate them into the underlying matrix of the global ecosystem. Such an approach will transcend by far anything attempted in today's sciences; in its ultimate nature it will be scientific and spiritual at the same time.

8 · The Dark Side of Growth

The mechanistic Cartesian world view has had a powerful influence on all our sciences and on the general Western way of thinking. The method of reducing complex phenomena to basic building blocks, and of looking for the mechanisms through which these interact, has become so deeply ingrained in our culture that it has often been identified with the scientific method. Views, concepts, or ideas that did not fit into the framework of classical science were not taken seriously and were generally disdained, if not ridiculed. As a consequence of this overwhelming emphasis on reductionist science our culture has become progressively fragmented and has developed technologies, institutions, and life styles that are profoundly unhealthy.

That a fragmented world view should also be unhealthy is not surprising in view of the close connection between "health" and "whole." Both these words, as well as "hale," "heal," and "holy," derive from the Old English root word *hal*, which means sound, whole, and healthy. Indeed, our experience of feeling healthy involves the feeling of physical, psychological and spiritual integrity, of a sense of balance among the various components of the organism and between the organism and its environment. This sense of integrity and balance has been lost in our culture. The fragmented, mechanistic world view that has become all-pervasive, and the one-sided, sensate and "yang-oriented" value system that is the basis of this world view, have led to a profound cultural imbalance and have generated numerous symptoms of ill health.

Excessive technological growth has created an environment in which life has become physically and mentally unhealthy. Polluted air, irritating noise, traffic congestion, chemical contaminants, radiation hazards, and many other sources of physical and psychological stress have become part of everyday life for most of us. These manifold health hazards are not just incidental byproducts of technological progress; they are integral features of an economic system obsessed with growth and expansion, continuing to intensify its high technology in an attempt to increase productivity.

In addition to the health hazards we can see, hear, and smell there are other threats to our well-being which may be far more dangerous because they will affect us on a much larger scale, both in space and in time. Human technology is severely disrupting and upsetting the ecological processes that sustain our natural environment and are the very basis of our existence. One of the most serious threats, almost totally ignored until recently, is the poisoning of water and air by toxic chemical wastes.

The American public became acutely aware of the hazards of chemical waste several years ago, when the tragedy of Love Canal was reported in front-page stories. Love Canal was an abandoned trench in a residential area of Niagara Falls, New York, that was used for many years as a dumpsite for toxic chemical wastes. These chemical poisons polluted surrounding bodies of water, filtered into adjacent back yards, and generated toxic fumes, causing high rates of birth defects, liver and kidney damage, respiratory ailments, and various forms of cancer among the residents of the area. Eventually an emergency was declared by the State of New York and the area was evacuated.

The story of Love Canal was first pieced together by Michael Brown, a reporter for the *Niagara Gazette*, who then went on to inspect similar hazardous-waste dumps throughout the United States.[1] Brown's extensive investigations have made it clear that Love Canal was only the first of many similar tragedies that are bound to unfold during the coming years and will seriously affect the health of millions of Americans. The U.S. Environmental Protection Agency estimated in 1979 that there are more than 50,000 known sites where hazardous materials are stored or buried, and less than 7 percent has received proper disposal.[2]

These enormous amounts of hazardous chemical waste are a result of the combined effects of technological and economic growth. Obsessed with expansion, increasing profits, and raising "productivity," the United States and other industrial countries have developed societies of competitive consumers who have been induced to buy, use, and throw away ever increasing quantities of products of marginal utility. To produce these goods—food additives, synthetic fibers, plastics, drugs, and pesticides, for example—resource-intensive technologies were developed, many of them heavily dependent on complex chemicals; and as production and consumption increased, so did the chemical wastes that are inevitable byproducts of these manufacturing processes. The United States produces a thousand new chemical compounds a year, many of them more complex than their predecessors and more alien to the human organism, while the amount of annual hazardous waste has increased from ten to thirty-five million tons over the past decade.

As production and consumption accelerated at this hectic pace, technologies appropriate for dealing with the unwanted byproducts were not developed. The reason for this negligence was simple: while the wasteful production of consumer goods was highly profitable for the manufacturers, the proper treatment and recycling of the residues was not. For many decades the chemical industry dumped its wastes into the ground without adequate safeguards, and this irresponsible practice has now resulted in thousands of dangerous chemical dumpsites, "toxic time bombs" that are likely to become one of the most serious environmental threats of the 1980s.

Faced with the grim consequences of its methods of production, the chemical industry has shown the typical corporate response. As Brown shows in case after case, the chemical companies have tried to conceal the dangers of their manufacturing processes and of the resulting chemical wastes; they have also covered up accidents and pressured politicians to avoid full inquiry. But thanks in part to the tragedy of Love Canal, public awareness has increased dramatically. While manufacturers proclaim in slick advertising campaigns that life would be impossible without chemicals, more and more people are realizing that the chemical industry has become life-destroying instead of life-sustaining. We can hope that public opinion will exert increasing pressure on the industry to develop proper technologies for treating and recycling waste products, as is already being done in several European countries. In the long run the problems generated by chemical waste will become manageable only if we can minimize the production of hazardous substances, which will involve radical changes in our attitudes as producers and consumers.

Our excessive consumption and our strong emphasis on high technology not only create massive quantities of waste but also require huge amounts of energy. Nonrenewable energy, derived from fossil fuels, powers most of our production processes, and with the decline of these natural resources energy itself has become a scarce and expensive resource. In their attempts to maintain, and even increase, their current levels of production, the world's industrialized countries have ferociously exploited the available resources of fossil fuels. These processes of energy production have the potential to cause unprecedented ecological disturbances and human suffering.

Exorbitant use of petroleum has led to heavy tanker traffic with frequent collisions, in which huge amounts of oil are spilled into the seas. These oil spills have not only polluted the most beautiful shores and beaches of Europe, but are also seriously disrupting the marine food cycles and thus creating ecological hazards that are still poorly understood. The generation of electricity from coal is even more hazardous and more polluting than energy production from oil. Underground mining causes severe damage to miners' health, and strip mining creates conspicuous environmental consequences, since the mines are generally abandoned once the coal is exhausted, with huge areas of land left devastated. The worst damage of all, both to the environment and to human health, comes from the burning of coal. Coal-burning plants emit vast quantities of smoke, ash, gases, and various organic compounds, many of which are known to be toxic or carcinogenic. The most dangerous of the gases is sulfur dioxide, which can severely impair the lungs. Another pollutant released in the burning of coal is nitrogen oxide, which is also the main ingredient in air pollution from automobiles. A single coal-burning plant can emit as much nitrogen oxide as several hundred thousand cars.

The sulfur and nitrogen oxides emanating from coal-fired plants not only are hazardous to the health of people living in the plant's vicinity but also generate one of the most insidious and completely invisible forms of air pollution, acid rain.[3] The gases thrown up by power plants mix with the oxygen and water vapor in the air and, through a series of chemical reactions, turn into sulfuric and nitric acids. These acids are then carried by the wind until they collect at various atmospheric gathering points and are washed down to the earth as acid rain or snow.

Eastern New England, eastern Canada, and southern Scandinavia are heavily affected by this type of pollution. When acid rain falls on lakes it kills fish, insects, plants, and other forms of life; eventually the lakes die completely from acidity they can no longer neutralize. Thousands of lakes in Canada and Scandinavia are dead or dying already; entire fabrics of life that took thousands of years to evolve are rapidly disappearing.

At the heart of the problem, as usual, lies ecological shortsightedness and corporate greed. Technologies to reduce the pollutants that cause acid rain have already been developed, but the corporations that own the coal plants are vigorously opposed to environmental regulation, and they have the political power to prevent stringent controls. Thus American utility companies have forced the Environmental Protection Agency to relax emission standards for the old coal-fired plants in the Midwest, which continue to send vast amounts of pollutants downwind and are expected to be the source of 80 percent of U.S. sulfur emissions by 1990. These actions are based on the same irresponsible attitudes that bring the hazards of chemical waste. Rather than neutralizing their polluting waste products, the industries simply dump them somewhere else, without caring that in a finite ecosystem there is no such place as "somewhere else."

During the 1970s the world became acutely aware of a global shortage of fossil fuels, and with the inevitable decline of these conventional sources of energy in sight, the leading industrialized countries embarked on a vigorous campaign for nuclear power as an alternative energy source. The debate about how to solve the energy crisis usually focuses on the costs and risks of nuclear power, as compared to those of energy production from petroleum, coal, and oil shale. The arguments advanced by government and corporate economists, and by other representatives of the energy industry, are usually heavily biased in two ways. Solar energy—the only energy source that is abundant, renewable, stable in price and environmentally benign—is said to be "uneconomical" or "not yet feasible," in spite of considerable evidence to the contrary,[4] and the need for more energy is assumed unquestioningly.

Any realistic discussion of the "energy crisis" has to start from a broader perspective than this, one that takes into account the roots of the present shortage of energy and their connections to the other critical problems facing us today. Such a perspective makes apparent what may at first sound paradoxical: What we need, to overcome the energy crisis, is not *more* energy but *less*. Our ever increasing energy needs reflect the general expansion of our economic and technological systems; they are caused by the patterns of undifferentiated growth that deplete our natural resources and contribute significantly to our multiple symptoms of individual and social illness. Energy, then, is a significant parameter of social and ecological balance. In our present, highly unbalanced state, more of it would not solve our problems but would make them worse. It would not only accelerate the depletion of our minerals and metals, forests and fish, but would also mean more pollution, more chemical poisoning, more social injustice, more cancer, more crime. What we need to overcome our multifaceted crisis is not more energy but a profound change of values, attitudes, and life styles.

Once these basic facts are perceived, it becomes evident that the use of nuclear power as an energy source is sheer folly. It surpasses the ecological impact of large-scale energy production from coal, which is already devastating, by several orders of magnitude, threatening not only to poison our natural environment for thousands of years but even to extinguish the entire human species. Nuclear power represents the most extreme case of a technology that has got out of hand, propelled by an obsession with self-assertion and control that has reached highly pathological levels.

In describing nuclear power in such terms, I am referring to both nuclear weapons and reactors. It is an intrinsic property of nuclear technology that these two applications cannot be separated. The term "nuclear power" itself has two linked meanings. "Power" has not only the technical meaning of "source of energy," but also the more general meaning of "possession of control or influence over others." In the case of nuclear power, these two kinds of power are inseparably connected, and both of them represent today the greatest threat to our survival and well-being.

Over the past two decades the U.S. Defense Department and the military industry have succeeded in creating a series of public hysterias about national defense in order to be granted regular increases in military spending. To do so military analysts have perpetuated the myth of an arms race in which the Russians are ahead of the United States. In reality the United States has been leading the Soviet Union in this insane competition ever since it began. Daniel Ellsberg has shown convincingly, by making available classified information, that the American military knew it was vastly superior to the Russians in strategic nuclear weapons throughout the 1950s and early 1960s.[5] American plans, based on this superiority, contemplated first use of nuclear weapons—in other words, initiating a nuclear war—and several American presidents made explicit nuclear threats, all of which were kept secret from the public.

In the meantime the Soviet Union has also built up a massive nuclear force, and today the Pentagon is again trying to brainwash the American people into believing that the Russians are ahead. Actually there is now a balance of power; equivalence in armament is a fair description of the current situation. The reason why the Pentagon is again distorting the truth is that it wants the American military to regain the kind of superiority it had from 1945 to about 1965, which would enable the United States to make the kinds of nuclear threats it was making then.

Officially the American nuclear policy is one of deterrence, but a closer look at the present American nuclear arsenal and at the new weapons being designed shows clearly that the Pentagon's current plans are not aimed at deterrence at all. Their only purpose is a nuclear first strike against the Soviet Union. To get an idea of the American force of deterrence it is sufficient to consider the nuclear submarines. In the words of President Jimmy Carter, "Just one of our relatively invulnerable Poseidon submarines—less than two percent of our total nuclear force of submarines, aircraft, and landbased missiles—carries enough warheads to destroy every large and medium-sized city in the Soviet Union. Our deterrent is overwhelming."[6] Twenty to thirty of these submarines are always at sea, where they are virtually invulnerable. Even if the Soviet Union sends its entire nuclear force against the United States, it cannot destroy a single American submarine; and each submarine can threaten every one of its cities. Thus the United States at all times has the power to destroy every Russian city twenty to thirty times over. Seen against this background, the current increase in armament clearly has nothing to do with deterrence.

What American military designers are now developing are high-precision weapons, such as the new cruise and MX missiles, which can strike their targets from a distance of 6,000 miles with highest accuracy. The purpose of these weapons is to destroy an enemy missile in its silo before it is fired; in other words, these weapons are to be used in a nuclear first strike. Since it would make no sense to aim laser-guided missiles at empty silos, they cannot be regarded as defensive weapons;

they are clearly weapons of aggression. One of the most detailed studies of the nuclear arms race which comes to that conclusion has been published by Robert Aldridge, an aeronautical engineer who formerly worked for the Lockheed corporation, America's largest producer of weapons.[7] For sixteen years Aldridge helped to design every submarine-launched ballistic missile bought by the U.S. Navy, but he quit Lockheed in 1973 when he became aware of a profound shift in American nuclear policy, a shift from retaliation to first strike. As an engineer he could see a clear discrepancy between the announced purposes of the programs he was working on and their intrinsic design. Since then Aldridge has found that the trend he detected has continued and accelerated. His deep concern about American military policy led him to write his detailed report, which ends with these words:

> I must reluctantly conclude from the evidence that the United States is ahead now and is rapidly approaching a first-strike capability—which it should start deploying by the mid-1980s. The Soviet Union, meanwhile, seems to be struggling for a second best. There is no available evidence that the U.S.S.R. has the combined missile lethality, anti-submarine warfare potential, ballistic missile defense, or space warfare technology to attain a disabling first-strike before the end of this century, if then.[8]

This study, like Ellsberg's, shows clearly that the military's new weapons, contrary to what the Pentagon would make us believe, no longer increase American national security. On the contrary, the likelihood of nuclear war becomes greater with every weapon that is added.

In 1960–61, according to Ellsberg, there were American plans for a nuclear first strike against the Soviet Union in case of any direct military confrontation with the Russians anywhere in the world. This was the only, and inevitable, American response to direct Russian involvement in some local conflict. We may be sure that such planning is still going on in the Pentagon. If it is, this means that in response to some local conflict in the Middle East, in Africa, or anywhere else in the world, the Defense Department intends to initiate an all-out nuclear war in which there would be half a billion dead after the first exchange. The entire war could be over in thirty to sixty minutes, and almost no living thing would survive its consequences. In other words, the Pentagon is planning to extinguish the human species as well as most others.

This concept is known in the Defense Department as "mutually assured destruction"; its acronym, very appropriately, is MAD.

The psychological background to this nuclear madness is an overemphasis on self-assertion, control and power, excessive competition, and an obsession with "winning"—the typical traits of patriarchal culture. The aggressive threats that have been made by men throughout human history are now being made with nuclear weapons, without recognition of the enormous difference in violence and destructive potential. Nuclear weapons, then, are the most tragic case of people holding onto an old paradigm that has long lost its usefulness.

Today the outbreak of a nuclear conflict depends no longer on the United States and the Soviet Union alone. American nuclear technology—and with it the raw materials to make nuclear bombs—is being exported all over the world. Only ten to twenty pounds of plutonium are required to make a bomb, and each nuclear reactor produces four hundred to five hundred pounds of plutonium annually, enough for twenty to fifty atomic bombs. Through plutonium, reactor technology and weapons technology have become inseparably linked.

Nuclear technology is now being promoted especially in the Third World. The aim of this promotion is not to satisfy the energy needs of Third World countries, but those of multinational corporations extracting the natural resources from these countries as fast as they can. Politicians in Third World countries often welcome nuclear technology, however, because it gives them a chance to use it for building nuclear weapons. Current American sales of reactor technology abroad guarantee that by the end of the century dozens of countries will possess enough nuclear material to manufacture bombs of their own, and we can expect those countries not only to acquire the American technology but also to copy the American patterns of behavior and use their nuclear power to make aggressive threats.

The potential of global destruction through nuclear war is the greatest environmental threat of nuclear power. If we are unable to prevent nuclear war, all other environmental concerns will become purely academic. But even without a nuclear holocaust, the environmental impact of nuclear power far exceeds all other hazards of our technology. At the beginning of the so-called peaceful use of atomic energy, nuclear power was advocated as cheap, clean, and safe. In the meantime we have become painfully aware that it is none of those. The construction and maintenance of nuclear power plants is becoming ever more expensive owing to the elaborate safety measures imposed on the nuclear industry by public protest; nuclear accidents have threatened the health and safety of hundreds of thousands of people; and radioactive substances continually poison our environment.

The health hazards of nuclear power are of an ecological nature and operate on an extremely large scale, both in space and in time. Nuclear power plants and military facilities release radioactive substances that contaminate the environment, thus affecting all living organisms, including humans. The effects are not immediate but gradual, and they are accumulating to more dangerous levels all the time. In the human organism these substances contaminate the internal environment with many medium- and long-term consequences. Cancer tends to develop after ten to forty years, and genetic diseases can appear in future generations.

Scientists and engineers very often do not fully grasp the dangers of nuclear power, partly because our science and technology have always had great difficulty dealing with ecological concepts. Another reason is the great complexity of nuclear technology. The people responsible for its development and applications—physicists, engineers, economists, politicians, and generals—are all used to a fragmented approach and each group concerns itself with narrowly defined problems. They often ignore how these problems interrelate and how they combine to produce a total impact on the global ecosystem. Besides, most nuclear scientists and engineers suffer from a profound conflict of interest. Most are employed by the military or by the nuclear industry, both of which exert powerful influences. Consequently the only experts who can provide a comprehensive assessment of the hazards of nuclear power are those independent of the military-industrial complex and able to adopt a broad ecological perspective. Not surprisingly, they all tend to be in the antinuclear movement.[9]

In the process of producing energy from nuclear power, both the workers in the nuclear industry and the whole natural environment are contaminated with radioactive substances at every step of the "fuel cycle." This cycle begins with the mining, milling, and enrichment of uranium, continues with the fabrication of fuel rods and the operation and maintenance of the reactor, and ends with the handling and storage or reprocessing of nuclear waste. The radioactive substances that escape into the environment at every stage of this process emit particles—alpha particles,* electrons, or photons—that can be highly energetic, penetrating the skin and damaging body cells. Radioactive substances can also be ingested with contaminated food or water and will then do their damage from within.

When considering the health hazards of radioactivity, it is important to note that there is no "safe" level of radiation, contrary to what the nuclear industry would have us believe. Medical scientists now generally agree that there is no evidence of a threshold below which radiation may be said to be harmless;[10] even the smallest amounts can produce mutations and diseases. In everyday life we are continually exposed to low-level background radiation, which has been impinging on the earth for billions of years and which also comes from natural sources present in rocks, in water, and in plants and animals. The risks from this natural background are unavoidable, but to increase them means to gamble with our health.

The nuclear reaction that takes place in a reactor is known as fission. It is a process in which uranium nuclei break into fragments—most of which are radioactive substances—plus heat, plus one or two free neutrons. These neutrons are absorbed by other nuclei which, in turn, break up, thus setting in motion a chain reaction. In an atomic bomb, this chain reaction ends in an explosion, but in a reactor it can be controlled with the help of control rods, which absorb some of the free neutrons. In this way the rate of fission can be regulated. The fission process releases a tremendous amount of heat that is used to boil water. The resulting steam drives a turbine that generates electricity. A nuclear reactor, then, is a highly sophisticated, expensive, and extremely dangerous device for boiling water.

The human factor involved in all stages of nuclear technology, military and nonmilitary, makes accidents unavoidable. These accidents result in the release of highly poisonous radioactive materials into the environment. One of the worst possibilities is the melt-down of a nuclear reactor, in which the whole mass of molten uranium would burn through the container of the reactor and into the earth, possibly triggering a steam explosion that would scatter deadly radioactive materials. The effects would be similar to those of an atomic bomb. Thousands of people would die from immediate radiation exposure; more deaths would occur after two or three weeks from acute radiation ill-

* Alpha particles are compounds of two protons and two neutrons.

ness; large areas of land would be contaminated and made uninhabitable for thousands of years.

Many nuclear accidents have already happened, and major catastrophes have often been narrowly avoided. The accident at the Three Mile Island nuclear power plant near Harrisburg, Pennsylvania, in which the health and safety of hundreds of thousands of people were threatened, is still vivid in our memory. Less known, but not less frightening, are accidents involving nuclear weapons, accidents that have become more and more frequent as the number and capacity of those weapons have increased.[11] By 1968 there were more than thirty major accidents involving American nuclear weapons that came close to an explosion. One of the most serious occurred in 1961, when an H-bomb was accidentally dropped over Goldsboro, North Carolina, and five of its six safety devices failed. That one safety device protected us from a thermonuclear explosion of twenty-four million tons TNT, an explosion one thousand times more powerful than that of the Nagasaki bomb and, in fact, more powerful than the combined explosions of all the wars in human history. Several of these twenty-four-million-ton bombs have been dropped accidentally over Europe, the United States, and other parts of the world, and these accidents are going to occur even more frequently as more and more countries build nuclear weapons, probably with much less sophisticated safety devices.

Another major problem of nuclear power is the disposal of nuclear waste. Each reactor annually produces tons of radioactive waste that remains toxic for thousands of years. Plutonium, the most dangerous of the radioactive byproducts, is also the most long-lived; it remains poisonous for at least 500,000 years.* It is difficult to grasp the enormous length of this time span, which far exceeds the length of time we are used to contemplating within our individual lifetimes, or within the lifetime of a society, nation, or civilization. Half a million years, as shown on the chart below, is more than one hundred times longer than all of recorded history. It is a time span fifty times longer than that from the end of the Ice Age to the present day, and more than ten times longer than our entire existence as humans with our present

* The half-life of plutonium (Pu-239)—the time after which one-half of a given quantity has decayed—is 24,400 years. This means that if one gram of plutonium is released into the environment, about one-millionth of a gram will be left after 500,000 years, a quantity which is minute but still toxic.

physical characteristics.* This is the length of time that plutonium must be isolated from the environment. What moral right do we have to leave such a deadly legacy to thousands and thousands of generations?

No human technology can create safe containers for such an enormous time span. Indeed, no permanent safe method of disposal or storage has been found for nuclear waste, in spite of millions of dollars spent during three decades of research. Numerous leaks and accidents have shown the shortcomings of all current devices. In the meantime, nuclear waste keeps piling up. Projections by the nuclear industry anticipate a total of 152 million gallons of intensely radioactive, "high-level" waste by the year 2000, and while the precise amounts of military radioactive waste are kept secret, they can be expected to be enormously larger than those from industrial reactors.

Plutonium, named after Pluto, the Greek god of the underworld, is by far the most deadly of all nuclear waste products. Less than one-millionth of a gram—an invisible dose—is carcinogenic. One pound, if

* The ancestors of the European races are usually identified with the Cro-Magnon race, which appeared about 30,000 years ago and possessed all modern skeletal characteristics, including the large brain.

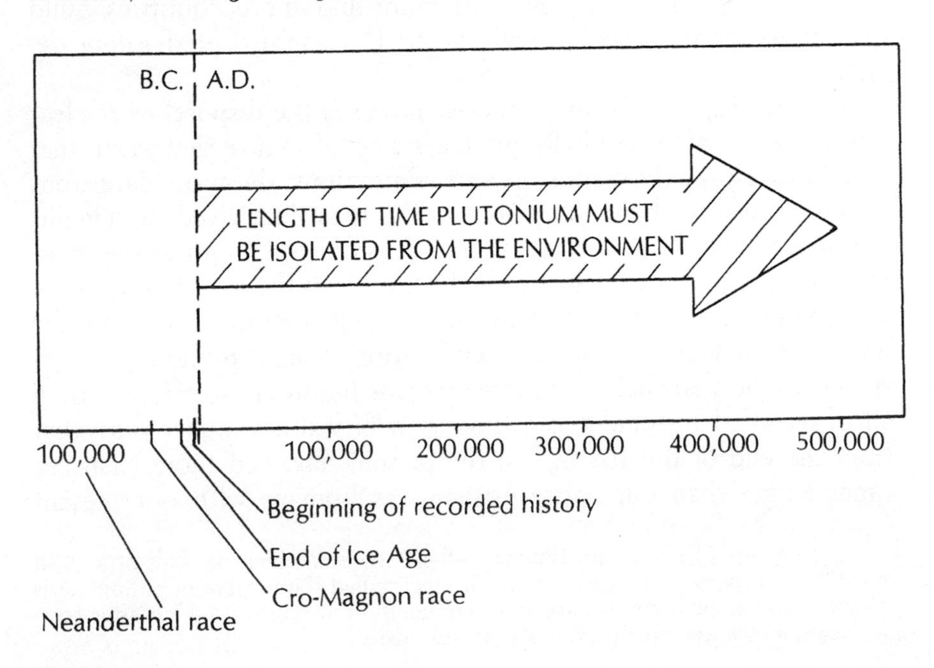

uniformly distributed, could potentially induce lung cancer in every person on earth. Given these facts, it is truly frightening to know that each commercial reactor produces four hundred to five hundred pounds of plutonium per year. Moreover, tons of plutonium are routinely transported along American highways and railroads and are flown into airports.

Once created, plutonium must be isolated from the environment virtually forever, since even the tiniest amounts will contaminate it for eons to come. It is important to realize that plutonium does not simply vanish with the death of a contaminated organism. A contaminated dead animal, for example, may be eaten by another animal, or it may decay and rot away, its dust scattered by the winds. In any case the plutonium will remain in the environment and will continue its lethal action, on and on, from organism to organism, for half a million years.

Since there is no hundred-percent-safe technology, some plutonium inevitably escapes when it is handled. It has been estimated that if the American nuclear industry expands according to projections made in 1975, and if it contains its plutonium with 99.99 percent perfection—which would be something of a miracle—it will be responsible for 500,000 fatal lung cancers per year for about fifty years following the year 2020. This will amount to a 25 percent increase in the total death rate in the United States.[12] In view of these estimates, it is difficult to understand how anybody can call nuclear power a safe source of energy.

Nuclear power also creates many other problems and hazards. They include the unsolved problem of disassembling, or "decommissioning" nuclear reactors at the end of their useful lives; the development of "fast breeder reactors," which use plutonium as a fuel and are far more dangerous than ordinary commercial reactors; the threat of nuclear terrorism and the ensuing loss of basic civil liberties in a totalitarian "plutonium economy"; and the disastrous economic consequences of the use of nuclear power as a capital- and technology-intensive, highly centralized source of energy.[13] The total impact of the unprecedented threats of nuclear technology should make it abundantly clear to anyone that it is unsafe, uneconomical, irresponsible, and immoral: totally unacceptable.

If the case against nuclear power is so convincing, why is nuclear technology still promoted so heavily? The deep reason is the obsession with power. Of all the available energy sources nuclear power is the one that leads to the highest concentration of political and economic power in the hands of a small elite. Because of its complex technology it requires highly centralized institutions, and because of its military aspects it lends itself to excessive secrecy and extensive use of police power. The various protagonists of the nuclear economy—the utilities, the manufacturers of reactors, and the energy corporations*—all benefit from a source of energy that is highly capital-intensive and centralized. They have invested billions of dollars in nuclear technology and continue to promote it vigorously in spite of its steadily growing problems and hazards. They are not prepared to abandon that technology, even if they are forced to ask for massive taxpayer subsidies and to use a large police force to protect it. As Ralph Nader says, nuclear power has become in many ways America's "technological Vietnam."[14]

Our obsession with economic growth and the value system underlying it have created a physical and mental environment in which life has become extremely unhealthy. Perhaps the most tragic aspect of this social dilemma is the fact that the health hazards created by the economic system are caused not only by the production process but by the consumption of many of the goods that are produced and heavily advertised to sustain economic expansion. To increase their profits in a saturated market, manufacturers have to produce their goods cheaper, and one way of doing this is to lower the quality of the products. To satisfy customers in spite of these low-quality products, vast sums of money are spent on conditioning consumer opinion and tastes through advertising. This practice, which has become an integral part of our economy, amounts to a serious health hazard because many of the goods produced and sold in this way have a direct effect on our health.

The food industry represents an outstanding example of health hazards generated by commercial interests. Although nutrition is one of the most important influences on our health, this is not emphasized in our system of health care and doctors are notoriously ignorant when it comes to questions of diet. Still, the basic features of a healthy diet are well known.[15] To be healthy and nutritious, our diet should be well balanced, low in animal protein and high in natural, nonrefined carbo-

* "Energy corporations" is an appropriate term used by Ralph Nader to describe the oil companies which have extended their business into all branches of the energy industry, including the supply of uranium and plutonium, in an attempt to monopolize energy production.

hydrates. This can be achieved by relying on three basic foods—whole grains, vegetables, and fruits. Even more important than the detailed composition of our diet are the following three requirements: our foods should be *natural*, consisting of organic food elements in their natural, unaltered state; they should be *whole*, complete and unfragmented, neither refined nor enriched; and they should be *poison-free*, organically grown, free from poisonous chemical residues and additives. These dietary requirements are extremely simple, and yet they are almost impossible to follow in today's world.

To improve their business food manufacturers add preservatives to food to increase its shelf life; they replace healthy organic food with synthetic products, and try to make up for the lack of nutritious content by adding artificial flavoring and coloring agents. Such overprocessed, artificial food is heavily advertised on billboards and television, together with alcohol and cigarettes, those other two major health hazards. We are subject to barrages of commercials for "junk food"—soft drinks, sweet snacks, high-fat foods—which have all been proved unhealthy. A recent study in Chicago that analyzed advertising by food companies on four television stations concluded that "on weekdays over 70 percent, and on weekends over 85 percent of food advertising is negatively related to the nation's health needs." Another study found that more than 50 percent of the money spent for food advertising on television is used to promote items closely linked to the most significant risk factors in the American diet.[16]

For a large number of people in our culture the problems of an unhealthy diet are compounded by excessive use of drugs, both medical and nonmedical. Although alcohol continues to cause more problems for individual and social health than all other drugs combined, other kinds of drug abuse have become a significant threat to public health. In the United States aspirin alone is now consumed at the rate of 20,000 tons per year, which amounts to almost 225 tablets per person.[17] But the biggest problem today is the excessive use of prescription drugs. Their sales have soared at an unprecedented rate, especially during the last twenty years, with the strongest increase occurring in the prescription of psychoactive drugs—tranquilizers, sedatives, stimulants, and antidepressants.[18]

Medical drugs can be extremely helpful if used intelligently. They have alleviated a great deal of pain and suffering and have helped many patients with degenerative diseases who, even ten years ago, would have been much more miserable. At the same time countless people have been victims of overuse and misuse. The overuse of drugs in contemporary medicine is based on a limited conceptual model of illness and is perpetuated by the powerful pharmaceutical industry. The biomedical model of illness and the economic model on which drug manufacturers base their business reinforce each other because they both reflect the same reductionist approach to reality. In both cases a complex pattern of phenomena and values is reduced to a single overriding aspect.

The pharmaceutical industry is one of the largest industries whose rate of profit has remained very high throughout the past two decades, outranking the rates of other manufacturing industries by significant margins. One of the outstanding characteristics of the drug industry is an excessive emphasis on differentiation of basically similar products. To a large extent research and marketing are devoted to developing drugs that are perceived as distinctive and superior, no matter how closely they resemble competing products, and huge sums are spent on advertising and promotion to establish a drug's distinctiveness far beyond any scientific justification.[19] As a consequence the market has been inundated with thousands of redundant medical drugs, many of them only marginally effective and all with harmful side effects.

To study the ways in which the drug industry sells its products is highly instructive.[20] In the United States the industry is controlled by the Pharmaceutical Manufacturers Association, a policy-making body that influences nearly every facet of the medical system. The PMA has close ties with the American Medical Association, and a large amount of the AMA's revenues comes from advertising in its medical journals. The largest of these periodicals is the *Journal of the American Medical Association*, whose apparent purpose is to keep physicians informed about new developments in medicine but which in fact is largely dominated by the interests of the pharmaceutical industry. The same is true for most other medical journals, which according to reliable estimates receive about half their income from advertising accounts with drug companies.[21]

This strong financial dependence of professional journals on an industry—a unique characteristic of the medical profession—is bound to affect the editorial policies of these journals. Indeed, numerous examples of conflict of interest have been observed. One involved a certain hormone called Norlutin, which was found to have harmful effects on some fetuses when taken during pregnancy.[22] According to a report in the March 1960 issue of the *JAMA*, these side effects of Norlutin occurred "with sufficient frequency to preclude its use or advertisement as a safe hormone to be taken during pregnancy." Yet in the same issue and for the next three months the journal continued to run a full-page advertisement for Norlutin with no reference to its possible side effects. Eventually the drug was taken off the market.

This was not an isolated event. The AMA has consistently neglected to inform physicians sufficiently about the adverse effects of antibiotics, which may well be the drugs most abused by physicians and most dangerous to patients. Unnecessary or negligent prescription of antibiotics has resulted in thousands of deaths, yet the AMA provides unlimited advertising space for antibiotics with no attempts at a disclaimer. This irresponsible advertising is certainly not unrelated to the fact that, next to sedatives and tranquilizers, antibiotics provide the largest advertising income for the American Medical Association.

Pharmaceutical advertising is designed specifically to induce doctors to prescribe ever more drugs. Naturally, then, these drugs are described as the ideal solution to a wide variety of everyday problems. Stressful life situations with physical, psychological, or social origins are presented as diseases amenable to drug treatment. Thus tranquilizers are advertised as remedies for "environmental depression" or "not fitting in," and other drugs are suggested as convenient means to "pacify" elderly patients or unruly school children. The tone of some of these ads, which are addressed to doctors, is absolutely horrifying to the lay person, especially when they advertise the treatment of women.[23] Women suffer disproportionately from drug treatments; they take over 60 percent of all psychoactive drugs prescribed and over 70 percent of all antidepressants. Advertisements often advise doctors in blatantly sexist language to get rid of female patients by giving them tranquilizers for vague complaints, or to prescribe drugs for women who are unhappy with their roles in society.

The influence of drug manufacturers on medical care extends far beyond journal advertising. In the United States the *Physician's Desk Reference* is the most popular reference book on drugs and is consulted regularly by more than 75 percent of the doctors. It lists every drug on the market, with its uses, recommended dosages, and side effects. Yet this standard work amounts to little more than straight advertising, since all its contents are prepared and paid for by the drug companies and it is distributed free to every physician in the country. For most doctors education about current drugs does not come from independent and objective pharmacologists but is provided almost exclusively by the highly media-conscious and manipulative producers. We can gauge the strength of this influence by noticing how rarely doctors use the proper technical terms when they refer to drugs; instead they use, and thus promote, the brand names made up by the drug companies.

Even more influential than its manual and journal advertising is the pharmaceutical industry's sales force. To sell their goods these "detail men" saturate doctors not only with smooth sales talk but also with briefcases full of drug samples, plus every imaginable promotional ploy. Many companies offer physicians prizes, gifts, and premiums in proportion to the amount of drugs prescribed—tape recorders, pocket calculators, dishwashers, refrigerators, and portable TV sets.[24] Others offer week-long "educational seminars" in the Bahamas with all expenses paid. It has been estimated that the drug companies collectively spend an average of $4,000 a year *per physician* on promotional ploys,[25] which amounts to 65 percent more than they spend on research and development.

The influence of the pharmaceutical industry on the practice of medicine has an interesting parallel in the influence of the petrochemical* industry on agriculture and farming. Farmers, like doctors, deal with living organisms that are severely affected by the mechanistic and reductionist approach of our science and technology. Like the human organism, the soil is a living system that has to remain in a state of dynamic balance to be healthy. When the balance is disturbed there will be pathological growth of certain components—bacteria or cancer cells in the human body, weeds or pests in the fields. Disease will occur, and eventually the whole organism may die and turn into inorganic matter. These effects have become major problems in modern agriculture because of the farming methods promoted by the petrochemical companies. As the pharmaceutical industry has conditioned doctors and patients to believe that the human body needs continual medical supervision and drug treatment to stay healthy, so the petrochemical industry has made farmers believe that soil needs massive in-

* Petrochemicals are chemicals isolated or derived from petroleum.

fusions of chemicals, supervised by agricultural scientists and technicians, to remain productive. In both cases these practices have seriously disrupted the natural balance of the living system and thus generated numerous diseases. Moreover, the two systems are directly connected, since any imbalance in the soil will affect the food that grows in it and thus the health of the people who eat that food.

A fertile soil is a living soil containing billions of living organisms in every cubic centimeter. It is a complex ecosystem in which the substances that are essential to life move in cycles from plants to animals, to soil bacteria, and back again to plants.[26] Carbon and nitrogen are two basic chemical elements that go through these ecological cycles, in addition to many other nutrient chemicals and minerals. Solar energy is the natural fuel that drives the soil cycles, and living organisms of all sizes are necessary to sustain the whole system and keep it in balance. Thus bacteria carry out various chemical transformations, such as the process of nitrogen fixation, which make nutrients accessible to plants; deep-rooted weeds bring trace minerals to the soil surface where crops can make use of them; earthworms break up the soil and loosen its texture; and all these activities are interdependent and combine harmoniously to provide the nourishment that sustains all life on earth.

The basic nature of living soil requires agriculture, first and foremost, to preserve the integrity of the great ecological cycles. This principle was embodied in traditional farming methods, which were based on a profound respect for life. Farmers used to plant different crops every year, rotating them so that the balance in the soil was preserved. No pesticides were needed, since insects attracted to one crop would disappear with the next. Instead of using chemical fertilizers, farmers would enrich their fields with manure, thus returning organic matter to the soil to reenter the biological cycle.

This age-old practice of ecological farming changed drastically about three decades ago, when farmers switched from organic to synthetic products, which opened up vast markets for the oil companies. While the drug companies manipulated doctors to prescribe ever more drugs, the oil companies manipulated farmers to use ever more chemicals. Both the pharmaceutical industry and the petrochemical industry became multibillion-dollar businesses. For the farmers the immediate effect of the new farming methods was a spectacular improvement in agricultural production, and the new era of chemical farming was hailed as the "Green Revolution." Soon, however, the dark side of the new technology became apparent, and today it is evident that the Green Revolution has helped neither the farmers nor the land nor the starving millions. The only ones to gain from it were the petrochemical corporations.

The massive use of chemical fertilizers and pesticides changed the whole fabric of agriculture and farming. The industry persuaded farmers that they could make money by planting large fields with a single highly profitable crop and controlling weeds and pests with chemicals. The results of this practice of single-crop monocultures were great losses of genetic variety in the fields and, consequently, high risks of large acreages being destroyed by a single pest. Monocultures also affected the health of the people living in the farming areas, who were no longer able to obtain a balanced diet from locally grown foods and thus became more disease-prone.

With the new chemicals, farming became mechanized and energy-intensive, with automated harvesters, feeders, waterers, and many other labor-saving machines performing the work that had previously been done by millions of people. Narrow notions of efficiency helped to conceal the drawbacks of these capital-intensive farming methods, as farmers were seduced by the wonders of modern technology. Even as late as 1970 an article in the *National Geographic* presented the following enthusiastic and utterly naïve vision of future agriculture:

> Fields will be larger, with fewer trees, hedges, and roadways. Machines will be bigger and more powerful . . . They'll be automated, even radio-controlled, with closed circuit TV to let an operator sitting on a front porch monitor what is going on . . . Weather control may tame hailstorm and tornado dangers . . . Atomic energy may supply power to level hills or provide irrigation water from the sea.[27]

The reality, of course, was far less encouraging. While American farmers were able to triple their corn yields per acre and, at the same time, cut their labor by two-thirds, the amount of energy used to produce one acre of corn increased fourfold. The new style of farming favored large corporate farmers with big capital and forced most of the traditional single-family farmers, who could not afford to mechanize, to leave their land. Three million American farms have been eliminated in this way since 1945, with large numbers of people forced to leave the rural areas and join the masses of the urban unemployed as victims of the Green Revolution.[28]

Those farmers who were able to remain on the land had to accept a profound transformation of their image, role, and activities. From growers of edible foods, taking pride in feeding the world's people, farmers have turned into producers of industrial raw materials to be processed into commodities designed for mass marketing. Thus corn is converted to starch or syrup; soybeans become oils, pet food, or protein concentrates; wheat flour is made into frozen dough or packaged mixes. For the consumer the tie of these products to the land has almost disappeared, and it is not surprising that many children today grow up believing that food comes from supermarket shelves.

Farming as a whole has been turned into a huge industry, in which key decisions are made by "agriscientists" and passed on to "agribusinessmen" or "farming technicians"—the former farmers—through a chain of agents and sales people. Thus farmers have lost most of their freedom and creativity, and have become, in effect, consumers of production techniques. These techniques are not based on ecological considerations but are determined by the commodity market. Farmers can no longer grow or breed what the land indicates, nor what people need; they have to grow and breed what the market dictates.

In this industrialized system, which treats living matter like dead substances and uses animals like machines, penned in feedlots and cages, the process of farming is almost totally controlled by the petrochemical industry. Farmers get virtually all their information about farming techniques from the industry's sales force, just as most doctors get their information about drug therapy from the drug industry's "detail men." The information about chemical farming is almost totally unrelated to the real needs of the land. As Barry Commoner has noted, "One can almost admire the enterprise and clever salesmanship of the petrochemical industry. Somehow it has managed to convince the farmer that he should give up the free solar energy that drives the natural cycles and, instead, buy the needed energy—in the form of fertilizer and fuel—from the petrochemical industry."[29]

In spite of this massive indoctrination by the energy corporations, many farmers have preserved their ecological intuition, passed down from generation to generation. These men and women know that the chemical way of farming is harmful to the land, but they are often forced to adopt it because the whole economy of farming—tax structure, credit system, real estate system, and so on—has been set up in a way that gives them no choice. To quote Commoner again, "The giant corporations have made a colony of rural America."[30]

Nevertheless, a growing number of farmers have become aware of the hazards of chemical farming and are turning back to organic, ecological methods. Just as there is a grass-roots movement in the health field, there is a grass-roots movement in farming. The new organic farmers grow their crops without synthetic fertilizers, rotating them carefully and controlling pests with new ecological methods. Their results have been most impressive. Their food is healthier and tastes better, and their operations have also been shown to be more productive than those of conventional farms.[31] The new organic farming has recently sparked serious interest in the United States and in many European countries.

The long-term effects of excessive "chemotherapy" in agriculture have proven disastrous for the health of the soil and the people, for our social relations, and for the entire ecosystem of the planet. As the same crops are planted and fertilized synthetically year after year, the balance in the soil is disrupted. The amount of organic matter diminishes, and with it the soil's ability to retain moisture. The humus content is depleted and the soil's porosity reduced. These changes in soil texture entail a multitude of interrelated consequences. The depletion of organic matter makes the soil dead and dry; water runs through it but does not wet it. The ground becomes hard-packed, which forces farmers to use more powerful machines. On the other hand, dead soil is more susceptible to wind and water erosion, which are taking an increasing toll. For example, half of the topsoil in Iowa has been washed away in the last twenty-five years, and in 1976 two-thirds of America's agricultural counties were designated drought disaster areas. What is often called "drought," "wind breaking down the land," or "winterkill," are all consequences of sterile soil.

The massive use of chemical fertilizers has seriously affected the natural process of nitrogen fixation by damaging soil bacteria involved in this process. As a consequence crops are losing their ability to take up nutrients from the soil and becoming more and more addicted to synthetic chemicals. Because their efficiency in absorbing nutrients this way is much lower, not all the chemicals are taken up by the crop but leach into the ground water or drain from the fields into rivers and lakes.

The ecological imbalance caused by monocropping and by excessive use of chemical fertilizers inevitably results in enormous increases in pests and crop diseases, which farmers counteract by spraying ever larger doses of pesticides, thus fighting the effects of their overuse of chemicals by using even more. However, pesticides often can no longer destroy the pests because they tend to become immune to the chemicals. Since World War II, when massive use of pesticides began, crop losses due to insects have not decreased; on the contrary, they have almost doubled. Moreover, many crops are now attacked by new insects that were never known as pests before, and these new pests are becoming increasingly resistant to all insecticides.[32]

Since 1945 there has been a sixfold increase in the use of chemical fertilizers and a twelvefold increase in the use of pesticides on American farms. At the same time increased mechanization and longer transport routes have contributed further to the energy dependence of modern agriculture. As a result, 60 percent of the costs of food are now costs of petroleum. As the farmer Wes Jackson puts it succinctly: "We have literally moved our agricultural base from soil to oil."[33] When energy was cheap, it was easy for the petrochemical industry to persuade farmers to change from organic to chemical farming, but since the costs of petroleum began their steady climb, many farmers have realized that they can no longer afford the chemicals they now depend upon. With every enlargement of farming technology the indebtedness of farmers increases as well. Even in the 1970s an Iowa banker remarked quite frankly, "I occasionally wonder whether the average farmer will ever get out of debt."[34]

If the Green Revolution has had disastrous consequences for the well-being of farmers and the health of the soil, the hazards for human health have been no less severe. Excessive use of fertilizers and pesticides has sent great quantities of toxic chemicals seeping through the soil, contaminating the water table and showing up in food. Perhaps half the pesticides on the market are mixed with petroleum distillates that may destroy the body's natural immune system. Others contain substances which are related specifically to cancer.[35] Yet these alarming results have barely affected the sale and use of fertilizers and pesticides. Some of the more dangerous chemicals have been outlawed in the United States, but the oil companies continue to sell them in the Third World, where legislation is less strict, as the drug companies sell dangerous prescription drugs. In the case of pesticides all populations are directly affected by this unethical practice because the toxic chemicals come back on fruits and vegetables imported from Third World countries.[36]

One of the principal justifications for the Green Revolution has been the argument that the new agricultural technology is needed to feed the world's hungry. In an age of scarcity, so the argument goes, only increased production will solve the problem of hunger, and only large-scale agribusiness is able to produce more food. This argument is still used, long after detailed research has made it quite clear that the problem of world hunger is not at all a technical problem; it is social and political. One of the most lucid discussions of the relation between agribusiness and world hunger can be found in the work of Frances Moore Lappé and Joseph Collins,[37] founders of the Institute for Food and Development Policy in San Francisco. Extensive research has led these authors to conclude that scarcity of food is a myth and that agribusiness does not solve the problem of hunger but, on the contrary, perpetuates and even aggravates it. They point out that the central question is not how production can be increased, but rather what is grown and who eats it, and that the answers are determined by those who control the food-producing resources. Merely to introduce new technologies into a system marred by social inequalities will never solve the hunger problem; on the contrary, it will make it worse. Indeed, studies of the impact of the Green Revolution on hunger in the Third World have confirmed the same paradoxical and tragic result again and again. More food is being produced, yet more people are hungry. As Moore Lappé and Collins write, "In the Third World, on the whole there is more food and less to eat."

Research codirected by Moore Lappé and Collins has shown that there is no country in the world in which people could not feed themselves from their own resources, and that the amount of food produced in the world at present is sufficient to provide about eight billion people—more than twice the world population—with an adequate diet. Nor can scarcity of agricultural land be considered a cause of hunger. For example, China has twice as many people per cultivated acre as India, yet in China there is no large-scale hunger. Inequality is the main stumbling block in all current attempts to fight world hunger.

Agricultural "modernization"—mechanized large-scale farming—is highly profitable for a small elite, the new corporate "farmers," and drives millions of people off the land. Thus fewer people are gaining control over more and more land, and once these large landholders are established they no longer grow food according to local needs but switch to more profitable crops for export, while the local population starves. Examples of this vicious practice abound in all countries of the Third World. In Central America at least half of the agricultural land—and precisely the most fertile land—is used to grow cash crops for export while up to 70 percent of the children are undernourished. In Senegal vegetables for export to Europe are grown on choice land while the country's rural majority goes hungry. Rich, fertile land in Mexico that previously produced a dozen local foods is now used to grow asparagus for European gourmets. Other landowners in Mexico are switching to grapes for brandy, while entrepreneurs in Colombia are changing from growing wheat to growing carnations for export to the United States.

World hunger can be overcome only by transforming social relations in such a way that inequality is reduced at every level. The primary problem is not the redistribution of food but the redistribution of control over agricultural resources. Only when this control is democratized will the hungry be able to eat what is produced. Many countries have proved that social changes of this kind can be successful. In fact, 40 percent of the Third World population now lives in countries where hunger has been eliminated through common struggle. These countries do not use agriculture as the means to export income but rather use it to produce food first for themselves. Such a "food first" policy requires, as Moore Lappé and Collins have emphasized, that industrial crops should be planted only after people have met their basic needs, and that trade should be seen as an extension of domestic need rather than being determined strictly by foreign demand.

At the same time, we who live in industrialized countries will have to realize that our own food security is not being threatened by the hungry masses in the Third World, but by the food and agricultural corporations that perpetuate this massive starvation. Multinational agribusiness corporations are now in the process of creating a single world agricultural system in which they will be able to control all stages of food production and to manipulate both food supply and prices through well-established monopoly practices. This process is now well under way. In the United States almost 90 percent of the vegetable production is controlled by major processing corporations, and many farmers have no choice but to sign up with them or go out of business.

World-wide corporate control of food production would make it impossible ever to eliminate hunger. It would, in effect, establish a Global Supermarket in which the world's poor would be in direct competition with the affluent and thus would never be able to feed themselves. This effect can already be observed in many Third World countries, where people go hungry although food is grown in abundance right where they live. Their own government may subsidize its production and they themselves may even grow and harvest it. Yet they will never eat any because they are unable to pay the prices resulting from international competition.

In its continual efforts to expand and increase its profits, agribusiness not only perpetuates world hunger but is extremely careless in the way it treats the natural environment, to the extent of creating serious threats to the global ecosystem. For example, giant multinational companies such as Goodyear, Volkswagen, and Nestlé are now bulldozing hundreds of millions of acres in the Amazon River basin in Brazil to raise cattle for export. The environmental consequences of clearing such vast areas of tropical forest are likely to be disastrous. Ecologists warn that the actions of the torrential tropical rains and the equatorial sun may set off chain reactions that could significantly alter the climate throughout the world.

Agribusiness, then, ruins the soil on which our very existence depends, perpetuates social injustice and world hunger, and seriously threatens global ecological balance. An enterprise that was originally nourishing and life-sustaining has become a major hazard to individual, social, and ecological health.

The more we study the social problems of our time, the more we realize that the mechanistic world view and the value system associated with it have generated technologies, institutions, and life styles that are profoundly unhealthy. Many of these health hazards are further aggravated by the fact that our health care system is unable to deal with them appropriately because of its adherence to the same paradigm that is perpetuating the causes of ill health. Current health care is reduced to medical care within the biomedical framework, that is, centered on acute, hospital-based, and drug-oriented medicine. Health care and the prevention of illness are perceived as two different problems and, accordingly, health professionals are not very active in supporting environmental and social policies that are directly related to public health.

The shortcomings of our present system of health care result from the subtle interplay of two tendencies, both examined in some detail in previous chapters. One is the adherence to the narrow biomedical framework in which the relevance of nonbiological aspects to the understanding of illness is systematically denied. The other, no less important, is the pursuit of economic and institutional growth and of political power by the health industry, which has heavily invested in the technologies that emerged from the reductionist view of illness. The American health care system consists of a vast conglomeration of powerful institutions, motivated by economic growth and lacking any effective incentives to hold down health costs.[38] The system is dominated by the same financial and corporate forces that have shaped the other sectors of the economy, forces that are not primarily interested in public health but that control virtually all facets of health care—the structure of health insurance, the management of hospitals, the manufacture and promotion of drugs, the orientation of medical research and education, and recognition and licensing of nonmedical therapists. The dominance of corporate values in this system is evident in the current debates on national health insurance, in which the basic patterns of power are never questioned. That is why none of the schemes currently under discussion is likely to satisfy the health needs of the American people. As one study of health care in the United States has noted, "Just as federal defense appropriations subsidize the military-industrial complex, national health insurance will subsidize the medical-industrial complex."[39]

The aim of the health industry has been to turn health care into a commodity that can be sold to consumers according to the rules of the "free market" economy. To achieve this purpose the "health care delivery" system has been structured and organized like the large manufacturing industries. Rather than encouraging health care in small community health centers, where it can be adapted to individual needs and practiced with an emphasis on prevention and health education, the current system favors a highly centralized and technology-intensive approach that is profitable for the industry but expensive and unhealthy for the patients.

The present "health establishment" has heavily invested in the status quo and is vigorously opposed to any fundamental revision of health care. By effectively controlling medical education, research, and practice, the health industry tries to suppress all incentives to change and to make the current approach intellectually and financially rewarding for the medical elite that directs the practice of health care. However, the problems of rising health costs, the diminishing returns from medical care, and the increasing evidence that environmental, occupational, and social factors are the primary causes of ill health will inevitably force change. In fact this change has already begun and is rapidly gaining momentum. The holistic health movement is active both within and outside the medical system, and is supported and complemented by other popular movements—environmental groups, antinuclear organizations, consumer groups, social liberation movements—that have perceived the environmental and social influences on health and are committed to opposing and preventing the creation of health hazards through political action. All these movements subscribe to a holistic and ecological view of life and reject the value system that dominates our culture and is perpetuated by our social and political institutions. The new rising culture shares a vision of reality that is still being discussed and explored but will eventually emerge as a new paradigm, destined to eclipse the Cartesian world view in our society.

In the following chapters I shall try to outline a coherent conceptual framework based on the new vision of reality. I hope it will help the various movements in the rising culture become aware of their common ground. The new framework will be profoundly ecological, compatible with the views of many traditional cultures and consistent with the concepts and theories of modern physics. As a physicist, I find it rewarding to observe that the world view of modern physics is not only having a strong impact on the other sciences but also has the potential of being therapeutic and culturally unifying.

IV

THE NEW VISION OF REALITY

9 · The Systems View of Life

The new vision of reality we have been talking about is based on awareness of the essential interrelatedness and interdependence of all phenomena—physical, biological, psychological, social, and cultural. It transcends current disciplinary and conceptual boundaries and will be pursued within new institutions. At present there is no well-established framework, either conceptual or institutional, that would accommodate the formulation of the new paradigm, but the outlines of such a framework are already being shaped by many individuals, communities, and networks that are developing new ways of thinking and organizing themselves according to new principles.

In this situation it would seem that a bootstrap approach, similar to the one that contemporary physics has developed, may be most fruitful. This will mean gradually formulating a network of interlocking concepts and models and, at the same time, developing the corresponding social organizations. None of the theories and models will be any more fundamental than the others, and all of them will have to be mutually consistent. They will go beyond the conventional disciplinary distinctions, using whatever language becomes appropriate to describe different aspects of the multileveled, interrelated fabric of reality. Similarly, none of the new social institutions will be superior to or more important than any of the others, and all of them will have to be aware of one another and communicate and cooperate with one another.

In the following chapters I shall discuss some concepts, models, and organizations of that kind which have recently emerged, and I shall try to show how they hang together conceptually. I want to concentrate especially on approaches that are relevant to dealing with individual and social health. Since the concept of health itself depends crucially on one's view of living organisms and their relation to the environment, this presentation of the new paradigm will begin with a discussion of the nature of living organisms.

Most of contemporary biology and medicine adheres to a mechanistic view of life and tries to reduce the functioning of living organisms to well-defined cellular and molecular mechanisms. The mechanistic view is justified to some extent because living organisms do act, in part, like machines. They have developed a wide variety of machinelike parts and mechanisms—bones, muscle action, blood circulation, and so on—probably because machinelike functioning was advantageous in their evolution. This does not mean that living organisms *are* machines. Biological mechanisms are merely special cases of much broader principles of organization; in fact, no operation of any organism consists entirely of such mechanisms. Biomedical science, following Descartes, has concentrated too much on the machinelike properties of living matter and has neglected to study its organismic, or systemic, nature. Although knowledge of the cellular and molecular aspects of biological structures will continue to be important, a fuller understanding of life will be achieved only by developing a "systems biology," a biology that sees an organism as a living system rather than a machine.

The systems view looks at the world in terms of relationships and integration.[1] Systems are integrated wholes whose properties cannot be reduced to those of smaller units. Instead of concentrating on basic building blocks or basic substances, the systems approach emphasizes basic principles of organization. Examples of systems abound in nature. Every organism—from the smallest bacterium through the wide range of plants and animals to humans—is an integrated whole and thus a living system. Cells are living systems, and so are the various tissues and organs of the body, the human brain being the most complex example. But systems are not confined to individual organisms and their parts. The same aspects of wholeness are exhibited by social systems—such as an anthill, a beehive, or a human family—and by ecosystems that consist of a variety of organisms and inanimate matter in mutual interaction. What is preserved in a wilderness area is not individual trees or organisms but the complex web of relationships between them.

All these natural systems are wholes whose specific structures arise from the interactions and interdependence of their parts. The activity of systems involves a process known as transaction—the simultaneous and mutually interdependent interaction between multiple components.[2] Systemic properties are destroyed when a system is dissected, either physically or theoretically, into isolated elements. Although we can discern individual parts in any system, the nature of the whole is always different from the mere sum of its parts.

Another important aspect of systems is their intrinsically dynamic nature. Their forms are not rigid structures but are flexible yet stable manifestations of underlying processes. In the words of Paul Weiss,

> The features of order, manifested in the particular form of a structure and the regular array and distribution of its substructures, are no more than the visible index of regularities of the underlying dynamics operating in its domain. . . . Living form must be regarded as essentially an overt indicator of, or clue to, dynamics of the underlying formative processes.[3]

This description of the systems approach sounds quite similar to the description of modern physics in a previous chapter. Indeed, the "new physics," especially its bootstrap approach, is very close to general systems theory. It emphasizes relationships rather than isolated entities and, like the systems view, perceives these relationships as being inherently dynamic. Systems thinking is process thinking; form becomes associated with process, interrelation with interaction, and opposites are unified through oscillation.

The emergence of organic patterns is fundamentally different from the consecutive stacking of building blocks, or the manufacture of a machine product in precisely programmed steps. Nevertheless, it is important to realize that these operations, too, take place in living systems. Although they are of a more specialized and secondary nature, machinelike operations occur throughout the living world. The reductionist description of organisms can therefore be useful and may in some cases be necessary. It is dangerous only when it is taken to be the complete explanation. Reductionism and holism, analysis and synthesis, are complementary approaches that, used in proper balance, help us obtain a deeper knowledge of life.

With this understanding we can now approach the question of the nature of living organisms, and here it will be useful to examine the essential differences between an organism and a machine. Let us begin by specifying what kind of machine we are talking about. There are modern cybernetic* machines that exhibit several properties characteristic of organisms, so that the distinction between machine and organism becomes quite subtle. But these were not the machines that served as models for the mechanistic philosophy of seventeenth-century science. In the views of Descartes and Newton, the world was a seventeenth-century machine, essentially a clockwork. This is the type of machine we have in mind when we compare its functioning to that of living organisms.

The first obvious difference between machines and organisms is the fact that machines are constructed, whereas organisms grow. This fundamental difference means that the understanding of organisms must be process-oriented. For example, it is impossible to convey an accurate picture of a cell by means of static drawings or by describing the cell in terms of static forms. Cells, like all living systems, have to be understood in terms of processes reflecting the system's dynamic organization. Whereas the activities of a machine are determined by its structure, the relation is reversed in organisms—organic structure is determined by processes.

Machines are constructed by assembling a well-defined number of parts in a precise and preestablished way. Organisms, on the other hand, show a high degree of internal flexibility and plasticity. The shape of their components may vary within certain limits and no two organisms will have identical parts. Although the organism as a whole exhibits well-defined regularities and behavior patterns, the relationships between its parts are not rigidly determined. As Weiss has shown with many impressive examples, the behavior of the individual parts can, in fact, be so unique and irregular that it bears no sign of relevance to the order of the whole system.[4] This order is achieved by coordinating activities that do not rigidly constrain the parts but leave room for variation and flexibility, and it is this flexibility that enables living organisms to adapt to new circumstances.

* Cybernetics, from the Greek *kybernan* ("to govern"), is the study of control and self-regulation in machines and living organisms.

Machines function according to linear chains of cause and effect, and when they break down a single cause for the breakdown can usually be identified. In contrast, the functioning of organisms is guided by cyclical patterns of information flow known as feedback loops. For example, component A may affect component B; B may affect C; and C may "feed back" the influence to A and thus close the loop. When such a system breaks down, the breakdown is usually caused by multiple factors that may amplify each other through interdependent feedback loops. Which of these factors was the initial cause of the breakdown is often irrelevant.

This nonlinear interconnectedness of living organisms indicates that the conventional attempts of biomedical science to associate diseases with single causes are highly problematic. Moreover, it shows the fallacy of "genetic determinism," the belief that various physical or mental features of an individual organism are "controlled" or "dictated" by its genetic makeup. The systems view makes it clear that genes do not uniquely determine the functioning of an organism as cogs and wheels determine the working of a clock. Rather, genes are integral parts of an ordered whole and thus conform to its systemic organization.

The internal plasticity and flexibility of living systems, whose functioning is controlled by dynamic relations rather than rigid mechanical structures, gives rise to a number of characteristic properties that can be seen as different aspects of the same dynamic principle—the principle of self-organization.[5] A living organism is a self-organizing system, which means that its order in structure and function is not imposed by the environment but is established by the system itself. Self-organizing systems exhibit a certain degree of autonomy; for example, they tend to establish their size according to internal principles of organization, independent of environmental influences. This does not mean that living systems are isolated from their environment; on the contrary, they interact with it continually, but this interaction does not determine their organization. The two principal dynamic phenomena of self-organization are self-renewal—the ability of living systems continuously to renew and recycle their components while maintaining the integrity of their overall structure—and self-transcendence—the ability to reach out creatively beyond physical and mental boundaries in the processes of learning, development, and evolution.

The relative autonomy of self-organizing systems sheds new light on the age-old philosophical question of free will. From the systems point of view, both determinism and freedom are relative concepts. To the extent that a system is autonomous from its environment it is free; to the extent that it depends on it through continuous interaction its activity will be shaped by environmental influences. The relative autonomy of organisms usually increases with their complexity, and it reaches its culmination in human beings.

This relative concept of free will seems to be consistent with the views of mystical traditions that exhort their followers to transcend the notion of an isolated self and become aware that we are inseparable parts of the cosmos in which we are embedded. The goal of these traditions is to shed all ego sensations completely and, in mystical experience, merge with the totality of the cosmos. Once such a state is reached, the question of free will seems to lose its meaning. If I *am* the universe, there can be no "outside" influences and all my actions will be spontaneous and free. From the point of view of mystics, therefore, the notion of free will is relative, limited and—as they would say—illusory, like all other concepts we use in our rational descriptions of reality.

To maintain their self-organization living organisms have to remain in a special state that is not easy to describe in conventional terms. The comparison with machines will again be helpful. A clockwork, for example, is a relatively isolated system that needs energy to run but does not necessarily need to interact with its environment to keep functioning. Like all isolated systems it will proceed according to the second law of thermodynamics, from order to disorder, until it has reached a state of equilibrium in which all processes—motion, heat exchange, and so on—have come to a standstill. Living organisms function quite differently. They are open systems, which means that they have to maintain a continuous exchange of energy and matter with their environment to stay alive. This exchange involves taking in ordered structures, such as food, breaking them down and using some of their components to maintain or even increase the order of the organism. This process is known as metabolism. It allows the system to remain in a state of nonequilibrium, in which it is always "at work." A high degree of nonequilibrium is absolutely necessary for self-organization; living organisms are open systems that continually operate far from equilibrium.

At the same time these self-organizing systems have a high degree of stability, and this is where we run into difficulties with conventional language. The dictionary meanings of the word "stable" include "fixed," "not fluctuating," "unvarying," and "steady," all of which are inaccurate to describe organisms. The stability of self-organizing systems is utterly dynamic and must not be confused with equilibrium. It consists in maintaining the same overall structure in spite of ongoing changes and replacements of its components. A cell, for example, according to Weiss, "retains its identity far more conservatively and remains far more similar to itself from moment to moment, as well as to any other cell of the same strain, than one could ever predict from knowing only about its inventory of molecules, macromolecules, and organelles which is subject to incessant change, reshuffling, and milling of its population."[6] The same is true for human organisms. We replace all our cells, except for those in the brain, within a few years, yet we have no trouble recognizing our friends even after long periods of separation. Such is the dynamic stability of self-organizing systems.

The phenomenon of self-organization is not limited to living matter but occurs also in certain chemical systems, which have been studied extensively by the physical chemist and Nobel laureate Ilya Prigogine, who developed a detailed dynamic theory to describe their behavior.[7] Prigogine has called these systems "dissipative structures" to express the fact that they maintain and develop structure by breaking down other structures in the process of metabolism, thus creating entropy—disorder—which is subsequently dissipated in the form of degraded waste products. Dissipative chemical structures display the dynamics of self-organization in its simplest form, exhibiting most of the phenomena characteristic of life—self-renewal, adaptation, evolution, and even primitive forms of "mental" processes. The only reason why they are not considered alive is that they do not reproduce or form cells. These intriguing systems thus represent a link between animate and inanimate matter. Whether they are called living organisms or not is, ultimately, a matter of convention.

Self-renewal is an essential aspect of self-organizing systems. Whereas a machine is constructed to produce a specific product or to carry out a specific task intended by its designer, an organism is primarily engaged in renewing itself; cells are breaking down and building up structures, tissues and organs are replacing their cells in continual cycles. Thus the pancreas replaces most of its cells every twenty-four hours, the stomach lining every three days; our white blood cells are renewed in ten days and 98 percent of the protein in the brain is turned over in less than one month. All these processes are regulated in such a way that the overall pattern of the organism is preserved, and this remarkable ability of self-maintenance persists under a variety of circumstances, including changing environmental conditions and many kinds of interference. A machine will fail if its parts do not work in the rigorously predetermined manner, but an organism will maintain its functioning in a changing environment, keeping itself in running condition and repairing itself through healing and regeneration. The power of regenerating organic structures diminishes with increasing complexity of the organism. Flatworms, polyps, and starfish can regenerate almost their entire body from a small fraction; lizards, salamanders, crabs, lobsters, and many insects are able to renew a lost organ or limb; and higher animals, including humans, can renew tissues and thus heal their injuries.

Even though they are capable of maintaining and repairing themselves, no complex organisms can function indefinitely. They gradually deteriorate in the process of aging and, eventually, succumb to exhaustion even when relatively undamaged. To survive, these species have developed a form of "super-repair."[8] Instead of replacing the damaged or worn-out parts they replace the whole organism. This, of course, is the phenomenon of reproduction, which is characteristic of all life.

Fluctuations play a central role in the dynamics of self-maintenance. Any living system can be described in terms of interdependent variables, each of which can vary over a wide range between an upper and a lower limit. All variables oscillate between these limits, so that the system is in a state of continual fluctuation, even when there is no disturbance. Such a state is known as homeostasis. It is a state of dynamic, transactional balance in which there is great flexibility; in other words, the system has a large number of options for interacting with its environment. When there is some disturbance, the organism tends to return to its original state, and it does so by adapting in various ways to environmental changes. Feedback mechanisms come into play and tend to reduce any deviation from the balanced state. Because of these regulatory mechanisms, also known as negative feedback, the body temperature, blood pressure, and many other important conditions of higher organisms remain relatively constant even when the environment changes considerably. However, negative feedback is only one aspect of self-organization through fluctuations. The other aspect is positive feedback, which consists in amplifying certain deviations rather than damping them. We shall see that this phenomenon plays a crucial role in the processes of development, learning, and evolution.

The ability to adapt to a changing environment is an essential characteristic of living organisms and of social systems. Higher organisms are usually capable of three kinds of adaptation, which come into play successively during prolonged environmental changes.[9] A person who goes from sea level to a high altitude may begin to pant and her heart may race. These changes are swiftly reversible; descending the same day will make them disappear immediately. Adaptive changes of this kind are part of the phenomenon of stress, which consists of pushing one or several variables of the organism to their extreme values. As a consequence the system as a whole will be rigid with respect to these variables and thus unable to adapt to further stress. For example, the person at high altitude will not be able to run up a staircase. Furthermore, since all variables in the system are interlinked, a rigidity in one will also affect the others, and the loss of flexibility will spread through the system.

If the environmental change persists, the organism will go through a further process of adaptation. Complex physiological changes take place among the more stable components of the system to absorb the environmental impact and restore flexibility. Thus the person at high altitude will be able to breathe normally again after a certain period of time and to use her panting mechanism for adjusting to other emergencies that might otherwise be lethal. This form of adaptation is known as somatic* change. Acclimatization, habit-forming, and addiction are special cases of this process.

Through somatic change the organism recaptures some of its flexibility by substituting a deeper and more enduring change for a more superficial and reversible one. Such an adaptation will be achieved comparatively slowly and will be slower to reverse. Yet somatic changes are still reversible. This means that various circuits of the biological system must be available for such a reversal for the entire time during which the change is maintained. Such a prolonged loading of circuits will limit the organism's freedom to control other functions and thus reduce its flexibility. Although the system is more flexible

* Somatic means "bodily," from the Greek *soma* ("body").

after the somatic change than it was before, when it was under stress, it is still less flexible than it was before the original stress occurred. Somatic change, then, internalizes stress, and the accumulation of such internalized stress may, eventually, lead to illness.

The third kind of adaptation available to living organisms is the adaptation of the species in the process of evolution. The changes brought about by mutation, also known as genotypic* changes, are totally different from somatic changes. Through genotypic change a species adapts to the environment by shifting the range of some of its variables, and notably of those which result in the most economical changes. For example, when the climate gets colder an animal will grow thicker fur rather than just running around more to keep warm. Genotypic change provides more flexibility than somatic change. Since every cell contains a copy of the new genetic information, it will behave in the changed manner without needing any messages from surrounding tissues and organs. Thus more circuits of the system will remain open and the overall flexibility is increased. On the other hand, genotypic change is irreversible within the lifetime of an individual.

The three modes of adaptation are characterized by increasing flexibility and decreasing reversibility. The quickly reversible stress reaction will be replaced by somatic change in order to increase flexibility under continuing stress, and evolutionary adaptation will be induced to further increase flexibility when the organism has accumulated so many somatic changes that it becomes too rigid for survival. Thus successive modes of adaptation restore as much as possible the flexibility that the organism has lost under environmental stress. The flexibility of an individual organism will depend on how many of its variables are kept fluctuating within their tolerance limits; the more fluctuations, the greater the stability of the organism. For populations of organisms the criterion corresponding to flexibility is variability. Maximum genetic variation within a population provides the maximum number of possibilities for evolutionary adaptation.

The ability of species to adapt to environmental changes through genetic mutations has been studied extensively and very successfully in our century, together with the mechanisms of reproduction and heredity. However, these aspects represent only one side of the phenomenon

* Genotype is a technical term for the genetic constitution of an organism; genotypic changes are changes in the genetic makeup.

of evolution. The other side is the creative development of new structures and functions without any environmental pressure, which is a manifestation of the potential for self-transcendence that is inherent in all living organisms. The Darwinian concepts, therefore, express only one of two complementary views that are both necessary in understanding evolution. Discussion of the view of evolution as an essential manifestation of self-organizing systems will be easier if we first take a closer look at the relation between organisms and their environment.

As the notion of an independent physical entity has become problematic in subatomic physics, so has the notion of an independent organism in biology. Living organisms, being open systems, keep themselves alive and functioning through intense transactions with their environment, which itself consists partially of organisms. Thus the whole biosphere—our planetary ecosystem—is a dynamic and highly integrated web of living and nonliving forms. Although this web is multileveled, transactions and interdependencies exist among all its levels.

Most organisms are not only embedded in ecosystems but are complex ecosystems themselves, containing a host of smaller organisms that have considerable autonomy and yet integrate themselves harmoniously into the functioning of the whole. The smallest of these living components show an astonishing uniformity, resembling one another quite closely throughout the living world, as vividly described by Lewis Thomas:

> There they are, moving about in my cytoplasm. . . . They are much less closely related to me than to each other and to the free-living bacteria out under the hill. They feel like strangers, but the thought comes that the same creatures, precisely the same, are out there in the cells of seagulls, and whales, and dune grass, and seaweed, and hermit crabs, and further inland in the leaves of the beech in my backyard, and in the family of skunks beneath the back fence, and even in that fly on the window. Through them, I am connected: I have close relatives, once removed, all over the place.[10]

Although all living organisms exhibit conspicuous individuality and are relatively autonomous in their functioning, the boundaries between organism and environment are often difficult to ascertain. Some organisms can be considered alive only when they are in a certain environment; others belong to larger systems that behave more like an autonomous organism than its individual members; still others collaborate to build large structures which become ecosystems supporting hundreds of species.

In the world of microorganisms, viruses are among the most intriguing creatures, existing on the borderline between living and nonliving matter. They are only partly self-sufficient, alive only in a limited sense. Viruses are unable to function and multiply outside of living cells. They are vastly simpler than any microorganism, the simplest among them consisting of just a nucleic acid, DNA or RNA. In fact, outside of cells viruses show no apparent signs of life. They are simply chemicals, exhibiting highly complex but completely regular molecular structures.[11] In some cases it has even been possible to take viruses apart, purify their components, and then put them back together again without destroying their capacity to function.

Although isolated virus particles are just assemblages of chemicals, they consist of chemical substances of a very special kind—the proteins and nucleic acids that are the essential constituents of living matter.[12] In viruses these substances can be studied in isolation, and it was such studies that led molecular biologists to some of their greatest discoveries in the 1950s and 1960s. Nucleic acids are chainlike macromolecules that carry information for self-replication and protein synthesis. When a virus enters a living cell it is able to use the cell's biochemical machinery to build new virus particles according to the instructions encoded in its DNA or RNA. A virus, therefore, is not an ordinary parasite which takes nourishment from its host to live and reproduce itself. Being essentially a chemical message, it does not provide its own metabolism, nor can it perform many other functions characteristic of living organisms. Its only function is to take over the cell's replication machinery and use it to replicate new virus particles. This activity takes place at a frantic rate. Within an hour an infected cell can produce thousands of new viruses and in many cases the cell will be destroyed in the process. Since so many virus particles are produced by a single cell, a virus infection of a multicelled organism can rapidly destroy a great number of cells and thus lead to disease.

Although the structure and functioning of viruses is now well known, their basic nature still remains intriguing. Outside living cells a virus particle cannot be called a living organism; inside a cell it forms a living system together with the cell, but one of a very special kind. It is self-organizing, but the purpose of its organization is not the stability and survival of the entire virus-cell system. Its only aim is the production of new viruses that will then go on to form living systems of this peculiar kind in the environments provided by other cells.

The special way in which viruses exploit their environment is an exception in the living world. Most organisms integrate themselves harmoniously into their surroundings, and some of them reshape their environment in such a way that it becomes an ecosystem capable of supporting large numbers of animals and plants. The outstanding example of such ecosystem-building organisms are corals, which for a long time were thought to be plants but are more appropriately classified as animals. Coral polyps are tiny multicellular organisms that join to form large colonies and, as such, can grow massive skeletons of limestone. Over long periods of geological time many of these colonies have grown into huge coral reefs, which represent by far the largest structures created by living organisms on earth. These massive structures support innumerable bacteria, plants, and animals: encrusting organisms living on top of the coral framework, fishes and invertebrates hiding in its nooks and crannies, and various other creatures that cover virtually all the available space on the reef.[13] To build these densely populated ecosystems the coral polyps function in a highly coordinated way, sharing nervous networks and reproductive capabilities to such an extent that it is often difficult to consider them individual organisms.

Similar patterns of coordination exist in tightly knit animal societies of higher complexity. Extreme examples are the social insects—bees, wasps, ants, termites, and others—that form colonies whose members are so interdependent and in such close contact that the whole system resembles a large, multicreatured organism.[14] Bees and ants are unable to survive in isolation, but in great numbers they act almost like the cells of a complex organism with a collective intelligence and capabilities for adaptation far superior to those of its individual members. This phenomenon of animals joining up to form larger organismic systems is not limited to insects but can also be observed in several other species, including, of course, the human species.

Close coordination of activities exists not only among individuals of the same species but also among different species, and again the resulting living systems have the characteristics of single organisms. Many types of organisms that were thought to represent well-defined biological species have turned out, upon close examination, to consist of two or more different species in intimate biological association. This phenomenon, known as symbiosis, is so widespread throughout the living world that it has to be considered a central aspect of life. Symbiotic relationships are mutually advantageous to the associated partners, and they involve animals, plants, and microorganisms in almost every imaginable combination.[15] Many of these may have formed their union in the distant past and evolved toward ever more interdependence and exquisite adaptation to one another.

Bacteria frequently live in symbiosis with other organisms in a way that makes both their own lives and the lives of their hosts dependent on the symbiotic relationship. Soil bacteria, for example, alter the configurations of organic molecules so that they become usable for the energy needs of plants. To do so the bacteria incorporate themselves so intimately into the roots of the plants that the two are almost indistinguishable. Other bacteria live in symbiotic relationships in the tissues of higher organisms, especially in the intestinal tracts of animals and humans. Some of these intestinal microorganisms are highly beneficial to their hosts, contributing to their nutrition and increasing their resistance to disease.

At an even smaller scale, symbiosis takes place within the cells of all higher organisms and is crucial to the organization of cellular activities. Most cells contain a number of organelles, which perform specific functions and until recently were thought to be molecular structures built by the cell. But it now appears that some organelles are organisms in their own right.[16] The mitochondria, for example, which are often called the powerhouses of the cell because they fuel almost all cellular energy systems, contain their own genetic material and can replicate independently of the replication of the cell. They are permanent residents in all higher organisms, passed on from generation to generation and living in intimate symbiosis within each cell. Similarly, the chloroplasts of green plants which contain the chlorophyll and the apparatus for photosynthesis are independent, self-replicating inhabitants in the plants' cells.

The more one studies the living world the more one comes to realize that the tendency to associate, establish links, live inside one another and cooperate is an essential characteristic of living organisms. As Lewis Thomas has observed, "We do not have solitary beings. Every creature is, in some sense, connected to and dependent on the rest."[17]

Larger networks of organisms form ecosystems, together with various inanimate components linked to the animals, plants, and microorganisms through an intricate web of relations involving the exchange of matter and energy in continual cycles. Like individual organisms, ecosystems are self-organizing and self-regulating systems in which particular populations of organisms undergo periodic fluctuations. Because of the nonlinear nature of the pathways and interconnections within an ecosystem, any serious disturbance will not be limited to a single effect but is likely to spread throughout the system and may even be amplified by its internal feedback mechanisms.

In a balanced ecosystem animals and plants live together in a combination of competition and mutual dependency. Every species has the potential of undergoing an exponential population growth but these tendencies are kept in check by various controls and interactions. When the system is disturbed, exponential "runaways" will start to appear. Some plants will turn into "weeds" and some animals into "pests," and other species will be exterminated. The balance, or health, of the whole system will be threatened. Explosive growth of this kind is not limited to ecosystems but occurs also in single organisms. Cancers and other tumors are dramatic examples of pathological growth.

Detailed study of ecosystems over the past decades has shown quite clearly that most relationships between living organisms are essentially cooperative ones, characterized by coexistence and interdependence, and symbiotic in various degrees. Although there is competition, it usually takes place within a wider context of cooperation, so that the larger system is kept in balance. Even predator-prey relationships that are destructive for the immediate prey are generally beneficient for both species. This insight is in sharp contrast to the views of the Social Darwinists, who saw life exclusively in terms of competition, struggle, and destruction. Their view of nature has helped create a philosophy that legitimates exploitation and the disastrous impact of our technology on the natural environment. But such a view has no scientific justification, because it fails to perceive the integrative and cooperative principles that are essential aspects of the ways in which living systems organize themselves at all levels.

As Thomas has emphasized, even in cases where there have to be winners and losers the transaction is not necessarily a combat. For example, when two individuals of a certain species of corals find themselves in a place where there is room for only one, the smaller of the two will always disintegrate, and it will do so by means of its own autonomous mechanisms: "He is not thrown out, not outgamed, not outgunned; he simply chooses to bow out."[18] Excessive aggression, competition, and destructive behavior are predominant only in the human species and have to be dealt with in terms of cultural values rather than being "explained" pseudoscientifically as inherently natural phenomena.

Many aspects of the relationships between organisms and their environment can be described very coherently with the help of the systems concept of stratified order, which has been touched upon earlier.[19] The tendency of living systems to form multileveled structures whose levels differ in their complexity is all-pervasive throughout nature and has to be seen as a basic principle of self-organization. At each level of complexity we encounter systems that are integrated, self-organizing wholes consisting of smaller parts and, at the same time, acting as parts of larger wholes. For example, the human organism contains organ systems composed of several organs, each organ being made up of tissues and each tissue made up of cells. The relations between these systems levels can be represented by a "systems tree."

As in a real tree, there are interconnections and interdependencies between all systems levels; each level interacts and communicates with its total environment. The trunk of the systems tree indicates that the individual organism is connected to larger social and ecological systems, which in turn have the same tree structure (p. 281).

At each level the system under consideration may constitute an individual organism. A cell may be part of a tissue but may also be a microorganism which is part of an ecosystem, and very often it is impossible to draw a clear-cut distinction between these descriptions. Every subsystem is a relatively autonomous organism while also being a component of a larger organism; it is a "holon," in Arthur Koestler's term, manifesting both the independent properties of wholes and the dependent properties of parts. Thus the pervasiveness of order in the universe takes on a new meaning: order at one systems level is the consequence of self-organization at a larger level.

From an evolutionary point of view it is easy to understand why stratified, or multileveled, systems are so widespread in nature.[20] They evolve much more rapidly and have much better chances of survival

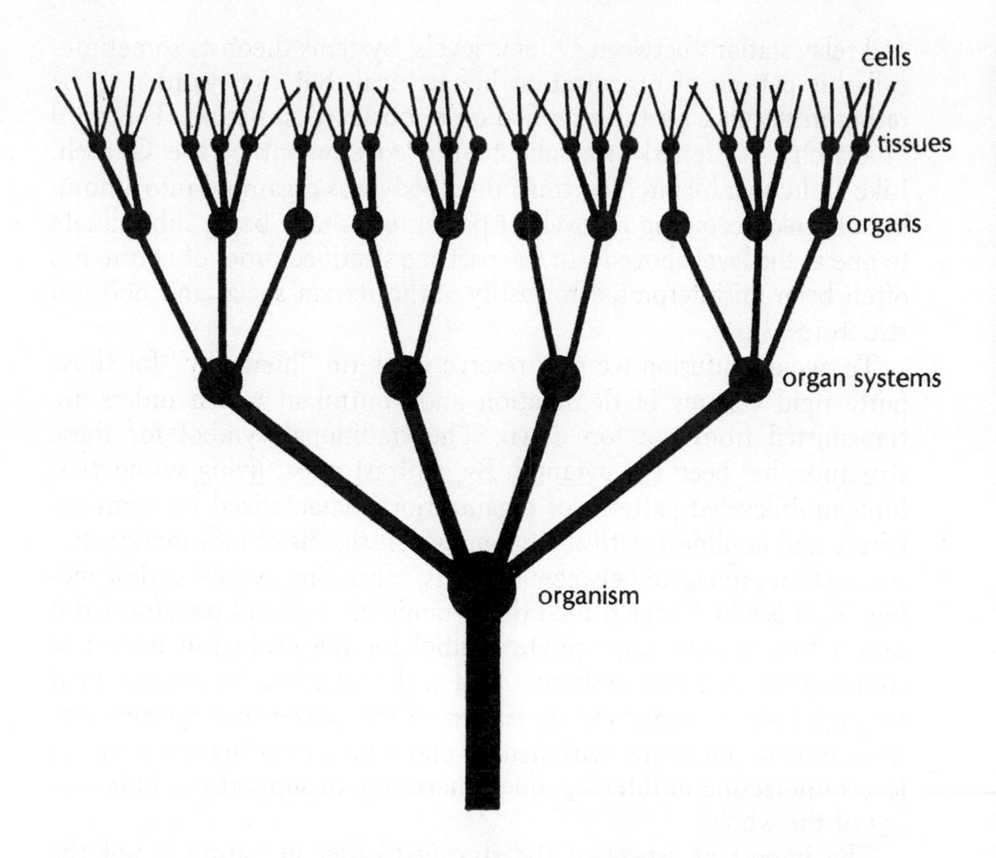

Systems tree representing various levels of complexity within an individual living organism.

than nonstratified systems, because in cases of severe disturbances they can decompose into their various subsystems without being completely destroyed. Nonstratified systems, on the other hand, would totally disintegrate and would have to start evolving again from scratch. Since living systems encounter many disturbances during their long history of evolution, nature has sensibly favored those which exhibit stratified order. As a matter of fact, there seem to be no records of survival of any others.

The multileveled structure of living organisms, like any other biological structure, is a visible manifestation of the underlying processes of self-organization. At each level there is a dynamic balance between self-assertive and integrative tendencies, and all holons act as interfaces and relay stations between systems levels. Systems theorists sometimes call this pattern of organization hierarchical, but that word may be rather misleading for the stratified order observed in nature. The word "hierarchy"* referred originally to the government of the Church. Like all human hierarchies, this ruling body was organized into a number of ranks according to levels of power, each rank being subordinate to one at the level above it. In the past the stratified order of nature has often been misinterpreted to justify authoritarian social and political structures.[21]

To avoid confusion we may reserve the term "hierarchy" for those fairly rigid systems of domination and control in which orders are transmitted from the top down. The traditional symbol for these structures has been the pyramid. By contrast, most living systems exhibit multileveled patterns of organization characterized by many intricate and nonlinear pathways along which signals of information and transaction propagate between all levels, ascending as well as descending. That is why I have turned the pyramid around and transformed it into a tree, a more appropriate symbol for the ecological nature of stratification in living systems. As a real tree takes its nourishment through both its roots and its leaves, so the power in a systems tree flows in both directions, with neither end dominating the other and all levels interacting in interdependent harmony to support the functioning of the whole.

The important aspect of the stratified order in nature is not the transfer of control but rather the organization of complexity. The various systems levels are stable levels of differing complexities, and this makes it possible to use different descriptions for each level. However, as Weiss has pointed out, any "level" under consideration is really the level of the observer's attention.[22] The new insight of subatomic physics also seems to hold for the study of living matter: the observed patterns of matter are reflections of patterns of mind.

The concept of stratified order also provides the proper perspective on the phenomenon of death. We have seen that self-renewal—the breaking down and building up of structures in continual cycles—is an essential aspect of living systems. But the structures that are continually being replaced are themselves living organisms. From their point of view the self-renewal of the larger system is their own cycle of birth

* From the Greek *hieros* ("sacred") and *arkhia* ("rule").

and death. Birth and death, therefore, now appear as a central aspect of self-organization, the very essence of life. Indeed, all living things around us renew themselves all the time, and this also means that everything around us dies all the time. "If you stand in a meadow," Thomas writes, "at the edge of a hillside and look around carefully, almost everything you can catch sight of is in the process of dying."[23] But for every organism that dies another one is born. Death, then, is not the opposite of life but an essential aspect of it.

Although death is a central aspect of life, not all organisms die. Simple one-celled organisms, such as bacteria and amoebae, reproduce by cell division and in doing so simply live on in their progeny. The bacteria around today are essentially the same that populated the earth billions of years ago, but they have branched into innumerable organisms. This kind of life without death was the only kind of life for the first two-thirds of evolutionary history. During that immense time span there was no aging and no death, but there was not much variety either—no higher life forms and no self-awareness. Then, about a billion years ago, the evolution of life went through an extraordinary acceleration and produced a great variety of forms. To do so, "life had to invent sex and death," as Leonard Shlain put it. "Without sex there could be no variety, without death no individuality."[24] From then on higher organisms would age and die and individuals would pair their chromosomes in sexual reproduction, thus generating enormous genetic variety which made evolution proceed several thousand times faster.

Stratified systems evolved along with these higher life forms, systems that renew themselves at all levels and thus maintain ongoing cycles of birth and death for all organisms throughout the tree structure. And this development brings us to questions about the place of human beings in the living world. Since we too are born and are bound to die, does this mean that we are parts of larger systems that continually renew themselves? Indeed, this seems to be the case. Like all other living creatures we belong to ecosystems and we also form our own social systems. Finally, at an even larger level, there is the biosphere, the ecosystem of the entire planet, upon which our survival is utterly dependent. We do not usually consider these larger systems as individual organisms like plants, animals, or people, but a new scientific hypothesis does just that at the largest accessible level. Detailed studies of the ways in which the biosphere seems to regulate the chemical composition of the air, the temperature on the surface of the earth, and many other aspects of the planetary environment have led the chemist James Lovelock and the microbiologist Lynn Margulis to suggest that these phenomena can be understood only if the planet as a whole is regarded as a single living organism. Recognizing that their hypothesis represents a renaissance of a powerful ancient myth, the two scientists have called it the Gaia hypothesis, after the Greek goddess of the earth.[25]

Awareness of the earth as alive, which played an important role in our cultural past, was dramatically revived when astronauts were able, for the first time in human history, to look at our planet from outer space. Their perception of the planet in all its shining beauty—a blue and white globe floating in the deep darkness of space—moved them deeply and, as many of them have since declared, was a profound spiritual experience that forever changed their relationship to the earth. The magnificent photographs of the "Whole Earth" which these astronauts brought back became a powerful new symbol for the ecology movement and may well be the most significant result of the whole space program.

What the astronauts, and countless men and women on earth before them, realized intuitively is now being confirmed by scientific investigations, as described in great detail in Lovelock's book. The planet is not only teeming with life but seems to be a living being in its own right. All the living matter on earth, together with the atmosphere, oceans, and soil, forms a complex system that has all the characteristic patterns of self-organization. It persists in a remarkable state of chemical and thermodynamic nonequilibrium and is able, through a huge variety of processes, to regulate the planetary environment so that optimal conditions for the evolution of life are maintained.

For example, the climate on earth has never been totally unfavorable for life since living forms first appeared, about four billion years ago. During that long period of time the radiation from the sun increased by at least 30 percent. If the earth were simply a solid inanimate object, its surface temperature would follow the sun's energy output, which means that the whole earth would have been a frozen sphere for more than a billion years. We know from geological records that such adverse conditions never existed. The planet maintained a fairly constant surface temperature throughout the evolution of life, much as a human organism maintains a constant body temperature in spite of varying environmental conditions.

Similar patterns of self-regulation can be observed for other environmental properties, such as the chemical composition of the atmosphere, the salt content of the oceans, and the distribution of trace elements among plants and animals. All these are regulated by intricate cooperative networks that exhibit the properties of self-organizing systems. The earth, then, is a living system; it functions not just *like* an organism but actually seems to *be* an organism—Gaia, a living planetary being. Her properties and activities cannot be predicted from the sum of her parts; every one of her tissues is linked to every other tissue and all of them are mutually interdependent; her many pathways of communication are highly complex and nonlinear; her form has evolved over billions of years and continues to evolve. These observations were made within a scientific context, but they go far beyond science. Like many other aspects of the new paradigm, they reflect a profound ecological awareness that is ultimately spiritual.

The systems view of living organisms is difficult to grasp from the perspective of classical science because it requires significant modifications of many classical concepts and ideas. The situation is not unlike the one encountered by physicists during the first three decades of this century, when they were forced to adopt drastic revisions of their basic concepts of reality to understand atomic phenomena. This parallel is further enforced by the fact that the notion of complementarity, which was so crucial in the development of atomic physics, also seems to play an important role in the new systems biology.

Besides the complementarity of self-assertive and integrative tendencies, which can be observed at all levels of nature's stratified systems, living organisms display another pair of complementary dynamic phenomena that are essential aspects of self-organization. One of them, which may be described loosely as self-maintenance, includes the processes of self-renewal, healing, homeostasis, and adaptation. The other, which seems to represent an opposing but complementary tendency, is that of self-transformation and self-transcendence, a phenomenon that expresses itself in the processes of learning, development, and evolution. Living organisms have an inherent potential for reaching out beyond themselves to create new structures and new patterns of behavior. This creative reaching out into novelty, which in time leads to an ordered unfolding of complexity, seems to be a fundamental property of life, a basic characteristic of the universe which is not—at least for the time being—amenable to further explanation. We can, however, explore the dynamics and mechanisms of self-transcendence in the evolution of individuals, species, ecosystems, societies, and cultures.

The two complementary tendencies of self-organizing systems are in continual dynamic interplay, and both of them contribute to the phenomenon of evolutionary adaptation. To understand this phenomenon, therefore, two complementary descriptions will be needed. One will have to include many aspects of neo-Darwinian theory, such as mutation, the structure of DNA, and the mechanisms of reproduction and heredity. The other description must deal not with the genetic mechanisms but with the underlying dynamics of evolution, whose central characteristic is not adaptation but creativity. If adaptation alone were the core of evolution, it would be hard to explain why living forms ever evolved beyond the blue-green algae, which are perfectly adapted to their environment, unsurpassed in their reproductive capacities, and have proved their fitness for survival over billions of years.

The creative unfolding of life toward forms of ever increasing complexity remained an unsolved mystery for more than a century after Darwin, but recent study has outlined the contours of a theory of evolution that promises to shed light on this striking characteristic of living organisms. This is a systems theory that focuses on the dynamics of self-transcendence and is based on the work of a number of scientists from various disciplines. Among the main contributors are the chemists Ilya Prigogine and Manfred Eigen, the biologists Conrad Waddington and Paul Weiss, the anthropologist Gregory Bateson, and the systems theorists Erich Jantsch and Ervin Laszlo. A comprehensive synthesis of the theory has recently been published by Erich Jantsch, who regards evolution as an essential aspect of the dynamics of self-organization.[26] This view makes it possible to begin to understand biological, social, cultural and cosmic evolution in terms of the same pattern of systems dynamics, even though the different kinds of evolution involve very different mechanisms. A basic complementarity of descriptions, which is still far from being understood, is manifest throughout the theory, examples being the interplay between adaptation and creation, the simultaneous action of chance and necessity, and the subtle interaction between macro- and microevolution.

The basic dynamics of evolution, according to the new systems view, begins with a system in homeostasis—a state of dynamic balance characterized by multiple, interdependent fluctuations. When the system is disturbed it has the tendency to maintain its stability by means of negative feedback mechanisms, which tend to reduce the deviation from the balanced state. However, this is not the only possibility. Deviations may also be reinforced internally through positive feedback, either in response to environmental changes or spontaneously without any external influence. The stability of a living system is continually tested by its fluctuations, and at certain moments one or several of them may become so strong that they drive the system over an instability into an entirely new structure, which will again be fluctuating and relatively stable. The stability of living systems is never absolute. It will persist as long as the fluctuations remain below a critical size, but any system is always ready to transform itself, always ready to evolve. This basic model of evolution, worked out for chemical dissipative structures by Prigogine and his collaborators, has since been applied successfully to describe the evolution of various biological, social, and ecological systems.

There are a number of fundamental differences between the new systems theory of evolution and the classical neo-Darwinian theory. The classical theory sees evolution as moving toward an equilibrium state, with organisms adapting themselves ever more perfectly to their environment. According to the systems view, evolution operates far from equilibrium and unfolds through an interplay of adaptation and creation. Moreover, the systems theory takes into account that the environment is, itself, a living system capable of adaptation and evolution. Thus the focus shifts from the evolution of an organism to the coevolution of organism plus environment. The consideration of such mutual adaptation and coevolution was neglected in the classical view, which has tended to concentrate on linear, sequential processes and to ignore transactional phenomena that are mutually conditioning and going on simultaneously.

Jacques Monod saw evolution as a strict sequence of chance and necessity, the chance of random mutations and the necessity of survival.[27] Chance and necessity are also aspects of the new theory, but their roles are quite different. The internal reinforcement of fluctuations and the way the system reaches a critical point may occur at random and are unpredictable, but once such a critical point has been reached the system is forced to evolve into a new structure. Thus chance and necessity come into play simultaneously and act as complementary principles. Moreover, the unpredictability of the whole process is not limited to the origin of the instability. When a system becomes unstable, there are always at least two new possible structures into which it can evolve. The further the system has moved from equilibrium, the more options will be available. Which of these options is chosen is impossible to predict; there is true freedom of choice. As the system approaches the critical point, it "decides" itself which way to go, and this decision will determine its evolution. The totality of possible evolutionary pathways may be imagined as a multiforked graph with free decisions at each branching point.[28]

This picture shows that evolution is basically open and indeterminate. There is no goal in it, or purpose, and yet there is a recognizable pattern of development. The details of this pattern are unpredictable because of the autonomy living systems possess in their evolution as in other aspects of their organization.[29] In the systems view the process of evolution is not dominated by "blind chance" but represents an unfolding of order and complexity that can be seen as a kind of learning process, involving autonomy and freedom of choice.

Since the days of Darwin, scientific and religious views about evolution have often been in opposition, the latter assuming that there was some general blueprint designed by a divine creator, the former reducing evolution to a cosmic game of dice. The new systems theory accepts neither of these views. Although it does not deny spirituality and can even be used to formulate the concept of a deity, as we shall see below, it does not allow for a preestablished evolutionary plan. Evolution is an ongoing and open adventure that continually creates its own purpose in a process whose detailed outcome is inherently unpredictable. Nevertheless, the general pattern of evolution can be recognized and is quite comprehensible. Its characteristics include the progressive increase of complexity, coordination, and interdependence; the integration of individuals into multileveled systems; and the continual refinement of certain functions and patterns of behavior. As Ervin Laszlo sums it up, "There is a progression from multiplicity and chaos to oneness and order."[30]

In classical science nature was seen as a mechanical system composed of basic building blocks. In accordance with this view, Darwin proposed a theory of evolution in which the unit of survival was the species, the subspecies, or some other building block of the biological world. But a century later it has become quite clear that the unit of survival is not any of these entities. What survives is the organism-in-its-environment.[31] An organism that thinks only in terms of its own survival will invariably destroy its environment and, as we are learning from bitter experience, will thus destroy itself. From the systems point of view the unit of survival is not an entity at all, but rather a pattern of organization adopted by an organism in its interactions with its environment; or, as neurologist Robert Livingston has expressed it, the evolutionary selection process acts on the basis of behavior.[32]

In the history of life on earth, the coevolution of microcosm and macrocosm is of particular importance. Conventional accounts of the origin of life usually describe the build-up of higher life forms in microevolution and neglect the macroevolutionary aspects. But these two are complementary aspects of the same evolutionary process, as Jantsch has emphasized.[33] From one perspective microscopic life creates the macroscopic conditions for its further evolution; from the other perspective the macroscopic biosphere creates its own microscopic life. The unfolding of complexity arises not from adaptation of organisms to a given environment but rather from the coevolution of organism and environment at all systems levels.

When the earliest life forms appeared on earth around four billion years ago—half a billion years after the formation of the planet—they were single-celled organisms without a cell nucleus that looked rather like some of today's bacteria. These so-called prokaryotes lived without oxygen, since there was little or no free oxygen in the atmosphere. But almost as soon as the microorganisms originated they began to modify their environment and create the macroscopic conditions for the further evolution of life. For the next two billion years some prokaryotes produced oxygen through photosynthesis, until it reached its present levels of concentration in the earth's atmosphere. Thus the stage was set for the emergence of more complex, oxygen-breathing cells that would be capable of forming cell tissues and multicellular organisms.

The next important evolutionary step was the emergence of eukaryotes, single-celled organisms with a nucleus containing the organism's genetic material in its chromosomes. It was these cells that later on formed multicellular organisms. According to Lynn Margulis, coauthor of the Gaia hypothesis, eukaryotic cells originated in a symbiosis between several prokaryotes that continued to live on as organelles within the new type of cell.[34] We have mentioned the two kinds of organelles—mitochondria and chloroplasts—that regulate the complementary respiration requirements of animals and plants. These are nothing but the former prokaryotes, which still continue to manage the energy household of the planetary Gaia system, as they have done for the past four billion years.

In the further evolution of life, two steps enormously accelerated the evolutionary process and produced an abundance of new forms. The first was the development of sexual reproduction, which introduced extraordinary genetic variety. The second step was the emergence of consciousness, which made it possible to replace the genetic mechanisms of evolution with more efficient social mechanisms, based upon conceptual thought and symbolic language.

To extend our systems view of life to a description of social and cultural evolution, we will deal first with the phenomena of mind and consciousness. Gregory Bateson proposed to define mind as a systems phenomenon characteristic of living organisms, societies, and ecosystems, and he listed a set of criteria which systems have to satisfy for mind to occur.[35] Any system that satisfies those criteria will be able to process information and develop the phenomena we associate with mind—thinking, learning, memory, for example. In Bateson's view, mind is a necessary and inevitable consequence of a certain complexity which begins long before organisms develop a brain and a higher nervous system.

Bateson's criteria for mind turn out to be closely related to those characteristics of self-organizing systems which I have listed above as the critical differences between machines and living organisms. Indeed, mind is an essential property of living systems. As Bateson said, "Mind is the essence of being alive."[36] From the systems point of view, life is not a substance or a force, and mind is not an entity interacting with matter. Both life and mind are manifestations of the same set of systemic properties, a set of processes that represent the dynamics of self-organization. This new concept will be of tremendous value in our attempts to overcome the Cartesian division. The description of mind as a pattern of organization, or a set of dynamic relationships, is related to the description of matter in modern physics. Mind and matter no longer appear to belong to two fundamentally separate categories, as Descartes believed, but can be seen to represent merely different aspects of the same universal process.

Bateson's concept of mind will be useful throughout our discussion, but to remain closer to conventional language I shall reserve the term "mind" for organisms of high complexity and will use "mentation," a term meaning mental activity, to describe the dynamics of self-organization at lower levels. This terminology was suggested some years ago by the biologist George Coghill, who developed a beautiful systemic view of living organisms and of mind well before the advent of systems theory.[37] Coghill distinguished three essential and closely interrelated patterns of organization in living organisms: structure, function, and mentation. He saw structure as organization in space, function as organization in time, and mentation as a kind of organization which is intimately interwoven with structure and function at low levels of complexity but goes beyond space and time at higher levels. From the modern systems perspective, we can say that mentation, being the dynamics of self-organization, represents the organization of all functions and is thus a meta-function. At lower levels it will often look like behavior, which can be defined as the totality of all functions, and thus the behaviorist approach is often successful at these levels. But at higher levels of complexity mentation can no longer be limited to behavior, as it takes on the distinctive nonspatial and nontemporal quality that we associate with mind.

In the systems concept of mind, mentation is characteristic not only of individual organisms but also of social and ecological systems. As Bateson has emphasized, mind is immanent not only in the body but also in the pathways and messages outside the body. There are larger manifestations of mind of which our individual minds are only subsystems. This recognition has very radical implications for our interactions with the natural environment. If we separate mental phenomena from the larger systems in which they are immanent and confine them to human individuals, we will see the environment as mindless and will tend to exploit it. Our attitudes will be very different when we realize that the environment is not only alive but also mindful, like ourselves.

The fact that the living world is organized in multileveled structures means that there are also levels of mind. In the organism, for example, there are various levels of "metabolic" mentation involving cells, tissues, and organs, and then there is the "neural" mentation of the brain, which itself consists of multiple levels corresponding to different stages of human evolution. The totality of these mentations constitutes what we would call the human mind. Such a notion of mind as a multileveled phenomenon, of which we are only partly aware in ordinary states of consciousness, is widespread in many non-Western cultures and has recently been studied extensively by some Western psychologists.[38]

In the stratified order of nature, individual human minds are embedded in the larger minds of social and ecological systems, and these are integrated into the planetary mental system—the mind of Gaia—which in turn must participate in some kind of universal or cosmic mind. The conceptual framework of the new systems approach is in no way restricted by associating this cosmic mind with the traditional idea of God. In the words of Jantsch, "God is not the creator, but the mind of the universe."[39] In this view the deity is, of course, neither male nor female, nor manifest in any personal form, but represents nothing less than the self-organizing dynamics of the entire cosmos.

The organ of neural mentation—the brain and its nervous system—is a highly complex, multileveled, and multidimensional living system that has remained deeply mysterious in many of its aspects in spite of several decades of intensive research in neuroscience.[40] The human brain is a living system par excellence. After the first year of growth no new neurons are produced, yet plastic changes will go on for the rest of its life. As the environment changes, the brain models itself in response to these changes, and any time it is injured the system makes very rapid adjustments. You can never wear it out; on the contrary, the more you use it, the more powerful it becomes.

The major function of neurons is to communicate with one another by receiving and transmitting electrical and chemical impulses. To do so each neuron has developed numerous fine filaments that branch out to make connections with other cells, thus establishing a vast and intricate network of communication which interweaves tightly with the muscular and skeletal systems. Most neurons are engaged in continual spontaneous activity, sending out a few pulses per second and modulating the patterns of their activity in various ways to transmit information. The entire brain is always active and alive, with billions of nervous impulses flashing through its pathways every second.

The nervous systems of higher animals and humans are so complex and display such a rich variety of phenomena that any attempt to understand their functioning in purely reductionistic terms seems quite hopeless. Indeed, neuroscientists have been able to map out the structure of the brain in some detail and have clarified many of its electrochemical processes, but they have remained almost completely ignorant about its integrative activities. As in the case of evolution, it would seem that two complementary approaches are needed: a reductionist approach to understand the detailed neural mechanisms, and a holistic approach to understand the integration of these mechanisms into the functioning of the whole system. So far there have been very few attempts to apply the dynamics of self-organizing systems to neural phenomena, but those currently undertaken have brought some encouraging results.[41] In particular, the significance of regular fluctuations in the process of perception, in the form of frequency patterns, has received considerable attention.

Another interesting development is the discovery that the two complementary modes of description which seem to be required to understand the nature of living systems are reflected in the very structure and functioning of our brains. Research over the past twenty years has shown consistently that the two hemispheres of the brain tend to be involved in opposite but complementary functions. The left hemisphere, which controls the right side of the body, seems to be more specialized in analytic, linear thinking, which involves processing information sequentially; the right hemisphere, controlling the left side of the body, seems to function predominantly in a holistic mode that is appropriate for synthesis and tends to process information more diffusely and simultaneously.

The two complementary modes of functioning have been demonstrated dramatically in a number of "split-brain" experiments involving epileptic patients whose corpus callosum, the band of fibers that normally connects the two hemispheres, had been cut. These patients showed some very striking anomalies. For example, with closed eyes they could describe an object they were holding in their right hand but could only make a guess if the object was held in the left hand. Similarly, the right hand could still write but could no longer draw pictures, whereas the opposite was the case for the left. Other experiments indicated that the different specializations of the two sides of the brain represented preferences rather than absolute distinctions, but the general picture was confirmed.[42]

In the past, brain researchers often referred to the left hemisphere as the major, and to the right as the minor hemisphere, thus expressing our culture's Cartesian bias in favor of rational thought, quantification, and analysis. Actually the preference for "left-brain" or "right-hand" values and activities is much older than the Cartesian world view. In most European languages the right side is associated with the good, the just, and the virtuous, the left side with evil, danger, and suspicion. The very word "right" also means "correct," "appropriate," "just," whereas "sinister," which is the Latin word for "left," conveys the idea of something evil and threatening. The German for "law" is *Recht,* and the French *droit,* both of which also mean "right." Examples of this kind can be found in virtually all Western languages and probably in many others as well. The deep-rooted preference for the right side—the one controlled by the left brain—in so many cultures makes one wonder whether it may not be related to the patriarchal value system. Whatever its origins may be, there have recently been attempts to promote more balanced views of brain functioning and to develop methods for increasing one's mental faculties by stimulating and integrating the functioning of both sides of the brain.[43]

The mental activities of living organisms from bacteria to primates can be discussed fairly consistently in terms of patterns of self-organization, without the need to modify one's language very much as one moves up the evolutionary ladder in the direction of increasing complexity. But with human organisms things become quite different. The human mind is able to create an inner world that mirrors the outer reality but has an existence of its own and can move an individual or a society to act upon the outer world. In human beings this inner world—the psychological realm—unfolds as an entirely new level and involves a number of phenomena that are characteristic of human nature.[44] They include self-awareness, conscious experience, conceptual thought, symbolic language, dreams, art, the creation of culture, a sense of values, interest in the remote past, and concern for the distant future. Most of these characteristics exist in rudimentary form in various animal species. In fact, there seems to be no single criterion that would allow us to distinguish humans from other animals. What is unique about human nature is a combination of characteristics foreshadowed in lower forms of evolution but integrated and developed to a high level of sophistication only in the human species.[45]

In our interactions with our environment there is a continual interplay and mutual influence between the outer world and our inner world. The patterns we perceive around us are based in a very fundamental way on the patterns within. Patterns of matter mirror patterns of mind, colored by subjective feelings and values. In the traditional Cartesian view it was assumed that every individual had basically the same biological apparatus and that each of us, therefore, had access to the same "screen" of sensory perception. The differences were assumed to arise from the subjective interpretation of the sensory data; they were due, in the well-known Cartesian metaphor, to the "little man looking at the screen." Recent neurophysiological studies have shown that this is not so. The modification of sensory perception by past experiences, expectations, and purposes occurs not only in the interpretation but begins at the very outset, at the "gates of perception." Numerous experiments have indicated that the registration of data by the sense organs will be different for different individuals *before* perception is experienced.[46] These studies show that the physiological aspects of perception cannot be separated from the psychological aspects of interpretation. Moreover, the new view of perception also blurs the conventional distinction between sensory and extrasensory perception—another vestige of Cartesian thinking—by showing that all perception is, to some extent, extrasensory.

Our responses to the environment, then, are determined not so much by the direct effect of external stimuli on our biological system but rather by our past experience, our expectations, our purposes, and the individual symbolic interpretation of our perceptual experience. The faint smell of a perfume may evoke joy or sorrow, pleasure or pain, through its association with past experience, and our response will vary accordingly. Thus the inner and outer worlds are always interlinked in the functioning of a human organism; they act upon each other and evolve together.

As human beings, we shape our environment very effectively because we are able to represent the outer world symbolically, to think conceptually, and to communicate our symbols, concepts, and ideas. We do so with the help of abstract language, but also nonverbally through paintings, music, and other forms of art. In our thinking and communication we not only deal with the present but can also refer to the past and anticipate the future, which gives us a degree of autonomy far beyond anything found in other species. The development of abstract thinking, symbolic language, and the various other human capabilities all depend crucially on a phenomenon that is characteristic of the human mind. Human beings possess consciousness; we are aware not only of our sensations but also of ourselves as thinking and experiencing individuals.

The nature of consciousness is a fundamental existential question that has fascinated men and women throughout the ages and has reemerged as a topic of intensive discussions among experts from various disciplines, including psychologists, physicists, philosophers, neuroscientists, artists, and representatives of mystical traditions. These discussions have often been very stimulating but have also created considerable confusion, because the term "consciousness" is being used in different senses by different people. It can mean subjective awareness, for example when conscious and unconscious activities are compared, but also self-awareness, which is the awareness of being aware. The term is also used by many in the sense of the totality of mind, with its many conscious and unconscious levels. And the discussion is further complicated by the recent strong interest in Eastern "psychologies" that have developed elaborate maps of the inner realm and use a dozen terms or more to describe its various aspects, all of them usually translated as "mind" or "consciousness."

In view of this situation, we need to specify carefully the sense in which the term consciousness is used. The human mind is a multileveled and integrated pattern of processes that represent the dynamics of human self-organization. Mind is a pattern of organization, and awareness is a property of mentation at any level, from single cells to human beings, although of course differing very widely in scope. Self-awareness, on the other hand, seems to manifest itself only in higher animals, unfolding fully in the human mind, and it is this property of mind that I mean by consciousness. The totality of the human mind, with its conscious and unconscious realms, I shall call, with Jung, the psyche.

Because the systems view of mind is not limited to individual organisms but can be extended to social and ecological systems, we may say that groups of people, societies, and cultures have a collective mind, and therefore also possess a collective consciousness. We may also follow Jung in the assumption that the collective mind, or collective psyche, also includes a collective unconscious.[47] As individuals we participate in these collective mental patterns, are influenced by them, and shape them in turn. In addition the concepts of a planetary mind and a cosmic mind may be associated with planetary and cosmic levels of consciousness.

Most theories about the nature of consciousness seem to be variations on either of two opposing views that may nevertheless be complementary and reconcilable in the systems approach. One of these views may be called the Western scientific view. It considers matter as primary and consciousness as a property of complex material patterns that emerges at a certain stage of biological evolution. Most neuroscientists today subscribe to this view.[48] The other view of concioussness may be called the mystical view, since it is generally held in mystical traditions. It regards consciousness as the primary reality and ground of all being. In its purest form consciousness, according to this view, is nonmaterial, formless, and void of all content; it is often described as "pure consciousness," "ultimate reality," "suchness," and the like.[49] This manifestation of pure consciousness is associated with the Divine in many spiritual traditions. It is said to be the essence of the universe and to manifest itself in all things; all forms of matter and all living beings are seen as patterns of divine consciousness.

The mystical view of consciousness is based on the experience of reality in nonordinary modes of awareness, which are traditionally achieved through meditation but may also occur spontaneously in the process of artistic creation and in various other contexts. Modern psychologists have come to call nonordinary experiences of this kind "transpersonal" because they seem to allow the individual mind to make contact with collective and even cosmic mental patterns. According to numerous testimonies, transpersonal experiences involve a strong, personal, and conscious relation to reality that goes far beyond the present scientific framework. We should therefore not expect science, at its present stage, to confirm or contradict the mystical view of consciousness.[50] Nevertheless, the systems view of mind seems perfectly consistent with both the scientific and the mystical views of consciousness, and thus to provide the ideal framework for unifying the two.

The systems view agrees with the conventional scientific view that consciousness is a manifestation of complex material patterns. To be more precise, it is a manifestation of living systems of a certain complexity. On the other hand, the biological structures of these systems are expressions of underlying processes that represent the system's self-organization, and hence its mind. In this sense material structures are no longer considered the primary reality. Extending this way of thinking to the universe as a whole, it is not too far-fetched to assume that all its structures—from subatomic particles to galaxies and from bacteria to human beings—are manifestations of the universe's self-organizing dynamics, which we have identified with the cosmic mind. But this is almost the mystical view, the only difference being that mystics emphasize the direct experience of cosmic consciousness that goes beyond the scientific approach. Still, the two approaches seem to be quite compatible. The systems view of nature at last seems to provide a meaningful scientific framework for approaching the age-old questions of the nature of life, mind, consciousness, and matter.

To understand human nature we study not only its physical and psychological dimensions but also its social and cultural manifestations. Human beings evolved as social animals and cannot keep well, physically or mentally, unless they remain in contact with other human beings. More than any other social species we engage in collective thinking, and in doing so we create a world of culture and values that becomes an integral part of our natural environment. Thus biological and cultural characteristics of human nature cannot be separated. Humankind emerged through the very process of creating culture and needs this culture for its survival and further evolution.

Human evolution, then, progresses through an interplay of inner and outer worlds, individuals and societies, nature and culture. All these realms are living systems in mutual interaction that display similar patterns of self-organization. Social institutions evolve toward increasing complexity and differentiation, not unlike organic structures, and mental patterns exhibit the creativity and urge for self-transcendence that is characteristic of all life. "It is the nature of the mind to be creative," observes the painter Gordon Onslow-Ford. "The more the depths of the mind are plumbed, the more abundantly they produce."[51]

According to generally accepted anthropological findings, the anatomical evolution of human nature was virtually completed some fifty thousand years ago. Since then the human body and brain have remained essentially the same in structure and size. On the other hand, the conditions of life have changed profoundly during this period and continue to change at a rapid pace. To adapt to these changes the human species used its faculties of consciousness, conceptual thought, and symbolic language to shift from genetic evolution to social evolution, which takes place much faster and provides far more variety. However, this new kind of adaptation was by no means perfect. We still carry around biological equipment from the very early stages of our evolution that often makes it difficult for us to meet the challenges of today's environment. The human brain, according to Paul Maclean's theory, consists of three structurally different parts, each endowed with its own intelligence and subjectivity, which stem from different periods of our evolutionary past.[52] Although the three parts are intimately linked, their activities are often contradictory and difficult to integrate, as MacLean shows in a picturesque metaphor: "Speaking allegorically of these three brains within a brain, we might imagine that when the psychiatrist bids the patient to lie on the couch, he is asking him to stretch out alongside a horse and a crocodile."[53]

The innermost part of the brain, known as the brain stem, is concerned with instinctive behavior patterns already exhibited by reptiles. It is responsible for biological drives and many kinds of compulsive behavior. Surrounding this part is the limbic system* which is well developed in all mammals and, in the human brain, is involved with emotional experience and expression. The two inner parts of the brain, also know as the subcortex, are strongly interconnected and express themselves nonverbally through a rich spectrum of body language. The outermost part, finally, is the neocortex† which facilitates higher-order abstract functions, such as thought and language. The neocortex originated in the earliest evolutionary phase of mammals and expanded in the human species at an explosive rate, unprecedented in the history of evolution, until it became stabilized about fifty thousand years ago.

By developing our capacity for abstract thinking at such a rapid pace, we seem to have lost the important ability to ritualize social conflicts. Throughout the animal world aggression rarely develops to the point where one of the two adversaries is killed. Instead, the fight is ritualized and usually ends with the loser conceding defeat but remaining relatively unharmed. This wisdom disappeared, or at least was deeply submerged, in the emergent human species. In the process of creating an abstract inner world we seem to have lost touch with the realities of life and have become the only creatures who often fail to cooperate with and even kill their own kind. The evolution of consciousness has given us not only the Cheops Pyramid, the Brandenburg Concertos,

* From the Latin *limbus* ("border").
† From the Latin *cortex* ("bark").

and the Theory of Relativity, but also the burning of witches, the Holocaust, and the bombing of Hiroshima. But that same evolution of consciousness gives us the potential to live peacefully and in harmony with the natural world in the future. Our evolution continues to offer us freedom of choice. We can consciously alter our behavior by changing our values and attitudes to regain the spirituality and ecological awareness we have lost.

In the future elaboration of the new holistic world view, the notion of rhythm is likely to play a very fundamental role. The systems approach has shown that living organisms are intrinsically dynamic, their visible forms being stable manifestations of underlying processes. Process and stability, however, are compatible only if the processes form rhythmic patterns—fluctuations, oscillations, vibrations, waves. The new systems biology shows that fluctuations are crucial in the dynamics of self-organization. They are the basis of order in the living world: ordered structures arise from rhythmic patterns.

The conceptual shift from structure to rhythm may be extremely useful in our attempts to find a unifying description of nature. Rhythmic patterns seem to be manifest at all levels. Atoms are patterns of probability waves, molecules are vibrating structures, and organisms are multidimensional, interdependent patterns of fluctuations. Plants, animals, and human beings undergo cycles of activity and rest, and all their physiological functions oscillate in rhythms of various periodicities. The components of ecosystems are interlinked through cyclical exchanges of matter and energy; civilizations rise and fall in evolutionary cycles, and the planet as a whole has its rhythms and recurrences as it spins around its axis and moves around the sun.

Rhythmic patterns, then, are a universal phenomenon, but at the same time they allow individuals to express their distinctive personalities. The manifestation of a unique personal identity is an important characteristic of human beings, and it appears that this identity may be, essentially, an identity of rhythm. Human individuals can be recognized by their characteristic speech patterns, body movements, gestures, breathing, all of which represent different kinds of rhythmic patterns. In addition, there are many "frozen" rhythms, like one's fingerprints or handwriting, that are uniquely associated with individuals. These observations indicate that the rhythmic patterns that characterize an individual human being are different manifestations of the same personal rhythm, an "inner pulse" which is the essence of personal identity.[54]

The crucial role of rhythm is not limited to self-organization and self-expression but extends to sensory perception and communication. When we see, our brain transforms the vibrations of light into rhythmic pulsations of its neurons. Similar transformations of rhythmic patterns occur in the process of hearing, and even the perception of odor seems to be based on "osmic frequencies." The Cartesian notion of separate objects and our experience with cameras have led us to assume that our senses create some kind of internal picture that is a faithful reproduction of reality. But this is not how sensory perception works. Pictures of separate objects exist only in our inner world of symbols, concepts, and ideas. The reality around us is an ongoing rhythmic dance, and our senses translate some of its vibrations into frequency patterns that can be processed by the brain.

The importance of frequencies in perception has been emphasized especially by the neuropsychologist Karl Pribram, who has developed a holographic* model of the brain in which visual perception is carried out through an analysis of frequency patterns and visual memory is organized like a hologram.[55] Pribram believes this explains why visual memory cannot be precisely localized within the brain. As in a hologram, the whole is encoded in each part. At present the validity of the hologram as a model for visual perception is not firmly established, but it is useful at least as a metaphor. Its main importance may be its emphasis on the fact that the brain does not store information locally but distributes it very widely, and, from a broader perspective, on the conceptual shift from structures to frequencies.

Another intriguing aspect of the holographic metaphor is a possible relation to two ideas in modern physics. One of them is Geoffrey Chew's idea of subatomic particles being dynamically composed of one another in such a way that each of them involves all the others;[56] the other idea is David Bohm's notion of implicate order, according to which all of reality is enfolded in each of its parts.[57] What all these approaches have in common is the idea that holonomy—the whole being somehow contained in each of its parts—may be a universal property of nature. This idea has also been expressed in many mystical tradi-

* Holography is a technique of lensless photography; see pp. 95–96 and reference note 29 for Chapter 3.

tions and seems to play an important role in mystical visions of reality.[58] The metaphor of the hologram has recently inspired a number of researchers and has been applied to various physical and psychological phenomena.[59] Unfortunately, this is not always done with the necessary caution, and the differences between a metaphor, a model, and the real world are sometimes overlooked in the general enthusiasm. The universe is definitely *not* a hologram, but it displays a multitude of vibrations of different frequencies, and thus the hologram may often be useful as an analogy to describe phenomena associated with these vibratory patterns.

As in the process of perception, rhythm also plays an important role in the many ways living organisms interact and communicate with one another. Human communication, for example, takes place to a significant extent through the synchronization and interlocking of individual rhythms. Recent film analyses have shown that every conversation involves a subtle and largely unseen dance in which the detailed sequence of speech patterns is precisely synchronized not only with minute movements of the speaker's body but also with corresponding movements by the listener.[60] Both partners are locked into an intricate and precisely synchronized sequence of rhythmic movements that lasts as long as they remain attentive and involved in their conversation. A similar interlocking of rhythms seems to be responsible for the strong bonding between infants and their mothers and, most likely, between lovers. On the other hand, opposition, antipathy, and disharmony will arise when the rhythms of two individuals are out of synchrony.

At rare moments in our lives we may feel that we are in synchrony with the whole universe. These moments may occur under many circumstances—hitting a perfect shot at tennis or finding the perfect run down a ski slope, in the midst of a fulfilling sexual experience, in contemplation of a great work of art, or in deep meditation. These moments of perfect rhythm, when everything feels exactly right and things are done with great ease, are high spiritual experiences in which every form of separateness or fragmentation is transcended.

In this discussion of the nature of living organisms we have seen that the systems view of life is spiritual in its deepest essence and thus consistent with many ideas held in mystical traditions. The parallels between science and mysticism are not confined to modern physics but can now be extended with equal justification to the new systems biology. Two basic themes emerge again and again from the study of living and nonliving matter and are also repeatedly emphasized in the teachings of mystics—the universal interconnectedness and interdependence of all phenomena, and the intrinsically dynamic nature of reality. We also find a number of ideas in mystical traditions that are less relevant, or not yet significant to, modern physics but are crucial to the systems view of living organisms.

The concept of stratified order plays a prominent role in many traditions. As in modern science, it involves the notion of multiple levels of reality which differ in their complexities and are mutually interacting and interdependent. These levels include, in particular, levels of mind, which are seen as different manifestations of cosmic consciousness. Although mystical views of consciousness go far beyond the framework of contemporary science, they are by no means inconsistent with the modern systems concepts of mind and matter. Similar considerations apply to the concept of free will, which is quite compatible with mystical views when associated with the relative autonomy of self-organizing systems.

The concepts of process, change, and fluctuation, which play such a crucial role in the systems view of living organisms, are emphasized in the Eastern mystical traditions, especially in Taoism. The idea of fluctuations as the basis of order, which Prigogine introduced into modern science, is one of the major themes in all Taoist texts. Because the Taoist sages recognized the importance of fluctuations in their observations of the living world, they also emphasized the opposite but complementary tendencies that seem to be an essential aspect of life. Among the Eastern traditions Taoism is the one with the most explicit ecological perspective, but the mutual interdependence of all aspects of reality and the nonlinear nature of its interconnections are emphasized throughout Eastern mysticism. For example, these are the ideas underlying the Indian concept of karma.

As in the systems view, birth and death are seen by many traditions as stages of endless cycles which represent the continual self-renewal that is characteristic of the dance of life. Other traditions emphasize vibratory patterns, often associated with "subtle energies," and many of them have described the holonomic nature of reality—the existence of "all in each and each in all"—in parables, metaphors, and poetic imagery.

Among Western mystics the one whose thought comes closest to that of the new systems biology is probably Pierre Teilhard de Chardin. Teilhard was not only a Jesuit priest but an eminent scientist who made major contributions to geology and paleontology.* He tried to integrate his scientific insights, mystical experiences, and theological doctrines into a coherent world view, which was dominated by process thinking and centered on the phenomenon of evolution.[61] Teilhard's theory of evolution is in sharp contrast to the neo-Darwinian theory but shows some remarkable similarities with the new systems theory. Its key concept is what he called the "Law of Complexity-Consciousness," which states that evolution proceeds in the direction of increasing complexity, and that this increase in complexity is accompanied by a corresponding rise of consciousness, culminating in human spirituality. Teilhard uses the term "consciousness" in the sense of awareness and defines it as "the specific effect of organized complexity," which is perfectly compatible with the systems view of mind.

Teilhard also postulated the manifestation of mind in larger systems and wrote that in human evolution the planet is covered with a web of ideas, for which he coined the term "mind-layer," or "noosphere."† Finally, he saw God as the source of all being, and in particular as the source of the evolutionary force. In view of the systems concept of God as the universal dynamics of self-organization, we can say that among the many images mystics have used to describe the Divine, Teilhard's concept of God, if liberated from its patriarchal connotations, may well be the one that comes closest to the views of modern science.

Teilhard de Chardin has often been ignored, disdained, or attacked by scientists unable to look beyond the reductionist Cartesian framework of their disciplines. However, with the new systems approach to the study of living organisms, his ideas will appear in a new light and are likely to contribute significantly to general recognition of the harmony between the views of scientists and mystics.

* Paleontology, from the Greek *palaios* ("ancient") and *onta* ("things"), is the study of past geological periods with the help of fossil remains.

† From the Greek *nóos* ("mind").

10·Wholeness and Health

To develop a holistic approach to health that will be consistent with the new physics and the systems view of living organisms, we do not need to break completely fresh ground but can learn from medical models existing in other cultures. Modern scientific thought—in physics, biology, and psychology—is leading to a view of reality that comes very close to the views of mystics and of many traditional cultures, in which knowledge of the human mind and body and the practice of healing are integral parts of natural philosophy and of spiritual discipline. A holistic approach to health and healing will therefore be in harmony with many traditional views, as well as consistent with modern scientific theories.

Comparisons of medical systems from different cultures should be made very carefully. Any system of health care, including modern Western medicine, is a product of its history and exists within a certain environmental and cultural context. As this context keeps changing, the health care system also changes, adapting itself continually to new situations and being modified by new economic, philosophical, and religious influences. Hence the usefulness of any medical system as a model for another society is quite limited. Nevertheless, it will be helpful to study traditional medical systems; not so much because they can serve as models for our society, but because cross-cultural studies will broaden our perspective and help us see current ideas about health and healing in a new light. In particular, we shall see that not all traditional cultures have approached health care in a holistic way. Throughout the ages cultures seem to have oscillated between reductionism and holism in their medical practices, probably in response to general fluctuations of value systems. However, when their approaches were fragmented and reductionistic, this reductionism was often very different from the one that dominates our current scientific medicine, and thus comparative studies may be very instructive.

In nonliterate cultures throughout the world, the origin of illness and the process of healing have been associated with forces belonging to the spirit world, and a great variety of healing rituals and practices have been developed to deal with illness accordingly. Among these, the phenomenon of shamanism offers a number of parallels to modern psychotherapies. The tradition of shamanism has existed since the dawn of history and still continues to be a vital force in many cultures throughout the world.[1] Its manifestations vary so much from culture to culture that it is almost impossible to make general statements about it, and there are probably many exceptions to each of the following generalizations.

A shaman is a man or woman who is able, at will, to enter into a nonordinary state of consciousness in order to make contact with the spirit world on behalf of members of his or her community. In nonliterate societies with little differentiation of roles and institutions, the shaman is usually the religious and political leader and also the doctor, and thus a very powerful and charismatic figure. As societies evolve, religion and politics become separate institutions, but religion and medicine generally stay together. The role of the shaman in those societies is to preside over religious rituals and to communicate with spirits for divination, diagnosis of illnesses, and healing. But it is also characteristic of traditional societies that most adults have some medical knowledge. Self-medication is very common and the shaman is needed only for difficult cases.

In addition to shamanistic traditions, the major cultures of the world have also developed secular medical systems that are not based on the use of trance but employ techniques passed on in written texts. These traditions usually establish themselves against the shamanistic systems. The shaman then loses his function as the leading ritual specialist and adviser to the people in power and becomes a peripheral figure often perceived as a potential threat to the power structure. In this situation the function of shamans is reduced to diagnosis, healing, and counseling at the local, village level. In spite of the widespread adoption of the Western and other secular medical systems, shamans have survived in this role throughout the world. In most countries with large rural areas shamanism is still the most important medical system, and it is also very much alive in the major cities of the world, especially in those with large populations of recent migrants.

The outstanding characteristic of the shamanistic conception of illness is the belief that human beings are integral parts of an ordered system and that all illness is the consequence of some disharmony with the cosmic order. Quite often illness is also interpreted as retribution for some immoral behavior. Accordingly, shamanistic therapies emphasize the restoration of harmony, or balance, within nature, in human relationships, and in relationships with the spirit world. Even minor illnesses and ailments such as sprains, fractures, or bites, are not seen as being due to bad luck but rather as inevitable manifestations of the larger order of things. Diagnosis and treatment of minor ailments rarely involve explanations beyond the immediate physical situation, however. Only when the patient does not get well quickly, or when the illness is more serious, are further explanations and causes sought.

Shamanistic ideas about disease causation are intimately linked with the patient's social and cultural environment. Whereas the focus of Western scientific medicine has been on the biological mechanisms and physiological processes that produce evidence of illness, the principal concern of shamanism is the sociocultural context in which the illness occurs. The disease process is either ignored altogether or is relegated to strictly secondary significance.[2] A Western doctor asked about the causes of an illness will talk about bacteria or physiological disorders; a shaman is likely to mention competition, jealousy and greed, witches and sorcerers, wrongdoing by a member of the patient's family, or some other way in which the patient or his kin failed to keep the moral order.

In shamanistic traditions human beings are viewed primarily in two ways: as part of a living social group and as part of a cultural belief system in which spirits and ghosts can actively intervene in human affairs. The individual psychological and spiritual state of the patient is less important. Men and women are not viewed predominantly as individuals; their life histories and personal experiences, including illnesses, are seen as the result of being part of a social group. In some traditions the social context is emphasized to such an extent that the organs, bodily functions, and symptoms of an individual are inseparably linked to social relations, plants, and other phenomena in the environment. For example, anthropologists observing the medical system of a village in

Zaire found it impossible to abstract a simple physical anatomy from the ideas about the body held in this culture, because the effective boundary of the person was consistently drawn much wider than in classical Western science and philosophy.[3]

In such cultures social circumstances are assigned overwhelmingly greater importance than are psychological or physical factors in determining the causes of an illness, and these medical systems are therefore often not holistic. Searching for a cause and pronouncing the diagnosis may sometimes be more important than the actual therapy. Diagnosis often takes place in front of the whole village and may involve disputes, arguments, and feuds between one family and another, with the patient barely noticed. The entire procedure is primarily a social event, the patient being merely a symbol of the conflict within the society.

Shamanistic therapies generally follow a psychosomatic* approach by applying psychological techniques to physical illnesses. The principal aim of these techniques is to reintegrate the patient's condition into the cosmic order. Claude Lévi-Strauss, in a classic article on shamanism, has given a detailed description of a complex Central American healing ritual in which the shaman cures a sick woman by calling upon the myths of her culture and using the appropriate symbolism to help her integrate her pain into a whole where everything is meaningful. Once the patient understands her condition in relation to this broader context, the healing takes place and she gets well.[4]

Shamanistic healing rituals often have the function of raising unconscious conflicts and resistances to a conscious level, where they can develop freely and find resolution. This, of course, is also the basic dynamic of modern psychotherapies, and indeed there are numerous similarities between shamanism and psychotherapy. Shamans used therapeutic techniques such as group sharing, psychodrama, dream analysis, suggestion, hypnosis, guided imagery, and psychedelic therapy for centuries before they were rediscovered by modern psychologists, but there is a significant difference between the two approaches. Whereas modern psychotherapists help their patients construct an individual myth with elements drawn from their past, shamans provide them with a social myth that is not limited to former personal experiences. Indeed, personal problems and needs are often ignored. The shaman does not work with the patient's individual unconscious, from which

* From the Greek *psyche* ("mind") and *soma* ("body").

these problems arise, but rather with the collective and social unconscious, which is shared by the whole community.

In spite of the difficulty of understanding shamanistic systems and comparing their concepts and techniques to those of our culture, such a comparison can be fruitful. The universal shamanistic view of human beings as integral parts of an ordered system is completely consistent with the modern systems view of nature, and the conception of illness as a consequence of disharmony and imbalance is likely to play a central role in the new holistic approach. Such an approach will have to go beyond the study of biological mechanisms and, like shamanism, find the causes of illness in environmental influences, psychological patterns, and social relations. Shamanism can teach us a lot about the social dimensions of illness, which are severely neglected not only in conventional medical care but also by many new organizations that claim to practice holistic medicine; and the great variety of psychological techniques that shamans use to integrate the patient's physical problems into a broader context offer many parallels to recently developed psychosomatic therapies.

Similar insights may be gained from the study of "high-tradition" medical systems, which were developed by the major civilizations of the world and passed on in written texts for hundreds and thousands of years. The wisdom and sophistication of these traditions are illustrated in two ancient medical systems—one Western and one Eastern—whose concepts of health and illness are still extremely relevant in our time and also resemble each other in several aspects. One of them is the tradition of Hippocratic medicine that lies at the roots of Western medical science, the other is the system of classical Chinese medicine that forms the basis of most East Asian medical traditions.

Hippocratic medicine emerged from an ancient Greek tradition of healing whose roots go far back into pre-Hellenic times. Throughout Greek antiquity healing was considered to be, essentially, a spiritual phenomenon and was associated with many deities. The most prominent among the early healing deities was Hygieia, one of the many manifestations of the Cretan goddess Athena, who was associated with snake symbolism and used the mistletoe as her all-heal.[5] Her curative rites were a secret guarded by priestesses. By the end of the second millennium B.C., patriarchal religion and social order had been imposed on Greece by three waves of barbarian invaders, and most of the earlier goddess myths were distorted and coopted into the new system, usually by portraying the goddess as the relative of a more powerful male god.[6] Thus Hygieia was made to be the daughter of Asclepius, who became the dominant healing god and was worshiped in temples all over Greece. In the cult of Asclepius, whose name is related etymologically to that of the mistletoe, snakes continued to play a prominent role, and the serpent, coiled around the Asclepian staff, has been the symbol of Western medicine ever since.

Hygieia, the goddess of health, continued to be associated with the Asclepian cult and was frequently portrayed with her father and with her sister Panakeia. In the new version of the myth the two goddesses associated with Asclepius represent two aspects of the healing arts that are as valid today as they were in ancient Greece—prevention and therapy.[7] Hygieia ("health") was concerned with the maintenance of health, personifying the wisdom that people would be healthy if they lived wisely. Panakeia ("all-healing") specialized in the knowledge of remedies, derived from plants or from the earth. The search for a panacea, or cure-all, has become a dominant theme in modern biomedical science, which often loses the balance between the two aspects of health care symbolized by the two goddesses. The Asclepian ritual involved a unique form of healing, based on dreams and known as temple incubation. Rooted in a firm belief in the healing powers of the god, it constituted an effective treatment procedure that Jungian psychotherapists have recently attempted to reinterpret in modern terms.[8]

The Asclepian ritual represented only one side of Greek medicine. Besides Asclepius the god there may also have existed a human physician of that name, said to be skilled in surgery and the use of drugs, and revered as the founder of medicine. Greek physicians called themselves Asclepiads ("sons of Asclepius") and formed medical guilds that promoted a form of medicine based on empirical knowledge. Although the Asclepiads had no connection with the dream therapy of the temple priests, the two schools were not in competition but complemented each other. Out of the lay Asclepiads emerged the tradition associated with the name of Hippocrates, which represents the culmination of Greek medicine and has had a lasting influence on Western medical science.[9] There is no doubt that a famous physician by that name lived in Greece around 400 B.C., practicing and teaching medicine as an Asclepiad on the island of Cos. The voluminous writings attributed to him, and known as the Hippocratic Corpus, were probably written by several authors at different times; they represent a compendium of the medical knowledge taught in various Asclepian guilds.

At the core of Hippocratic medicine is the conviction that illnesses are not caused by demons or other supernatural forces, but are natural phenomena that can be studied scientifically and influenced by therapeutic procedures and by wise management of one's life. Thus medicine should be practiced as a scientific discipline, based on the natural sciences and encompassing the prevention of illnesses, as well as their diagnosis and therapy. This attitude has formed the basis of scientific medicine to the present day, even though the successors of Hippocrates have rarely reached the breadth of vision and the depth of philosophical thought manifest in the Hippocratic writings.

Airs, Waters and Places, one of the most significant books of the Hippocratic Corpus, represents what we would now call a treatise on human ecology. It shows in great detail how the well-being of individuals is influenced by environmental factors—the quality of air, water, and food, the topography of the land, general living habits. The correlation between sudden changes in these factors and the appearance of disease is emphasized, and the understanding of environmental effects is seen as the essential basis of the physician's art. This aspect of Hippocratic medicine has been severely neglected with the rise of Cartesian science and is only now being appreciated again. According to René Dubos, "The relevance of environmental forces to the problems of human biology, medicine and sociology has never been formulated with greater breadth and sharper vision than it was at the dawn of scientific history!"[10]

Health, according to the Hippocratic writings, requires a state of balance among environmental influences, ways of life, and the various components of human nature. These components are described in terms of "humors" and "passions," which have to be in equilibrium. The Hippocratic doctrine of the humors can be restated in terms of chemical and hormonal balance, and the relevance of the passions refers to the interdependence of mind and body, which is strongly emphasized in the texts. Hippocrates was not only a shrewd observer of physical symptoms but also left excellent descriptions of many mental disorders that still occur in our time.

As far as healing was concerned, Hippocrates recognized the healing forces inherent in living organisms, forces he called "nature's healing power." The role of the physician was to assist these natural forces by creating the most favorable conditions for the healing process. This is the original meaning of the word "therapy," which comes from the Greek *therapeuin* ("to attend"). In addition to defining the role of the therapist as that of an attendant, or assistant, to the natural healing process, the Hippocratic writings also contain a strict code of medical ethics, known as the Hippocratic Oath, which has remained the ideal of physicians to the present day.

The Hippocratic tradition, with its emphasis on the fundamental interrelation of body, mind, and environment, represents a high point of Western medical philosophy that is as strong in its appeal for our time as it was twenty-five hundred years ago. As Dubos writes, paraphrasing Whitehead's remark on the debt of European philosophy to Plato, "modern medicine is but a series of commentaries and elaborations on the Hippocratic writings."[11]

The main themes of Hippocratic medicine—health as a state of balance, the importance of environmental influences, the interdependence of mind and body, and nature's inherent healing power—were developed in ancient China in a very different cultural context. Classical Chinese medicine had its roots in shamanistic traditions and was shaped by both Taoism and Confucianism, the two principal philosophical schools of the classical period.[12] During the Han period (206 B.C.–A.D. 220) Chinese medicine was formalized as a system of ideas and written down in the classical medical texts. The most important among the early medical classics is the *Nei Ching*, the Classic of Internal Medicine, which develops in a lucid and attractive way a theory of the human organism in health and illness, together with a theory of medicine.[13]

As in every other theoretical tradition developed in early China, the concepts of yin and yang are central. The entire universe, both natural and social, is in a state of dynamic balance, with all its components oscillating between the two archetypal poles. The human organism is a microcosm of the universe; its parts are assigned yin and yang qualities, and thus the individual's place in the great cosmic order is firmly established. Unlike the early Greek scholars, the Chinese were not deeply interested in causal relations but rather in the synchronic patterning of things and events. Joseph Needham has aptly called this attitude "correlative thinking." For the Chinese,

> Things behaved in particular ways not necessarily because of prior actions or impulsions of other things, but because their position in the ever-moving cyclical universe was such that they were endowed with intrinsic natures which made that behavior inevitable for them. If they did not behave in those particular ways they would lose their relational positions in the whole (which made them what they were) and turn into something other than themselves.[14]

This correlative and dynamic way of thinking is basic to the conceptual system of Chinese medicine.[15] The healthy individual and the healthy society are integral parts of a great patterned order, and illness is disharmony at the individual or social level. The cosmic patterns were mapped out by means of a complex system of correspondences and associations that was elaborated in great detail in the classical texts. In addition to the yin/yang symbolism, the Chinese used a system called *Wu Hsing*, which is usually translated as the Five Elements, but that interpretation is much too static. *Hsing* means "to act" or "to do," and the five concepts associated with wood, fire, earth, metal, and water represent qualities that succeed and influence one another in a well-defined cyclical order. Manfred Porkert has translated *Wu Hsing* as the Five Evolutive Phases,[16] which seems to be much better suited to describing the dynamic connotation of the Chinese term. From these Five Phases the Chinese derived a correspondence system extending to the entire universe. The seasons, atmospheric influences, colors, sounds, the parts of the body, emotional states, social relations, and numerous other phenomena were all classified into five types related to the Five Phases.[17] When the Five Phase theory was fused with the yin/yang cycles, the result was an elaborate system in which every aspect of the universe was described as a well-defined part of a dynamically patterned whole. This system formed the theoretical base for the diagnosis and treatment of illness.

The Chinese idea of the body has always been predominantly functional and concerned with the interrelations of its parts rather than with anatomical accuracy. Accordingly, the Chinese concept of a physical organ refers to a whole functional system, which has to be considered in its totality, along with the relevant parts of the correspondence system. For example, the idea of the lungs includes not only the lungs themselves but the entire respiratory tract, the nose, the skin, and the secretions associated with these organs. In the correspondence system the lungs are associated with metal, the color white, a piquant taste, grief and negativism, and various other qualities and phenomena.

The Chinese notion of the body as an indivisible system of interrelated components is obviously much closer to the modern systems approach than to the classical Cartesian model, and the similarity is reinforced by the fact that the Chinese saw the network of relationships they were studying as intrinsically dynamic. The individual organism, like the cosmos as a whole, was seen as being in a state of continual, multiple, and interdependent fluctuations whose patterns were described in terms of the flow of *ch'i.* The concept of *ch'i,* which played an important role in almost every Chinese school of natural philosophy, implies a thoroughly dynamic conception of reality. The word literally means "gas" or "ether" and was used in ancient China to denote the vital breath or energy animating the cosmos. But, neither of these Western terms describes the concept adequately. *Ch'i* is not a substance, nor does it have the purely quantitative meaning of our scientific concept of energy. It is used in Chinese medicine in a very subtle way to describe the various patterns of flow and fluctuation in the human organism, as well as the continual exchanges between organism and environment. *Ch'i* does not refer to the flow of any particular substance but rather seems to represent the principle of flow as such, which, in the Chinese view, is always cyclical.

The flow of *ch'i* keeps a person alive; imbalances, and hence illnesses, occur when the *ch'i* does not circulate properly. There are definite pathways of *ch'i,* called *ching-mo* and usually translated as "meridians," which are associated with the primary organs and are assigned yin and yang qualities. Along the meridians lie series of pressure points that can be used to stimulate the various flow processes in the body. From the Western scientific point of view, there is now considerable documentation to show that the pressure points have distinct electrical resistance and thermosensitivity, unlike other areas at the body surface, but there has been no scientific demonstration of the existence of meridians.

A key concept in the Chinese view of health is that of balance. The classics state that diseases become manifest when the body gets out of balance and the *ch'i* does not circulate properly. The causes for such imbalances are multiple. Through poor diet, lack of sleep, lack of exercise, or by being in a state of disharmony with one's family or society, the body can lose its balance, and it is at times like this that illnesses occur. Among the external causes, seasonal changes are given special attention and their influences on the body are described in great detail. Internal causes are attributed to an imbalance of one's emotional states, which are classified and associated with specific internal organs according to the correspondence system.

Illness is not thought of as an intruding agent but as due to a pattern of causes leading to disharmony and imbalance. However, the nature of all things, including the human organism, is such that there is a natural tendency to return to a dynamic state of balance. Going in and out of balance is seen as a natural process that happens constantly throughout the life cycle. Accordingly, the traditional texts draw no sharp line between health and illness. Both health and ill health are seen as natural and as being part of a continuum. They are aspects of the same process in which the individual organism changes continually in relation to the changing environment.

Since illness will at times be inevitable in the ongoing process of life, perfect health is not the ultimate goal of either patient or doctor. The aim of Chinese medicine, rather, is to achieve the best possible adaptation to the individual's total environment. In pursuing this aim the patient plays an important and active role. In the Chinese view the individual is responsible for the maintenance of her own health, and to a great extent even for the restoration of health when the organism gets out of balance. The doctor takes part in this process but the main responsibility is the patient's. It is the individual's duty to keep healthy, and this is done by living according to the rules of society and taking care of one's body in a highly practical way.

It is easy to see that a system of medicine which regards balance and harmony with the environment as the basis of health will be likely to emphasize preventive measures. Indeed, to prevent any imbalance in their patients has always been an important role of Chinese doctors. It is said that doctors in China used to be paid only while their patients stayed well and that payments stopped when they became ill. That is probably an exaggeration, but Chinese doctors did refuse patients once their condition reached a certain point of severity. As the *Nei Ching* explains,

> To administer medicines to diseases which have already developed . . . is comparable to the behavior of those persons who begin to

> dig a well after they have become thirsty, and of those who begin to cast weapons after they have already engaged in battle. Would these actions not be too late?[18]

These concepts and attitudes imply a role of the doctor quite different from that in the West. In Western medicine the doctor with the highest reputation is a specialist who has detailed knowledge about a specific part of the body. In Chinese medicine the ideal doctor is a sage who knows how all the patterns of the universe work together; who treats each patient on an individual basis; whose diagnosis does not categorize the patient as having a specific disease but records as fully as possible the individual's total state of mind and body and its relation to the natural and social environment.

To arrive at such a complete picture the Chinese developed not only highly refined diagnostic methods of observing and questioning the patient but also a unique art of pulse taking that allows them to determine the detailed flow of patterns of *ch'i* along the meridians, and thus the dynamic state of the entire organism.[19] Traditional Chinese practitioners believe that these methods allow them to recognize imbalances and hence potential problems before they manifest themselves in symptoms that can be detected with Western diagnostic techniques.

The traditional Chinese diagnosis is necessarily a lengthy process in which the patient must actively participate by contributing considerable information about her personal way of life. Ideally each patient is a unique case presenting a vast array of variables to be taken into account. In actual practice there was probably always a tendency to classify according to patterns of symptoms, but the desire for precise classification and labeling was never present. The entire diagnosis relies heavily on subjective judgments by doctor and patient and is based on a body of qualitative data obtained by the doctor through the use of his own senses—touch, hearing, and vision—and through close interaction with the patient.

Having determined the dynamic state of the patient in relation to the environment, the Chinese doctor then attempts to restore balance and harmony. Several therapeutic techniques are used, all designed to stimulate the patient's organism in such a way that it will follow its own natural tendency to return to a balanced state. Accordingly, one of the most important principles of Chinese medicine is always to give as mild a therapy as possible. The whole process, ideally, is one of ongoing interaction between doctor and patient, with the doctor continually modifying the therapy according to the patient's responses.

Herbal medicines are classified according to the yin/yang system and are associated with five basic flavors that, according to the Five Phase theory, will affect the corresponding internal organs. In actual practice herbal medicines are rarely given singly but are usually prescribed in mixtures reflecting the patient's pattern of *ch'i*. Massage therapy, moxibustion, and acupuncture all make use of the pressure points along the meridians to influence the flow of *ch'i*. Moxibustion consists of burning small cones of the powdered herb moxa (mugwort) on the body at the pressure points; in the case of acupuncture solid needles of varying gauges and lengths are inserted at these points. The needles can be used either to stimulate or to sedate the body, depending on how they are inserted and manipulated. What all these therapies have in common is that they are not aimed at treating the symptoms of the patient's illness. They work at a more fundamental level to counteract imbalances that are regarded as the source of ill health.

To apply our study of the Chinese medical model to developing a holistic approach to health in our culture, we need to deal with two questions: To what extent is the Chinese model holistic? And which of its aspects, if any, can be adapted to our cultural context? For the first question it is useful to distinguish two kinds of holism.[20] In a somewhat narrow sense, holism in medicine means that the human organism is seen as a living system whose components are all interconnected and interdependent. In a broader sense, the holistic view recognizes also that this system is an integral part of larger systems, which implies that the individual organism is in continual interaction with its physical and social environment, that it is constantly affected by the environment but can also act upon it and modify it.

The Chinese medical system is certainly holistic in the first sense. Its practitioners believe that their therapies will not just remove the principal symptoms of the patient's illness but will affect the entire organism, which they treat as a dynamic whole. In the broader sense, however, the Chinese system is holistic only in theory. The interdependence of organism and environment is acknowledged in the diagnosis of illness and is discussed extensively in the medical classics, but is generally neglected as far as therapy is concerned. The classical texts give equal weight to environmental influences, family relationships, emotional problems, and so on, but most practitioners today make no practical attempt to deal with the psychological and social aspects of illness therapeutically. When they make their diagnosis doctors spend considerable time talking to the patients about their work situation, their families, and their emotional states, but when it comes to therapy they concentrate on dietary counseling, herbal medicine, and acupuncture, restricting themselves to techniques that manipulate processes inside the body. There is no psychotherapy and there is no attempt to give patients advice on how they could change their life situation. The role of stressful events in the psychological and social realms is clearly acknowledged as a source of illness, but doctors do not feel that it is part of the therapeutic process to bring about changes at those levels.

As far as we can tell, this attitude was characteristic of Chinese doctors in the past also. The medical classics are rich documents expounding a broad holistic view of human nature and of medicine, but these are theoretical works written by doctors who were primarily scholars and were not very involved in curing patients. In practice the Chinese system was probably never very holistic as far as the psychological and social aspects of illness are concerned. The reluctance to act therapeutically by affecting the patient's social situation was certainly a result of the strong influence of Confucianism on all aspects of Chinese life. The Confucian system was mainly concerned with maintaining the social order. Illness, in the Confucian view, could arise from inadequate adjustment to the rules and customs of society, but the only way for an individual to get well was to change himself to fit the given social order. This attitude is so deeply ingrained in East Asian culture that it still underlies modern medical therapy in both China and Japan.

Which aspects of traditional Chinese medical philosophy and practice, then, can or should we incorporate into our own framework of health care? To answer this question the study of medical practice in contemporary Japan is extremely helpful. It provides a unique opportunity to see how modern Japanese doctors use traditional East Asian medical concepts and practices to deal with diseases that are not too different from those in our society. The Japanese adopted Western medicine voluntarily about a hundred years ago but are now increasingly reevaluating their own traditional practices, which, they believe, can fulfill many functions beyond the capacities of the biomedical model. Margaret Lock has made a detailed study of traditional East Asian medicine* in modern urban Japan and found that there is an increasing number of Japanese doctors, known as *kanpō* doctors,† who combine Eastern and Western techniques into an efficient system of medical care.[21] Although many aspects of *kanpō* medicine are effective only in the cultural context of Japan, others may well be adapted for our own culture.

One striking difference between Eastern and Western approaches to health is that in East Asian society in general, subjective knowledge is highly valued. Even in modern scientific Japan the value of subjective experience is strongly acknowledged, and subjective knowledge is considered as valuable as rational deductive thinking. Thus Japanese doctors can accept subjective judgments—both their own and their patients'—without seeing them as threats to their medical competence or their personal integrity. One consequence of this attitude is a distinctive lack of concern about quantification among East Asian doctors, supported by the doctors' awareness that they are dealing with living systems in continuous flux for which qualitative measurements are considered to be sufficient. For example, *kanpō* doctors would not measure patients' temperatures but would note their subjective feelings about having a fever; herbal medicines are measured very roughly in little boxes without the use of scales, and are then mixed together. Nor is the duration of acupuncture therapy measured—it is simply determined by asking the patient how it feels.

The proper valuation of subjective knowledge is surely something we could learn from the East. Ever since Galileo, Descartes, and Newton our culture has been so obsessed with rational knowledge, objectivity, and quantification, that we have become very insecure in dealing with human values and human experience. In medicine, intuition and subjective knowledge are used by every good physician, but this is not acknowledged in the professional literature, nor is it taught in our med-

* Lock and others use the term "East Asian medicine" for the medical system that was dominant until the nineteenth century among the literate populations of China, Korea, and Japan and is often called "classical Chinese medicine" or "Oriental medicine."

† *Kanpō* literally means the "Chinese method"; it refers to the entire medical system brought to Japan from China in the sixth century.

ical schools. On the contrary, the criteria for admission to most medical schools screen out those who have the greatest talents for practicing medicine intuitively.

Once we adopt a more balanced attitude toward rational and intuitive knowledge, it will be easier to incorporate into our system of health care some of the aspects characteristic both of East Asian medicine and of our own Hippocratic tradition. The main difference between such a new model of health and the East Asian approach will be the integration of psychological and social measures into our system of health care. Psychological counseling and psychotherapy are not part of the East Asian tradition but play an important role in our culture; nor are East Asian doctors concerned about changing the social situation, although they recognize the importance of social problems in the development of illness. In our society, however, a truly holistic approach will recognize that the environment created by our social and economic system, based on the fragmented and reductionist Cartesian world view, has become a major threat to our health. An ecological approach to health will therefore make sense only if it leads to profound changes in our technology and our social and economic structures.

Health care in Europe and North America is practiced by a large number of people and organizations, including physicians, nurses, psychotherapists, psychiatrists, public-health professionals, social workers, chiropractors, homeopaths, acupuncturists, and various "holistic" practitioners. These individuals and groups use a great variety of approaches that are based on different concepts of health and illness. To integrate them into an effective system of health care, based on holistic and ecological views, it will be crucial to establish a common conceptual basis for talking about health, so that all these groups can communicate and coordinate their efforts.

It will also be necessary to define health at least approximately. Although everybody knows what it feels like to be healthy, it is impossible to give a precise definition; health is a subjective experience whose quality can be known intuitively but can never be exhaustively described or quantified. Nevertheless, we may begin our definition by saying that health is a state of well-being that arises when the organism functions in a certain way. The description of this way of functioning will depend on how we describe the organism and its interactions with its environment. Different models of living organisms will lead to different definitions of health. The concept of health, therefore, and the related concepts of illness, disease, and pathology, do not refer to well-defined entities but are integral parts of limited and approximate models that mirror a web of relationships among multiple aspects of the complex and fluid phenomenon of life.

Once the relativity and subjective nature of the concept of health is perceived, it also becomes clear that the experience of health and illness is strongly influenced by the cultural context in which it occurs. What is healthy and sick, normal and abnormal, sane and insane, varies from culture to culture. Moreover, cultural context influences the specific ways people behave when they get sick. How we communicate our health problems, the manner in which we present our symptoms, when and to whom we go for care, the explanations and therapeutic measures offered by the doctor, therapist, or healer—all that is strongly affected by our society and culture.[22] It would seem, therefore, that a new framework for health can be effective only if it is based on concepts and ideas rooted in our own culture and evolving according to the dynamics of our social and cultural evolution.

For the past three hundred years our culture has been dominated by the view of the human body as a machine, to be analyzed in terms of its parts. The mind is separated from the body, disease is seen as a malfunctioning of biological mechanisms, and health is defined as the absence of disease. This view is now slowly being eclipsed by a holistic and ecological conception of the world which sees the universe not as a machine but rather as a living system, a view that emphasizes the essential interrelatedness and interdependence of all phenomena and tries to understand nature not only in terms of fundamental structures but in terms of underlying dynamic processes. It would seem that the systems view of living organisms can provide the ideal basis for a new approach to health and health care that is fully consistent with the new paradigm and is rooted in our cultural heritage. The systems view of health is profoundly ecological and thus in harmony with the Hippocratic tradition which lies at the roots of Western medicine. It is a view based on scientific notions and expressed in terms of concepts and symbols which are part of our everyday language. At the same time the new framework naturally takes into account the spiritual dimensions of health and is thus in harmony with the views of many spiritual traditions.

Systems thinking is process thinking, and hence the systems view sees health in terms of an ongoing process. Whereas most definitions, including some proposed recently by holistic practitioners, picture health as a static state of perfect well-being, the systems concept of health implies continual activity and change, reflecting the organism's creative response to environmental challenges. Since a person's condition will always depend importantly on the natural and social environment, there can be no absolute level of health independent of this environment. The continual changes of one's organism in relation to the changing environment will naturally include temporary phases of ill health, and it will often be impossible to draw a sharp line between health and illness.

Health is really a multidimensional phenomenon involving interdependent physical, psychological, and social aspects. The common representation of health and illness as opposite ends of a one-dimensional continuum is quite misleading. Physical disease may be balanced by a positive mental attitude and social support, so that the overall state is one of well-being. On the other hand, emotional problems or social isolation can make a person feel sick in spite of physical fitness. These multiple dimensions of health will generally affect one another, and the strongest feeling of being healthy will occur when they are well balanced and integrated. The experience of illness, from the systems point of view, results from patterns of disorders that may become manifest at various levels of the organism, as well as in the various interactions between the organism and the larger systems in which it is embedded. An important characteristic of the systems approach is the notion of stratified order involving levels of differing complexities, both within individual organisms and in social and ecological systems. Accordingly, the systems view of health can be applied to different systems levels, with the corresponding levels of health mutually interconnected. In particular we can discern three interdependent levels of health—individual, social, and ecological. What is unhealthy for the individual is generally also unhealthy for the society and for the embedding ecosystem.

The systems view of health is based on the systems view of life. Living organisms, as we have seen, are self-organizing systems that display a high degree of stability. This stability is utterly dynamic and is characterized by continual, multiple, and interdependent fluctuations. To be healthy such a system needs to be flexible, to have a large number of options for interacting with its environment. The flexibility of a system depends on how many of its variables are kept fluctuating within their tolerance limits: the more dynamic the state of the organism, the greater its flexibility. Whatever the nature of the flexibility—physical, mental, social, technological, or economic—it is essential to the system's ability to adapt to environmental changes. Loss of flexibility means loss of health.

This notion of dynamic balance is a useful concept for defining health. "Dynamic" is of crucial importance here, indicating that the necessary balance is not a static equilibrium but rather a flexible pattern of fluctuations of the kind described above. Health, then, is an experience of well-being resulting from a dynamic balance that involves the physical and psychological aspects of the organism, as well as its interactions with its natural and social environment.

The concept of health as dynamic balance is consistent both with the systems view of life and with many traditional models of health and healing, among them the Hippocratic tradition and the tradition of East Asian medicine. As in these traditional models, "dynamic balance" acknowledges the healing forces inherent in every living organism, the organism's innate tendency to reestablish itself in a balanced state when it has been disturbed. It may do so by returning, more or less, to the original state through various processes of self-maintenance, including homeostasis, adaptation, regeneration, and self-renewal. Examples of this phenomenon would be the minor illnesses which are part of our everyday life and usually cure themselves. On the other hand, the organism may also undergo a process of self-transformation and self-transcendence, involving stages of crisis and transition and resulting in an entirely new state of balance. Major changes in a person's life style, induced by a severe illness, are examples of such creative responses that often leave the person at a higher level of health than the one enjoyed before the challenge. This suggests that periods of ill health are natural stages in the ongoing interaction between the individual and the environment. To be in dynamic balance means to go through temporary phases of illness that can be used to learn and to grow.

The natural balance of living organisms includes a balance between their self-assertive and integrative tendencies. To be healthy an organism has to preserve its individual autonomy, but at the same time it has to be able to integrate itself harmoniously into larger systems. This capacity for integration is closely related to the organism's flexibility and to the concept of dynamic balance. Integration at one systems level will manifest itself as balance at a larger level, as the harmonious integration of individual components into larger systems results in the balance of those systems. Illness, then, is a consequence of imbalance and disharmony, and may very often be seen as stemming from a lack of integration. This is particularly true for mental illness, which often arises from a failure to evaluate and integrate sensory experience.

The notion of illness as originating in a lack of integration seems to be especially relevant to approaches that try to understand living organisms in terms of rhythmic patterns. From this perspective synchrony becomes an important measure of health. Individual organisms interact and communicate with one another by synchronizing their rhythms and thus integrating themselves into the larger rhythms of their environment. To be healthy, then, means to be in synchrony with oneself—physically and mentally—and also with the surrounding world. When a person is out of synchrony, illness is likely to occur. Many esoteric traditions associate health with the synchrony of rhythms and healing with a certain resonance between healer and patient.

To describe an organism's imbalance the concept of stress seems to be extremely useful. Although it is relatively new in medical research,[23] it has taken a firm hold in the collective consciousness and language of our culture. The stress concept is also completely consistent with the systems view of life and can be fully grasped only when the subtle interplay between mind and body is perceived.

Stress is an imbalance of the organism in response to environmental influences. Temporary stress is an essential aspect of life, since the ongoing interaction between organism and environment often involves temporary losses of flexibility. These will occur when the individual perceives a sudden threat, or when it has to adapt to sudden changes in the environment or is being strongly stimulated in some other way. These transitory phases of imbalance are an integral part of the way healthy organisms cope with their environment, but prolonged or chronic stress can be harmful and plays a significant role in the development of many illnesses.[24]

From the systems point of view, the phenomenon of stress occurs when one or several variables of an organism are pushed to their extreme values, which induces increased rigidity throughout the system. In a healthy organism the other variables will conspire to bring the whole system back into balance and restore its flexibility. The remarkable thing about this response is that it is fairly stereotyped. The physiological stress symptoms—tight throat, tense neck, shallow respiration, accelerated heart rates, and so on—are virtually identical in animals and humans and are quite independent of the source of stress. Because they constitute the organism's preparation to cope with the challenge by either fighting or fleeing, the whole phenomenon is known as the fight-or-flight response. Once the individual has taken action by fighting or fleeing, it will rebound into a state of relaxation and ultimately will return to homeostasis. The well-known "sigh of relief" is an example of such a relaxation rebound.

When the fight-or-flight response is prolonged, however, and when an individual cannot take action by fighting or fleeing to release the organism from the stressful state, the consequences are likely to be detrimental to health. The continual imbalance created by prolonged unabated stress can generate physical and psychological symptoms—muscle tension, anxiety, indigestion, insomnia—which will eventually lead to illness. The prolongation of stress often results from our failure to integrate the responses of our bodies with our cultural habits and social rules of behavior. Like most animals we react to any kind of challenge by arousing our organism in preparation for either physical fight or physical escape, but in most cases these reactions are no longer useful. In an intense business meeting we cannot win an argument by physically assaulting our opponent, nor can we run away from the situation. Being civilized, we try to deal with the challenge in socially acceptable ways, but "old" parts of our brain often continue mobilizing the organism for inappropriate physical responses. If this happens repeatedly we are likely to get sick; we may develop a peptic ulcer or have a heart attack.

A key element in the link between stress and illness, which is not yet known in all its details but has been verified by numerous studies, is the fact that prolonged stress suppresses the body's immune system, its natural defenses against infections and other diseases. Full recognition of this fact will bring about a major shift in medical research from the preoccupation with microorganisms to a careful study of the host organism and its environment. Such a shift is now urgently needed, since the chronic and degenerative diseases that are characteristic of our time and constitute the major causes of death and disability are closely connected with excessive stress.

The sources of this overload of stress are manifold. They may originate within an individual, may be generated collectively by our society and culture, or may be present in the physical environment. Stressful situations arise not only from personal emotional traumas, anxieties, and frustrations, but also from the hazardous environment created by our social and economic system. Stress, however, comes not only from negative experiences. All events—positive or negative, joyous or sad—that require a person to adapt to profound or rapid changes will be highly stressful. It is very unfortunate for our health that our culture has produced an accelerating rate of change in all areas, together with numerous physical health hazards, but has failed to teach us how to cope with the increasing amount of stress we encounter.

Recognition of the role of stress in the development of illness leads to the important idea of illness as a "problem solver." Because of social and cultural conditioning people often find it impossible to release their stresses in healthy ways and therefore choose—consciously or unconsciously—to get sick as a way out. Their illness may be physical or mental, or it may manifest itself as violent and reckless behavior, including crime, drug abuse, accidents, and suicides, which may appropriately be called social illnesses. All these "escape routes" are forms of ill health, physical disease being only one of several unhealthy ways of dealing with stressful life situations. Hence curing the disease will not necessarily make the patient healthy. If the escape into a particular disease is blocked effectively by medical intervention while the stressful situation persists, this may merely shift the person's response to a different mode, such as mental illness or antisocial behavior, which will be just as unhealthy. A holistic approach will have to look at health from this broad perspective, distinguishing clearly between the origins of illness and its manifestation. Otherwise it will not mean much to talk about successful therapies. As a doctor friend of mine put it forcefully, "If you are able to reduce physical illness, but at the same time increase mental illness or crime, what the hell have you done?"

The idea of illness as a way to cope with stressful life situations naturally leads to the notion of the meaning of illness, or of the "message" transmitted by a particular disease. To understand this message ill health should be taken as an opportunity for introspection, so that the original problem and the reasons for choosing a particular escape route can be brought to a conscious level where the problem can be resolved. This is where psychological counseling and psychotherapy can play an important role, even in the treatment of physical illnesses. To integrate physical and psychological therapies will amount to a major revolution in health care, since it will require full recognition of the interdependence of mind and body in health and illness.

When the systems view of mind is adopted, it becomes obvious that any illness has mental aspects. Getting sick and healing are both integral parts of an organism's self-organization, and since mind represents the dynamics of this self-organization, the processes of getting sick and of healing are essentially mental phenomena. Because mentation is a multileveled pattern of processes, most of them taking place in the unconscious realm, we are not always fully aware of how we move in and out of illness, but this does not alter the fact that illness is a mental phenomenon in its very essence.

The intimate interplay between physical and mental processes has been recognized throughout the ages. We all know that we express emotions through gestures, inflections, breathing patterns, and minute movements imperceptible to the untrained eye. The precise ways in which physical and psychological patterns interlink are still little understood, and thus most physicians tend to restrict themselves to the biomedical model and neglect the psychological aspects of illness. However, there have been significant attempts to develop a unified approach to the mind/body system throughout the history of Western medicine. Several decades ago these attempts culminated in the foundation of psychosomatic medicine as a scientific discipline, concerned specifically with the study of the relationships between the biological and psychological aspects of health.[25] This new branch of medicine is now rapidly gaining acceptance, especially with the growing awareness of the relevance of stress, and is likely to play an important role in a future holistic system of health care.

The term "psychosomatic" needs some clarification. In conventional medicine it was used to refer to a disorder without a clearly diagnosed organic basis. Owing to the strong biomedical bias, such "psychosomatic disorders" tended to be regarded as imagined, not real. The modern use of the term is quite different; it derives from the recognition of a fundamental interdependence between mind and body at all stages of illness and health. To single out any disorder as psychologically caused would be as reductionistic as the belief that there are purely organic diseases without any psychological components. Researchers and clinicians today are increasingly aware that virtually all disorders are psychosomatic in the sense that they involve the continual interplay of mind and body in their origin, development, and cure. In the words of René Dubos, "Whatever its precipitating cause and its manifestations, almost every disease involves both the body and the mind, and these two aspects are so interrelated that they cannot be separated one from the other."[26] So the term "psychosomatic disorder" has become redundant, although it is meaningful to speak of psychosomatic medicine.*

The manifestations of illness will vary from case to case, from almost purely psychological to almost exclusively physical symptoms. When psychological aspects dominate, the illness is usually referred to as mental illness. Mental illnesses involve physical symptoms, however, and in some cases biological and genetic factors may even be dominant in the causation of the disorder. Furthermore, the origin and development of many mental illnesses depend crucially on the individual's ability to interact with his family, friends, and other social groups. Those illnesses can be fully understood only by seeing how the individual organism is embedded in its social environment.[27]

Moreover, it is becoming apparent that the role of the patient's personality is a crucial element in the generation of many illnesses. Prolonged stress somehow seems to be channeled through a particular personality configuration to give rise to a specific disorder. The most convincing link between personality and illness has been found for heart disease, and links are being tentatively established for other major diseases, most notably cancer.[28] These results are extremely significant because as soon as the patient's personality enters into the clinical picture the illness becomes inseparably linked to his entire psyche, which suggests the unification of physical and psychological therapies.

In spite of the extensive literature on the role of psychological influences in the development of illness, very little work has been done to explore the methods of altering these influences. The key to any such attempt is the idea that mental attitudes and processes not only play a significant part in getting sick; they can also play a significant part in getting well. The psychosomatic nature of illness implies the possibil-

* The term "holistic health," which has recently become very popular, is similarly redundant, since health already implies wholeness, yet it makes sense to speak of holistic health care.

ity of psychosomatic self-healing. This idea is strongly supported by the recent discovery of the biofeedback phenomenon, which showed that a wide range of physical processes can be influenced by a person's mental efforts.[29]

The first step in this kind of self-healing will be the patients' recognition that they have participated consciously or unconsciously in the origin and development of their illness, and hence will also be able to participate in the healing process. In practice, this notion of patient participation, which implies the idea of patient responsibility, is extremely problematic and is vigorously denied by most patients. Conditioned as they are by the Cartesian framework, they refuse to consider the possibility that they may have participated in their illness, associating the idea with blame and moral judgment. It will be important to clarify exactly what is meant by patient participation and responsibility.

In the context of a psychosomatic approach, our participation in the development of an illness means that we make certain choices to expose ourselves to stressful situations and, furthermore, to react to these stresses in certain ways. These choices are influenced by the same factors that influence all the choices we make in life. They are made unconsciously more often than consciously, and will depend on our personality, on various external constraints, and on social and cultural conditioning. Any responsibility, therefore, can be only partial. Like the concept of free will, the notion of personal responsibility must necessarily be limited and relative, and neither of them can be associated with absolute moral values. The purpose of recognizing our participation in our illness is not to feel guilty about it but to adopt the necessary changes and to realize that we can also participate in the healing process.

Mental attitudes and psychological techniques are important means for both the prevention and healing of illness. A positive attitude combined with specific stress-reduction techniques will have a strong positive impact on the mind/body system and will often be able to reverse the disease process, even to heal severe biological disorders. The same techniques can be used to prevent illness by applying them to cope with excessive stress before any serious damage occurs.

An impressive proof of the healing power of positive expectations alone is provided by the well-known placebo effect. A placebo is an imitation medicine, dressed up like an authentic pill and given to patients who think they are receiving the real thing. Studies have shown that 35 percent of patients consistently experience "satisfactory relief" when placebos are used instead of regular medication for a wide range of medical problems.[30] Placebos have been strikingly successful in reducing or eliminating physical symptoms, and have produced dramatic recoveries from illnesses for which there are no known medical cures. The only active ingredient in these treatments appears to be the power of the patient's positive expectations, supported by interaction with the therapist.

The placebo effect is not limited to the administration of pills but can be associated with any form of treatment. Indeed, it is likely to play a significant part in all therapy. In medical jargon "placebo" has been used for any aspect of the healing process that is not based on physical or pharmacological intervention, and, like the term "psychosomatic," often has a pejorative connotation. Doctors have tended to classify illnesses whose origins and development could not be understood within the biomedical framework as "psychosomatic" and to label any healing process induced by the patient's positive expectations and faith in doctor and treatment as a "placebo effect," while self-healings without any medical intervention were described as "spontaneous remissions." The real meaning of these three terms is very similar; they all refer to the healing powers of the patient's mental attitude.

The patient's will to get well and confidence in the treatment are crucial aspects of any therapy, from shamanistic healing rituals to modern medical procedures. As the writer and editor Norman Cousins has noted, "Many medical scholars have believed that the history of medicine is actually the history of the placebo effect."[31] On the other hand, negative attitudes of the patient, the doctor, or the family may produce an "inverse placebo effect." Experience has shown repeatedly that patients who are told they have only six to nine months more to live will, indeed, not live longer. Statements of this kind have a powerful impact on the patient's mind/body system—they seem to act almost as a magic spell—and should therefore never be made.

In the past, psychosomatic self-healing has always been associated with faith in some treatment—a drug, the power of a healer, perhaps a miracle. In a future approach to health and healing, based on the new holistic paradigm, it should be possible to acknowledge the individual's potential for self-healing directly, with no need for any conceptual crutches, and to develop psychological techniques that will facilitate the healing process.

We have been building a model of illness that is both holistic and dynamic. In it illness is a consequence of imbalance and disharmony, often stemming from a lack of integration, which may arise at various levels of the organism and, accordingly, may generate symptoms of a physical, psychological, or social nature. Disease is the biological manifestation of illness, and the model clearly distinguishes between disease origins and disease processes. Excessive stress is believed to contribute significantly to the origin and development of most diseases, manifesting itself in the organism's initial imbalance and, subsequently, being channeled through a particular personality configuration to give rise to specific disorders. An important aspect of this process is the fact that illness is often perceived, consciously or unconsciously, as a way out of the stressful situation, various kinds of illness representing different escape routes. Curing the disease will not necessarily make the patient healthy, but illness may be an opportunity for introspection that will resolve the root problems.

The development of illness involves the continual interplay between physical and mental processes that reinforce one another through a complex network of feedback loops. Disease patterns at any stage appear as manifestations of underlying psychosomatic processes that should be dealt with in the course of therapy. This dynamic view of illness specifically acknowledges the organism's innate tendency to heal itself—to reestablish itself in a balanced state—which may include stages of crisis and major life transitions. Periods of ill health involving minor symptoms are normal and natural stages representing the organism's means of restoring balance by interrupting one's usual activities and forcing a change of pace. As a consequence, the symptoms associated with these minor illnesses usually disappear after a few days, whether or not any treatment is received. More serious illnesses will require greater efforts of regaining one's balance, generally including the help of a doctor or therapist, and the outcome will depend crucially on the patient's mental attitudes and expectations. Severe illnesses, finally, will require a therapeutic approach dealing not only with the physical and psychological aspects of the disorder, but also with the changes in the patient's life style and world view that will be an integral part of the healing process.

These views of health and illness imply a number of guidelines for health care and make it possible to sketch out the basic framework for a new holistic approach. Health care will consist of restoring and maintaining the dynamic balance of individuals, families, and other social groups. It will mean people taking care of their own health individually, as a society, and with the help of therapists. This kind of health care cannot just be "provided," or "delivered"—it has to be practiced. Furthermore, it will be important to consider the interdependence of our individual health and that of the social and ecological systems in which we are embedded. If you live in a stressful neighborhood the situation will not be improved if you move out and let somebody else take the stress, although your own health may improve. Similarly, an unhealthy economy is not improved by raising the level of unemployment. Such actions amount to managing stress by simply pushing it around—from one family to another, from individuals to the society and back to other individuals, or from society onto the ecosystem, whence it may come back forty years later, as in the case of Love Canal. Health care at all levels will consist of balancing and resolving the stressful situations by individual and social action.

A future system of health care will consist, first and foremost, of a comprehensive, effective, and well-integrated system of preventive care. Health maintenance will be partly an individual matter and partly a collective matter, and most of the time the two will be closely interrelated. Individual health care is based on the recognition that the health of human beings is determined, above all, by their behavior, their food, and the nature of their environment.[32] As individuals, we have the power and the responsibility to keep our organism in balance by observing a number of simple rules of behavior relating to sleep, food, exercise, and drugs. The role of therapists and health professionals will be merely to assist us in doing so. In the past this kind of preventive health care has been severely neglected in our society, but recently there has been a significant shift in attitudes that has generated a powerful grass-roots movement promoting healthy living habits—whole foods, physical exercise, home births, relaxation and meditation techniques—and emphasizing personal responsibility for health.

If acceptance of individual responsibility will be crucial to a future system of holistic health care, it will be equally crucial to recognize that this responsibility is subject to severe constraints. Individuals can be held responsible only to the extent that they have the freedom to look after themselves, and this freedom is often curtailed by heavy social and cultural conditioning. Moreover, many health problems arise from economic and political factors that can be modified only by collective action. Individual responsibility has to be accompanied by social responsibility, and individual health care by social actions and policies. "Social health care" seems an appropriate term for policies and collective activities dedicated to the maintenance and improvement of health.

Social health care will have two basic parts—health education and health policies—to be pursued simultaneously and in close coordination. The aim of health education will be to make people understand how their behavior and their environment affect their health, and to teach them how to cope with stress in their daily lives. Comprehensive programs of health education with this emphasis can be integrated into the school system and given central importance. At the same time they can be accompanied by public health education through the media to counteract the effects of advertising of unhealthy products and life styles. An important aim of health education will be to foster corporate responsibility. The business community needs to learn much more about the health hazards of its production and its products. It should develop and demonstrate concern about public health, become aware of the health costs generated by its activities, and formulate corporate policies accordingly.[33]

Health policies, to be established by government at various levels of administration, will consist of legislation to prevent health hazards from being generated, and also of social policies that provide for people's basic needs. The following suggestions include a few of the many measures needed to provide an environment that would encourage and make it possible for people to adopt healthy ways of living:

- Restrictions on all advertising of unhealthy products.
- "Health care taxes" on individuals and corporations who generate health hazards, to offset the medical costs which inevitably arise from these hazards; for example, there could be corporate taxes for creating various kinds of pollution, and graduated health care taxes on alcohol, on the tar content of cigarettes, and on junk foods.

- Social policies to improve education, employment, civil rights, and economic levels of large numbers of impoverished people; these social policies are also health policies, which affect not only the individuals concerned but also the health of society as a whole.
- Increased development of family-planning services, family counseling, day-care centers, etc.; this can be seen as preventive mental health care.
- Development of nutritional policies that provide incentives for industry to produce more nutritious foods, including restrictions on items offered in vending machines and nutritional specifications for food in schools, hospitals, prisons, cafeterias of government agencies, etc.
- Legislation to support and develop organic methods of farming.[34]

A careful study of these suggested policies shows that any one of them, ultimately, requires a different social and economic system if it is to be successful. There is no way to avoid the conclusion that the present system itself has become a fundamental threat to our health. We shall not be able to increase, or even maintain, our health unless we adopt profound changes in our value system and our social organization. One physician who has recognized this quite clearly is Leon Eisenberg:

> Our daily practice with human ailments makes us aware of the extent to which problems of ill health flow from failures in our political, economic and social institutions. The redesign of these institutions is the central challenge for the coming century, and gives the greatest promise for improving public health.[35]

The restructuring of social institutions required by the new holistic view of health will apply, first of all, to the health care system itself. Our present institutions of health care are based on the narrow biomedical approach to the treatment of diseases and are organized in such a fragmented way that they have become highly ineffective and inflationary. As Kerr White has pointed out, "It is difficult to overemphasize the negative impact that our fragmented, disorganized, and unbalanced health care arrangements have on the adequacy of the care provided in this country and the inflationary impact this confusion has on their costs."[36] What we need is a responsive and well-integrated system of health care that fulfills the needs of individuals and populations.

The first and most important step in a holistic approach to therapy will be to make the patient aware, as fully as possible, of the nature and extent of her imbalance. This means that her problems will have to be put into the broad context from which they arose, which will involve a careful examination of the multiple aspects of the particular illness by therapist and patient. The recognition of this context alone—of the web of interrelated patterns which led to the disorder—is highly therapeutic, as it relieves anxiety and gives hope and self-confidence, thus initiating the process of self-healing. Psychological counseling will play an important role in this process, and those administering primary health care should possess basic therapeutic skills at both the physical and the psychological level. The main purpose of the first encounter between patient and general practitioner, apart from emergency measures, will be to educate the patient about the nature and meaning of the illness, and about the possibilities of changing the patterns in the patient's life that have led to it. This, in fact, is the original role of the "doctor," which comes from the Latin *docere* ("to teach").

To assess the relative contribution of biological, psychological, and social factors to the illness of a particular person is the essence of the science and art of general practice. It requires not only some basic knowledge of human biology, psychology, and social science, but also experience, wisdom, compassion, and concern for the patient as a human being. General practitioners administering primary care of this kind need not be medical doctors, nor experts in any of the scientific disciplines concerned, but they will have to be sensitive to the multiple influences affecting health and illness, and able to decide which of these are the most relevant, best known, and most manageable in a particular case. If necessary, they will refer the patient to specialists in the relevant areas, but even when such special treatments are needed, the object of the therapy will still be the whole person.

The basic aim of any therapy will be to restore the patient's balance, and since the underlying model of health acknowledges the organism's innate tendency to heal itself, the therapist will try to intrude only minimally and keep the treatments as mild as possible. The healing will always be done by the mind/body system itself; the therapist will merely reduce the excessive stress, strengthen the body, encourage the patient to develop self-confidence and a positive mental attitude, and generally create the environment most conducive to the healing.

Such an approach to therapy will be multidimensional, involving treatments at several levels of the mind/body system, which will often require a multidisciplinary team effort. The members of the health team will be specialists in various fields but will share the same holistic view of health and a common conceptual framework that will allow them to communicate effectively and integrate their efforts systematically. Health care of this kind will require many new skills in disciplines not previously associated with medicine and is likely to be intellectually richer, more stimulating, and more challenging than a medical practice that adheres exclusively to the biomedical model.

The kind of primary patient care outlined above is being forcefully advocated today by nurses who find themselves at the forefront of the holistic health movement. Increasing numbers of nurses are deciding that they want to be independent therapists rather than assistants to doctors, and are in the process of applying a holistic approach to their practice. These highly educated and motivated nurses will be best qualified to take on the responsibilities of general practitioners. They will be able to provide the necessary health education and counseling and to assess the patients' life dynamics as a basis for preventive health care. They will keep regular contacts with their clients so that problems can be detected before serious symptoms develop, and will go out into the community to see and understand patients within the context of their work and family situations.

In such a system medical doctors will act as specialists. They will prescribe drugs and perform surgery in emergency cases, treat fractured bones, and practice the full range of medical care for which the biomedical approach is appropriate and successful. Even in those cases, however, the nurse practitioner will still play an important role, keeping the personal contact with the patient and integrating the special treatments into a meaningful whole. For example, if surgery is necessary, the nurse will stay with the patient, choose the appropriate hospital, cooperate with the hospital nurses, support the patient psychologically, and give him postoperative care. Ideally, she would know her patient well from previous consultations and would be available throughout the entire procedure, somewhat like a lawyer who guides a client through a trial.

The new holistic kind of primary care can, of course, also be practiced by physicians, and it seems that medical students have recently become more and more interested in such a career. On the other hand, a nurse may well specialize—in massage therapy, herbal medicine, midwifery, public health, or social work—in addition to her (or his) general practice. The important fact is that we now have a large number of highly qualified nurses who cannot use their full potential in the present system but are ready to carry out a holistic and humanistic approach to primary care. To incorporate nursing into a holistic framework of health care will mean to expand what already exists, and should therefore be the ideal strategy for the period of transition to the new system.

The reorganization of health care will also mean discouraging the construction and use of facilities that are inefficient and incompatible with the new view of health.[37] To change the present technology-intensive, hospital-based system, a first useful step may be, as Victor Fuchs has suggested, to impose a moratorium on all hospital construction and expansion to bring our escalating hospital costs under control.[38] At the same time hospitals will gradually be transformed into more efficient and more humane institutions, comfortable and therapeutic environments modeled after hotels rather than factories or machine shops, with good and nourishing food, family members included in patient care, and other such sensible improvements.

Drugs will be used only in emergency cases and then as sparingly and as specifically as possible. Thus health care will be liberated from the pharmaceutical industry and physicians and pharmacists will collaborate in selecting from the many thousands of pharmaceutical products the few dozen basic drugs that, according to experienced clinicians, are fully adequate for effective medical care.

These changes will be possible only with a thorough reorganization of medical education. To prepare medical students and other health professionals for the new holistic approach will require a considerable broadening of their scientific basis and much greater emphasis on the behavioral sciences and on human ecology. As Howard Rasmussen, professor of biochemistry and medicine at the University of Pennsylvania School of Medicine, has suggested, an educational program presenting a multidisciplinary study of human nature would be an ideal introductory course for health professionals.[39] Such a course, to deal with various levels of individual and social health, should be based on general systems theory and study the human condition in health and illness within an ecological context. It would be the foundation for more detailed medical studies and would provide all health professionals with a common language for their future collaboration in health teams. At the same time there would be a corresponding reorientation of research priorities from an overemphasis on cellular and molecular biology to a more balanced approach.

Undergraduate education in medical schools will concentrate much more on family practice and ambulatory medicine—that is, on understanding the patient as a walking, living person. It will prepare students for working in health teams by helping them understand the multifaceted nature of health and, accordingly, the interrelated roles played by the team members. This means radical changes. Indeed, Rasmussen thinks "nothing short of revolution can restore educational balance and relevance."[40]

An effective and well-integrated system of health care must be facilitated by financial incentives that will induce health professionals, health care institutions, and the general public to make appropriate choices and pursue appropriate policies. In the United States this will include, first and foremost, a system of national health insurance that is not dominated by corporate interests and provides economic incentives for holistic health care, including health education and other preventive measures.[41] In conjunction with such a system licensing laws for health care professionals will have to be revised to reflect the new approach to health and give the public greater freedom of choice.[42]

The paradigm shift in health care will involve the formulation of new conceptual models, the creation of new institutions, and the implementation of new policies. As far as organization and policies are concerned, there are a number of measures that can be adopted right away. With regard to therapeutic models and techniques, the situation is somewhat more complicated. As yet there is no well-established system of therapies that corresponds to the new view of illness as a multidimensional and multileveled phenomenon. However, there are at present a number of models and procedures that seem to deal successfully with various aspects of ill health. It would seem, therefore, that here too a "bootstrap" approach may be the most appropriate strategy. It would consist of developing a mosaic of therapeutic models and techniques of limited scope that are mutually consistent. It would be the role of the general practitioner or the health team to find out which model, or which approach, is most suitable and most efficient for a particular patient. At the same time researchers and clinicians will be exploring these models further and, eventually, integrating them into a coherent system.

A number of therapeutic models and techniques are already being developed that go beyond the biomedical framework and are consistent with the systems view of health. Some of them are based on well-established Western healing traditions, others are of more recent origin, and most of them are not taken very seriously by the medical establishment because they are difficult to understand in terms of classical scientific concepts.

To begin with, numerous unorthodox approaches to health share a belief in the existence of patterns of "subtle energies," or "life energies," and see illness as resulting from changes in these patterns. Although the therapies practiced in these traditions, which are sometimes referred to as "energy medicine," involve a variety of techniques, all of them are believed to influence the organism at a more fundamental level than the physical or psychological symptoms of illness. This view is quite similar to that of the Chinese medical tradition, and so are many of the concepts used in the various healing traditions. For example, when homeopaths speak about the "vital force," or Reichian therapists about "bioenergy," they use these terms in a sense that comes very close to the Chinese concept of *ch'i.* The three concepts are not identical, but they seem to refer to the same reality—a reality much more complex than any of them. The main purpose of these terminologies is to describe the patterns of flow and fluctuation in the human organism. It is also believed that "life energy" is exchanged between an organism and its environment, and many traditions hold that this energy can be transferred between human beings by the laying-on of hands and other techniques of "psychic" healing.[43]

Most approaches to "energy medicine" were developed when science was formulated almost exclusively in terms of mechanistic concepts, and their originators cannot be faulted for using terminologies that now seem vague, simplistic, or out of date. The founders and practitioners of these healing traditions often had a remarkable intuition for the nature of life, health, and illness, and many of their concepts are likely to be extremely useful when reformulated in the language of the new systems approach. When self-organization is seen as the essence of living organisms, one of the main tasks of the life sciences will be to study the patterned processes of self-organizing systems, as well as the energies involved in these processes. The processes of physical and chemical systems have been studied extensively, and the energies associated with them are well understood. By contrast, the processes of self-organizing systems, together with their associated energies, are just beginning to be explored and may well reveal phenomena that have so far not been considered by orthodox science.

However, the term "energy," as used in the unorthodox healing traditions, is rather problematic from the scientific point of view. "Life energy" is often thought of as some kind of substance which flows through the organism and is transferred between organisms. According to modern science, energy is not a substance but rather a measure of activity, of dynamic patterns.[44] To understand the models of "energy medicine" scientifically it would seem, therefore, that one should concentrate on the concepts of flow, fluctuation, vibration, rhythm, synchrony, and resonance, which are fully consistent with the modern systems view. Concepts like "subtle bodies," or "subtle energies" should not be taken to refer to underlying substances but as metaphors describing the dynamic patterns of self-organization.

One of the most intriguing approaches to the fundamental dynamic patterns of the human organism is that of homeopathy. The roots of homeopathic philosophy can be traced back to the teachings of Paracelsus and Hippocrates, but the formal therapeutic system was founded at the end of the eighteenth century by the German physician Samuel Hahnemann. Although vigorously opposed by the medical establishment, homeopathy spread steadily during the nineteenth century and became especially popular in the United States, where 15 percent of all doctors were homeopaths around 1900. In the twentieth century the movement could not hold its ground against modern biomedical science and has only very recently found a certain renaissance.

In the homeopathic view, illness results from changes in an energy pattern or "vital force," which is the basis of all physical, emotional, and mental phenomena and is characteristic of each individual. The aim of homeopathic therapy, like that of acupuncture, is to stimulate a person's energy levels. The traditional homeopathic approach is purely phenomenological and, unlike Chinese medicine, does not have a detailed theory of energy patterns, but in recent years George Vithoulkas, perhaps the most articulate leader of the modern homeopathic movement, has formulated the beginnings of a theoretical framework.[45] Vithoulkas has tentatively identified Hahnemann's vital force with the body's electromagnetic field and uses the term "dynamic plane" for the fundamental level at which illness originates. In his theory the dynamic plane is characterized by a pattern of vibrations which is unique for each individual. External or internal stimuli affect the organism's rate of vibration, and these changes generate physical, emotional, or mental symptoms.

Homeopaths claim to be able to detect imbalances of the organism before any serious disturbances occur by observing a variety of subtle symptoms: changes in behavior patterns such as sensitivity to cold, desire for salt or sugar, sleeping habits, and so on. These subtle symptoms represent the organism's reaction to imbalances on the dynamic plane. Homeopathic diagnosis aims to establish a total pattern, or gestalt, of symptoms that mirrors the personality of the patient and is a reflection of that person's vibrational pattern. This is consistent with a key idea in modern psychosomatic medicine, the idea that an initial imbalance of the organism is channeled through a particular personality configuration to produce specific symptoms.

Homeopathic therapy consists of matching the pattern of symptoms that is characteristic of the patient with a similar pattern characteristic of the remedy. Vithoulkas believes that each remedy is associated with a certain vibrational pattern that constitutes its very essence. When the remedy is taken its energy pattern resonates with the energy pattern of the patient and thereby induces the healing process. The resonance phenomenon seems to be central to homeopathic therapy, but what exactly resonates and how this resonance is brought about is not well understood. Homeopathic remedies are substances derived from animals, plants and minerals, and are taken in highly diluted form. The selection of the correct remedy is based on Hahnemann's Law of Similars—"Like Cures Like"—which gave homeopathy* its name. According to Hahnemann, any substance that can produce a total pattern of symptoms in a healthy human being can cure those same symptoms in a sick person. Homeopaths claim that literally any sub-

* From the Greek *homeo* ("similar") and *pathos* ("suffering").

stance can produce, and cure, a wide spectrum of highly individualized symptoms known as the "personality" of the remedy.

The first and perhaps most important part of homeopathic practice is "taking the homeopathic case." Each interview is a unique process demanding of the interviewer a high degree of intuition and sensitivity. The purpose is to experience the personality of the patient as an integrated, living entity, and to match its very essence with that of the remedy. Vithoulkas says this experience should emerge from an intimate human encounter between therapist and patient, which will deeply affect both of them:

> The encounter between a patient and a homeopath is an intimate interaction for both . . . The prescriber . . . is not merely a passive observer, protected behind a wall of objectivity. Each patient engages the homeopath in a deep and meaningful way. Because of the very nature of homeopathy, the prescriber becomes an intimate participant in the life of the patient, involved in every aspect of it, and being at once sympathetic and sensitive as well as objective and accepting . . . When homeopathy is practiced with this degree of involvement, it stimulates growth in the prescriber just as it does for the patient.[46]

This description of the homeopathic interview, with its strong emphasis on the mutual interaction between therapist and patient, is quite reminiscent of an intense session of psychotherapy as described, for example, by Jung.[47] Indeed, one is tempted to wonder whether the crucial resonance in homeopathic therapy is not the one between the patient and the homeopath, with the remedy merely a crutch.

The lack of any scientific explanation of homeopathic therapy is one of the main reasons why it has remained a highly controversial healing art. However, further development of psychosomatic medicine and the systems approach to health will help to clarify many of the homeopathic principles and may well encourage the medical profession to reexamine its position. Homeopathic philosophy, with its general view of illness, its emphasis on individualized treatment, and its basic trust in the human organism, exemplifies many important aspects of holistic health care.

A school of "energy medicine" that is of more recent origin than homeopathy and has had a strong influence on a variety of therapies is

Reichian therapy.[48] Wilhelm Reich started out as a psychoanalyst and disciple of Freud, but while Freud and the other analysts concentrated on the psychological contents of mental disorders, Reich became interested in the ways these disorders manifest themselves physically. As the emphasis of his treatment shifted from the psyche to the body he developed therapeutic techniques that involved physical contact between therapist and patient—a sharp break with traditional psychoanalytic practice. From the very beginning of his medical research, Reich was keenly interested in the role of energy in the functioning of living organisms, and one of the main goals of his psychoanalytic work was to associate the sexual drive, or libido, which Freud saw as an abstract psychological force, with concrete energy flowing through the physical organism. This approach led Reich to the concept of bioenergy, a fundamental form of energy that permeates and governs the entire organism and manifests itself in the emotions as well as in the flow of bodily fluids and other biophysical movements. Bioenergy, according to Reich, flows in wave movements and its basic dynamic characteristic is pulsation. Every mobilization of flow processes and emotions in the organism is based on a mobilization of bioenergy.

One of Reich's key discoveries was that attitudes and emotional experiences can give rise to certain muscular patterns which block the free flow of energy. These muscular blocks, which Reich called "character armor," are developed in nearly every adult individual. They reflect our personality and enclose key elements of our emotional history, locked up in the structure and tissue of our muscles. The central task of Reichian therapy is to destroy the muscular armor in order to reestablish the organism's full capacity for the pulsation of bioenergy. This is done with the help of deep breathing and a variety of other physical techniques aimed at helping patients express themselves through their bodies rather than with words. In this process past traumatic experiences will emerge into conscious awareness and will be resolved, together with the corresponding muscular blocks. The ideal result is the appearance of a phenomenon that Reich called the orgasm reflex, and which he regarded as central to the dynamics of living organisms, far transcending the usual sexual connotation of the term. "In the orgasm," writes Reich, "the living organism is nothing but a part of pulsating nature."[49]

It is evident that Reich's concept of bioenergy comes very close to the Chinese concept of *ch'i*. Like the Chinese, Reich emphasized the cyclical nature of the organism's flow processes and, like the Chinese, he also saw the energy flow in the body as the reflection of a process that goes on in the universe at large. To him bioenergy was a special manifestation of a form of cosmic energy that he called "orgone energy." Reich saw this orgone energy as some kind of primordial substance, present everywhere in the atmosphere and extending through all space, like the ether of nineteenth-century physics. Inanimate as well as living matter, according to Reich, derives from orgone energy through a complicated process of differentiation.

This concept of orgone energy is clearly the most controversial part of Reich's thinking and was the reason for his isolation from the scientific community, his persecution, and his tragic death.[50] From the perspective of the 1980s it seems that Wilhelm Reich was a pioneer of the paradigm shift. He had brilliant ideas, a cosmic perspective, and a holistic and dynamic world view that far surpassed the science of his time and was not appreciated by his contemporaries. Reich's way of thinking, which he called "orgonomic functionalism," is in perfect agreement with the process thinking of our modern systems theory, as this passage shows:

> Functional thinking does not tolerate any static conditions. For it, all natural processes are in motion, even in the case of rigidified structures and immobile forms . . . Nature, too, "flows" in every single one of its diverse functions as well as in its totality . . . Nature is functional in all areas, and not only in those of organic matter. Of course, there are mechanical laws, but the mechanics of nature are in themselves a special variant of functional processes.[51]

Unfortunately, the language of modern systems biology was not available to Reich, so he sometimes expressed his theory of living matter and his cosmology in terms which were rooted in the old paradigm and rather inappropriate. He could not conceive of orgone energy as a measure of organic activity, but had to see it as a substance that could be detected and accumulated, and in his attempts to verify such a notion he cited numerous atmospheric phenomena that are more likely to be explained in terms of conventional processes, such as ionization or ultraviolet radiation.[52] In spite of these conceptual problems, Reich's basic ideas about the dynamics of life have had a tremendous influence and have inspired therapists to develop a variety of new psychosomatic approaches. If Reichian theory were reformulated in modern systems language, its relevance to contemporary research and therapeutic practice would become even clearer.

The therapeutic models discussed in the rest of this chapter do not necessarily subscribe to the notion of fundamental energy patterns, but they all see the organism as a dynamic system with interrelated physical, biochemical, and psychological aspects that must be in balance for the human being to be in good health. Some of the therapies address themselves to the physical aspects of this balance by dealing with the body's muscular system or with other structural elements; others influence the organism's metabolism; and still others concentrate on establishing balance through psychological techniques. Whatever their approach, they all recognize the fundamental interdependence of the organism's biological, mental, and emotional manifestations, and are therefore mutually consistent.

The therapies that try to facilitate harmony, balance, and integration through physical methods have recently become known collectively as bodywork. They deal with the nervous system, the muscle system, and various other tissues, and with the interplay and coordinated movement of all these components. Bodywork therapy is based on the belief that all our activities, thoughts, and feelings are reflected in the physical organism, manifesting themselves in our posture and movements, in tensions, and in many other signs of "body language." The body as a whole is a reflection of the psyche, and work on either one will also change the other.

Since the Eastern philosophical and religious traditions have always tended to see mind and body as a unity, it is not surprising that a number of techniques to approach consciousness from the physical level have been developed in the East. The therapeutic significance of these meditative approaches is being noticed increasingly in the West, and many Western therapists are incorporating Eastern bodywork techniques such as Yoga, T'ai Chi, and Aikido in their treatments. A major aspect of these Eastern techniques, which is also strongly emphasized in Reichian therapy, is the fundamental role of breath as a link between the conscious and unconscious levels of mind. Our breathing patterns reflect the dynamics of our entire mind/body system, and breath is the key to our emotional memories. The practice of proper breathing and the use of various breathing techniques as therapeutic tools is therefore central to many schools of bodywork, in the West as well as in the East.

The dynamic manifestations of the human organism—its continual movements and the various processes of flow and fluctuation—all involve the muscle system. Work on the body's muscular system is ideally suited for studying and influencing physiological and psychological balance. Detailed studies of the physical organism from this perspective show that the conventional distinctions between nerves, muscles, skin, and bones are often quite artificial and do not reflect the physical reality. The entire muscle system of the organism is covered by loose connective tissues that integrate the muscles into a functional whole and cannot be separated either physically or conceptually from the muscle tissue, the nerve fibers, and the skin. Segments of this connective tissue are associated with different organs, and a variety of physiological disorders can be detected and cured by special techniques of connective-tissue massage.

Since the muscle system is an integrated whole, a disturbance of any part of it will propagate through the entire system, and since all bodily functions are supported by muscles, every weakening of the organism's balance will be reflected in the muscular system in a specific way. An important aspect of this balance is the regular flow of nerve current throughout the body, which is the focus of chiropractic. Chiropractors concentrate on the structural support of the nervous system along the spine. By means of manual adjustments involving gentle manipulations of joints and soft tissues, they are able to realign dislocated vertebras and thereby eliminate obstructions in the nervous flow that may cause many different disorders. Out of chiropractic emerged a special technique of muscle testing, known as applied kinesiology,* which has been developed into a valuable therapeutic tool that enables therapists to use the muscle system as a source of information about various aspects of the organism's state of balance.[53]

Influenced by the pioneering ideas of Wilhelm Reich, by Eastern concepts, and by the modern dance movement, a number of therapists have combined various elements from these traditions to develop bodywork techniques that have recently become very popular. The principal founders of these new approaches are Alexander Lowen ("bioen-

* Kinesiology, from the Greek *kinesis* ("motion"), is the study of human anatomy in relation to movement.

ergetics"), Frederick Alexander ("Alexander technique"), Moshe Feldenkrais ("functional integration"), Ida Rolf ("structural integration"), and Judith Aston ("structural patterning"). In addition, various massage therapies have been developed, many of them inspired by Eastern techniques like shiatsu and acupressure. All these approaches are based on the Reichian notion that emotional stress manifests itself in the form of blocks in the muscle structure and tissue, but they differ in the methods employed for releasing these psychosomatic blocks.[54] Some bodywork schools are based on a single idea that is translated into a single set of prescriptions and manipulations, but ideally a bodywork therapist should be familiar with each of these techniques and not use any one of them exclusively. Another problem is that many schools tend to treat muscular blocks as static entities and associate emotions with body postures in a rather rigid way, without perceiving the body as it moves through space and relates to its environment.

One of the most subtle approaches to bodywork, and one which focuses precisely on that aspect—the body moving through space and interacting with its environment—is practiced by dance and movement therapists, and in particular by a school of movement therapy based on the work of Rudolf Laban and further elaborated by Irmgard Bartenieff.[55] Laban developed a method and terminology for analyzing human movement that is relevant to many disciplines besides therapy, including anthropology, architecture, industry, theater, and dance. The therapeutic significance of this method derives from Laban's perception that all movement is functional and expressive at the same time. Whatever tasks people pursue, they will also express something about themselves through their movement. Laban's system deals explicitly with this expressive quality of movement and thus allows movement therapists to recognize many fine details of their patients' emotional and physical states by carefully watching how they move.

The Laban-Bartenieff school of movement therapy pays special attention to the ways in which individuals interact and communicate with their environment. This interplay is seen in terms of complex rhythmic patterns, flowing in and out of one another in various ways, with ill health arising from a lack of synchrony and integration. Healing, in this view, is induced by a special process of interaction between therapist and patient in which the rhythms of both are continually synchronizing. By communicating with their patients through movement, and by establishing a kind of resonance, movement therapists help these individuals to better integrate themselves, physically and emotionally, into their environment.

Another important approach to balance is through the organism's metabolism. Biochemical balance can be influenced by changing one's diet and by taking various medicines in the form of either herbs or synthetic drugs. In most medical traditions these three forms of treatment are not sharply separated, and it would seem most appropriate to adopt this view also in the new system of holistic health care. Nutritional therapy, herbal medicine, and the prescription of drugs all affect the body's biochemical balance and are variations of one and the same therapeutic approach. Recognizing the organism's inherent tendency to regain its balance, the holistic therapist will always use the mildest possible remedy, beginning with a change of diet, moving on to herbal medicines if needed to bring the desired effect, and using synthetic drugs only as a last resort and in emergencies.

Although nutrition has always been a major factor in the development of disease patterns, it is severely neglected in today's medical education and practice. Most doctors are not qualified to give sound nutritional advice, and the articles on nutrition published in popular magazines are often extremely confusing. Yet the basic principles of nutritional counseling are relatively simple and should be known to all general practitioners.[56]

Nutritional counseling and therapy are closely related to a new branch of medicine known as clinical ecology, which in the late 1940s grew out of the study of allergies and is concerned with the impact of foods and chemicals on our health and our mental state.[57] Clinical ecologists have found that common foods and seemingly harmless chemical products used every day in our homes, offices, and work places can cause mental, emotional, and physical problems, from headaches and depressions to aches and pains in muscles and joints. Patients who come to see their doctors with multiple symptoms, both physical and psychological, are frequently suffering from such allergies. The treatment of these patients by clinical ecologists is a highly individualized procedure, involving nutritional therapy and various other techniques, aimed at identifying and eliminating the environmental causes of the patients' illness.

Like nutritional counseling, the art of herbal medicine has been almost totally forgotten with the rise of the biomedical model, and only very recently has there been a certain revival of the therapeutic use of natural herbs. This development is encouraging, since natural, unrefined plant material seems to be the best type of oral medication, but herbal medicine can be successful only if the purpose of the treatment is to deal with the whole organism rather than trying to cure a specific disease. Otherwise there will invariably be a tendency to refine herbal mixtures in order to isolate their "active ingredients," which would significantly reduce the therapeutic effect. Pharmaceutical drugs, which are often the end products of such refinements, act much faster on the body's biochemistry than herbal mixtures, but they also cause a much greater shock to the organism and thus generate numerous harmful side effects that generally do not occur when the unrefined herbal remedies are used.[58]

More careful use of medical drugs illustrates the future role of biomedical therapy as a whole. The achievements of modern medical science will by no means be abandoned, but in the future holistic approach biomedical techniques will play a much more restricted role. They will be used to deal with the physical and biological aspects of illness, especially in emergencies, but always very judiciously and in conjunction with psychological counseling, stress-reduction techniques, and other methods of holistic patient care. The transition to the new system will have to be made slowly and carefully because of the tremendous symbolic power of biomedical therapy in our culture. The reductionist approach to illness, with its strong emphasis on drugs and surgery, will be supplemented and eventually replaced by the new holistic therapies in a gradual process, as our collective views of health and illness change and evolve.

The last group of therapeutic techniques to be reviewed here approach psychosomatic balance through the mind. Embracing various methods of relaxation and stress reduction, they are likely to play an important role in all future therapies.[59] Current attitudes toward relaxation in our culture are quite naïve. Many activities thought of as relaxing—watching television, reading, having a few drinks—do not reduce stress or mental anxiety. Deep relaxation is a psycho-physiological process that requires as much diligent practice as any other skill, and that has to be practiced regularly to be fully effective. Correct breathing is one of the most important aspects of relaxation and thus one of the most vital elements in all stress-reduction techniques.

Deep, regular breathing and profound relaxation are characteristic of the techniques of meditation developed in many cultures, but especially those of the Far East, for thousands of years. Recent interest in mystical traditions has brought an increasing number of Westerners to regular meditation, and there have been several empirical studies of the health benefits of meditative practice.[60] Because these studies indicate that the response of the human organism to meditation is opposite to its reaction to stress, meditative techniques will probably have important clinical applications in the future.

Over the past fifty years various deep-relaxation techniques have also been developed in the West and have been used successfully as therapeutic tools for stress management. They can be regarded as Western forms of meditation, not connected with any spiritual tradition but growing out of the need for dealing with stress. One of the most comprehensive and successful of these techniques is a method known as autogenic training, which Johannes Schultz, a German psychiatrist, developed in the 1930s. It is a form of self-hypnosis combined with certain specific exercises designed to integrate mental and physical functions and to induce deep states of relaxation. During the initial stages autogenic training emphasizes exercises dealing with the physical aspects of relaxation, but once these have been mastered it progresses to more subtle psychological aspects that, like meditation, involve the experience of nonordinary states of consciousness.

When the organism is fully relaxed, one can make contact with one's unconscious to obtain important information about one's problems or about the psychological aspects of one's illness. The communication with the unconscious takes place through a highly personal, visual, and symbolic language similar to that of dreams. Mental imagery and visualization, therefore, play a central role in the advanced stages of autogenic training, as they do in many traditional techniques of meditation. Visualization techniques have recently been applied directly to specific illnesses too, and have often yielded excellent results.

The psychological approach to stress reduction and healing has received dramatic support from a new technology known as biofeedback.[61] It is a technique that helps a person to achieve voluntary control over normally unconscious bodily functions by monitoring them, amplifying the results electronically, and displaying them ("feeding them back"). Numerous applications of this technique over the past decade have shown that a wide range of autonomic, or involuntary, physiological functions—heart rate, body temperature, muscle tension, blood pressure, brain-wave activity, and others—can be brought under conscious control in this way. Many clinicians now believe that it may be possible to achieve some degree of voluntary control over any biological process that can be continuously monitored, amplified, and displayed.

The term "voluntary control" is actually somewhat inappropriate to describe the regulation of autonomic functions through biofeedback. The idea of the mind controlling the body is based on the Cartesian division and does not correspond to the observations made in biofeedback practice. What is required for this subtle form of self-regulation is not control but, on the contrary, a meditative state of deep relaxation in which all control is relinquished. In such a state, channels of communication between the conscious and the unconscious mind open up and facilitate the integration of psychological and biological functions. This process of communication often takes place through visual imagery and symbolic language, and it was this role of visual imagery in biofeedback which led a number of therapists to use visualization techniques for the treatment of illnesses.

Clinical biofeedback can be used in conjunction with many therapeutic techniques, both physical and psychological, to teach patients relaxation and the management of stress. It is more likely to convince Westerners of the unity and interdependence of mind and body than Eastern techniques of meditation, and it facilitates the important shift of responsibility for health and illness from the therapist to the patient. The fact that individuals are able to correct a particular symptom by themselves through biofeedback will often dramatically reduce their feeling of helplessness and encourage the positive mental attitude that is so important in healing.

These experiences have shown the great value of biofeedback as a therapeutic tool, but it should not be used in a reductionist way. Since it concentrates on the single physiological function that is monitored, biofeedback is not an alternative to more traditional meditation and relaxation techniques. Stress involves various patterns of psychosomatic functions, and the regulation of any one of them is generally not enough. Hence biofeedback has to be supplemented by more general methods of relaxation if it is to be fully effective. To establish the appropriate combination of self-regulation and relaxation techniques is quite difficult and requires a great deal of experience.

To conclude our discussion of holistic health care, it is appropriate to talk about a new way of treating cancer known as the Simonton approach, which I consider a holistic therapy par excellence. Cancer is an exemplary phenomenon, an illness characteristic of our age that forcefully illustrates many of the points made in this chapter. The imbalance and fragmentation that pervade our culture play an important role in the development of cancer and, at the same time, prevent medical researchers and clinicians from understanding the disease or treating it successfully. The conceptual framework and therapy that Carl Simonton, a radiation oncologist,* and Stephanie Matthews-Simonton, a psychotherapist, have developed are fully consistent with the views of health and illness we have been discussing and have far-reaching implications for many areas of health and healing.[62] At present the Simontons see their work as a pilot study. They select their patients very carefully because they want to see how far they can go with a small number of highly motivated individuals to understand the basic dynamics of cancer. Once they have reached such an understanding, they will apply their knowledge and skills to larger numbers of patients. So far, the average survival time of their patients is twice that of the best institutions for cancer therapy and three times the national average in the United States. Moreover, the quality of life and levels of activity of these men and women, all of whom were considered medically incurable, are absolutely extraordinary.

The popular image of cancer has been conditioned by the fragmented world view of our culture, the reductionist approach of our science, and the technology-oriented practice of medicine. Cancer is seen as a strong and powerful invader that strikes the body from outside. There seems to be no hope of controlling it, and for most people cancer is synonymous with death. Medical treatment—whether radiation, chemotherapy, surgery, or a combination of these—is drastic, negative, and further injures the body. Physicians are increasingly coming to see cancer as a systemic disorder; a disease that has a localized appearance but has the ability to spread, and that really involves

* Oncology, from the Greek *onkos* ("mass"), is the study of tumors.

the entire body, the original tumor being merely the tip of the iceberg. Patients, however, often insist on viewing their cancer as a localized problem, especially during its initial phase. They see the tumor as a foreign object and want to get rid of it as quickly as possible and forget the whole episode. Most patients are so thoroughly conditioned in their views that they refuse to consider the broader context of their illness and do not perceive the interdependence of its psychological and physical aspects. For many cancer patients the body has become their enemy, one they mistrust and from which they feel thoroughly alienated.

One of the main aims of the Simonton approach is to reverse the popular image of cancer, which does not correspond to the findings of current research. Modern cellular biology has shown that cancer cells are not strong and powerful but, on the contrary, are weak and confused. They do not invade, attack, or destroy, but simply overproduce. A cancer begins with a cell that contains incorrect genetic information because it has been damaged by harmful substances or other environmental influences, or simply because the organism will occasionally produce an imperfect cell. The faulty information will prevent the cell from functioning normally, and if this cell reproduces others with the same incorrect genetic makeup, the result will be a tumor composed of a mass of these imperfect cells. Whereas normal cells communicate effectively with their environment to determine their optimal size and rate of reproduction, the communication and self-organization of malignant cells are impaired. As a consequence they grow larger than healthy cells and reproduce recklessly. Moreover, the normal cohesion between cells may weaken and malignant cells may break loose from the original mass and travel to other parts of the body to form new tumors—which is known as metastasis. In a healthy organism the immune system will recognize abnormal cells and destroy them, or at least wall them off so they cannot spread. But if for some reason the immune system is not strong enough, the mass of faulty cells will continue to grow. Cancer, then, is not an attack from without but a breakdown within.

The biological mechanisms of cancer growth make it clear that the search for its causes has to go in two directions. On the one hand we need to know what causes the formation of cancerous cells; on the other we need to understand what causes the weakening of the body's immune system. Many researchers have come to realize over the years that the answers to both these questions consist of a complex network of interdependent genetic, biochemical, environmental, and psychological factors. With cancer, more than with any other illness, the traditional biomedical practice of associating a physical disease with a specific physical cause is not appropriate. But since most researchers still operate within the biomedical framework, they find the phenomenon of cancer extremely bewildering. As Carl Simonton has noted, "Cancer management today is in a state of confusion. It looks almost like the disease itself—fragmented and confused."[63]

The Simontons fully recognize the role of carcinogenic substances and environmental influences in the formation of cancer cells, and they strongly advocate the implementation of appropriate social policies to eliminate these health hazards. However, they have also come to realize that neither carcinogenic substances, nor radiation, nor genetic predisposition alone will provide an adequate explanation of what causes cancer. No understanding of cancer will be complete without addressing the crucial question: What inhibits a person's immune system, at a particular time, from recognizing and destroying abnormal cells and thus allows them to grow into a life-threatening tumor? This is the question on which the Simontons have concentrated in their research and therapeutic practice, and they have found that it can be answered only by carefully considering the mental and emotional aspects of health and illness.

The emerging picture of cancer is consistent with the general model of illness we have been developing. A state of imbalance is generated by prolonged stress which is channeled through a particular personality configuration to give rise to specific disorders. In cancer the crucial stresses appear to be those that threaten some role or relationship that is central to the person's identity, or set up a situation from which there is apparently no escape. Several studies suggest that these critical stresses typically occur six to eighteen months before the diagnosis of cancer.[64] They are likely to generate feelings of despair, helplessness, and hopelessness. Because of these feelings, serious illness, and even death, may become consciously or unconsciously acceptable as a potential solution.

The Simontons and other researchers have developed a psychosomatic model of cancer that shows how psychological and physical states work together in the onset of the disease. Although many details of this process still need to be clarified, it has become clear that the emotional stress has two principal effects. It suppresses the body's immune system and, at the same time, leads to hormonal imbalances that result in an increased production of abnormal cells. Thus optimal conditions for cancer growth are created. The production of malignant cells is enhanced precisely at a time when the body is least capable of destroying them.

As far as the personality configuration is concerned, the individual's emotional states seem to be the crucial element in the development of cancer. The connection between cancer and emotions has been observed for hundreds of years, and today there is substantial evidence for the significance of specific emotional states. These are the result of a particular life history that seems to be characteristic of cancer patients. Psychological profiles of such patients have been established by a number of researchers, some of whom were even able to predict the incidence of cancer with remarkable accuracy on the basis of these profiles.

Lawrence LeShan studied more than five hundred cancer patients and identified the following significant components in their life histories:[65] feelings of isolation, neglect, and despair during youth, with intense interpersonal relationships appearing difficult or dangerous; a strong relationship with a person or great satisfaction with a role in early adulthood, which becomes the center of the individual's life; loss of the relationship or role, resulting in despair; internalizing of the despair to the extent that individuals are unable to let other people know when they feel hurt, angry, or hostile. This basic pattern has been confirmed as typical of cancer patients by a number of researchers.

The basic philosophy of the Simonton approach affirms that the development of cancer involves a number of interdependent psychological and biological processes, that these processes can be recognized and understood, and that the sequence of events which leads to illness can be reversed to lead the organism back into a healthy state. As in any holistic therapy, the first step toward initiating the healing cycle consists of making patients aware of the wider context of their illness. Establishing the context of cancer begins by asking patients to identify the major stresses occurring in their lives six to eighteen months prior to their diagnosis. The list of these stresses is then used as a basis for discussing the patients' participation in the onset of their disease. The purpose of the concept of patient participation is not to evoke guilt, but rather to create the basis for reversing the cycle of psychosomatic processes that led to the state of ill health.

While the Simontons are establishing the context of a patient's illness, they are also strengthening his belief in the effectiveness of the treatment and the potency of the body's defenses. The development of such a positive attitude is crucial for the whole treatment. Studies have shown that the patient's response to treatment depends more on his attitude than on the severity of the disease. Once feelings of hope and anticipation are generated, the organism translates them into biological processes that begin to restore balance and to revitalize the immune system, using the same pathways that were used in the development of the illness. The production of cancerous cells decreases, and at the same time the immune system becomes stronger and more efficient in dealing with them. While this strengthening takes place, physical therapy is used in conjunction with the psychological approach to help the organism destroy the malignant cells.

The Simontons see cancer not merely as a physical problem but as a problem of the whole person. Accordingly, their therapy does not focus on the disease alone but deals with the total human being. It is a multidimensional approach involving various treatment strategies designed to initiate and support the psychosomatic process of healing. At the biological level the aim is twofold: to destroy cancer cells and to revitalize the immune system. In addition, regular physical exercise is used to reduce stress, to alleviate depression, and to help patients get more in touch with their bodies. Experience has shown that cancer patients are capable of far more physical activity than most people would assume.

The principal technique of strengthening the immune system is a method of relaxation and mental imagery that the Simontons developed when they learned about the important role of visual imagery and symbolic language in biofeedback. The Simonton technique consists of the regular practice of relaxation and visualization during which the cancer and the action of the immune system are pictured in the patient's own symbolic imagery. This technique has turned out to be an extremely efficient tool for strengthening the immune system and has often resulted in dramatic reductions or eliminations of malignant tumors. Moreover, the visualization method is also an excellent way for patients to communicate with their unconscious. The Simontons work very closely with their patients' imagery and have learned that it tells them much more about the patients' feelings than any rational explanations.

Although the visualization technique plays a central role in the Simonton therapy, it is important to emphasize that visualization and physical therapy alone are not sufficient to heal cancer patients. According to the Simontons, the physical disease is a manifestation of underlying psychosomatic processes that may be generated by various psychological and social problems. As long as these problems are not solved the patient will not get well, even though the cancer may temporarily disappear. To help patients solve the problems that are at the roots of their illnesses, the Simontons have made psychological counseling and psychotherapy essential elements of their approach. The therapy usually takes place in group sessions, in which patients find mutual support and encouragement. It concentrates on their emotional problems but does not separate these from the larger patterns of their lives, and thus generally includes social, cultural, philosophical, and spiritual aspects.

For most cancer patients the impasse created by the accumulation of stressful events can be overcome only if they change part of their belief system. The Simonton therapy shows them that their situation seems hopeless only because they interpret it in ways that limit their responses. Patients are encouraged to explore alternative interpretations and responses in order to find healthy ways of resolving the stressful situation. Thus the therapy involves a continual examination of their belief system and world view.

Dealing with death is an integral part of the Simonton therapy. Patients are made aware of the possibility that, sometime in the future, they may come to a decision that it is time for them to move toward death. They are assured that they have the right to make such a decision and are promised that the therapists will be as supportive and caring through their dying as they are in their struggle to regain health. In dealing with death in such a way, a major task is often getting the family to give the patient permission to die. Once this permission is given and expressed—not just verbally but through the family's behavior—the whole perspective of that death is changed. As the Simontons point out to their patients, whether or not one recovers from cancer, one can succeed in improving the quality of one's living or of one's dying.

The confrontation with death that cancer patients have to face touches on the fundamental existential problem that is characteristic of the human condition. Cancer patients are thus naturally led to consider their goals in life, their reasons for living, and their relation to the cosmos as a whole. The Simontons do not avoid any of these issues in their therapy, and this is why their approach is of such exemplary value for health care as a whole.

11 · Journeys Beyond Space and Time

In the systems view of health, every illness is in essence a mental phenomenon, and in many cases the process of getting sick is reversed most effectively through an approach that integrates both physical and psychological therapies. The conceptual framework underlying such an approach will include not only the new systems biology but also a new systems psychology, a science of human experience and behavior that perceives the human organism as a dynamic system involving interdependent physiological and psychological patterns, and as being embedded in interacting larger systems of physical, social, and cultural dimensions.

Carl Gustav Jung was perhaps the first to extend classical psychology into these new realms. In breaking with Freud he abandoned the Newtonian models of psychoanalysis and developed a number of concepts that are quite consistent with those of modern physics and with systems theory. Jung, who was in close contact with several of the leading physicists of his time, was well aware of these similarities. In one of his major works, *Aion,* we find the following prophetic passage:

> Sooner or later, nuclear physics and the psychology of the unconscious will draw closer together as both of them, independently of one another and from opposite directions, push forward into transcendental territory . . . Psyche cannot be totally different from matter, for how otherwise could it move matter? And matter cannot be alien to psyche, for how else could matter produce psyche? Psyche and matter exist in the same

> world, and each partakes of the other, otherwise any reciprocal action would be impossible. If research could only advance far enough, therefore, we should arrive at an ultimate agreement between physical and psychological concepts. Our present attempts may be bold, but I believe they are on the right lines.[1]

Indeed, it seems that Jung's approach was very much on the right lines and, in fact, many of the differences between Freud and Jung parallel those between classical and modern physics, between the mechanistic and the holistic paradigm.[2]

Freud's theory of the mind was based on the concept of the human organism as a complex biological machine. Psychological processes were deeply rooted in the body's physiology and biochemistry and followed the principles of Newtonian mechanics.[3] Mental life in health and illness reflected the interplay of instinctual forces within the organism and their clashes with the external world. While Freud's views about the detailed dynamics of these phenomena changed over time, he never abandoned the basic Cartesian orientation of his theory. Jung, by contrast, was not so much interested in explaining psychological phenomena in terms of specific mechanisms, but rather attempted to understand the psyche in its totality and was particularly concerned with its relations to the wider environment.

Jung's ideas about the dynamics of mental phenomena came quite close to the systems view. He saw the psyche as a self-regulating dynamic system, characterized by fluctuations between opposite poles. To describe its dynamics he used the Freudian term "libido" but gave it a very different meaning. Whereas Freud saw the libido as an instinctual drive that was closely connected with sexuality and had properties similar to those of a force in Newtonian mechanics, Jung conceived it as a general "psychic energy" which he saw as a manifestation of the basic dynamics of life. Jung was quite aware that he was using the term "libido" much in the sense in which Reich used "bioenergy"; Jung, however, concentrated exclusively on the psychological aspects of the phenomenon:

> We would probably do best to regard the psychic process simply as a life-process. In this way we enlarge the narrower concept of psychic energy to a broader one of life-energy, which includes "psychic energy" as a specific part. We thus gain the advantage of being able to follow

> quantitative relations beyond the narrow confines of the psychic into the sphere of biological functions in general . . . In view of the psychological use we intend to make of it, we call our hypothetical life-energy "libido" . . . In adopting this usage, I do not in any way wish to forestall workers in the field of bioenergetics, but freely admit that I have adopted the term libido with the intention of using it for *our* purposes: for theirs, some such term as "bio-energy" or "vital energy" may be preferred.[4]

As in the case of Reich, it is unfortunate that the language of modern systems theory was not available to Jung. Instead, as Freud had done before him, he used the framework of classical physics, which is much less appropriate to describe the functioning of living organisms.[5] Consequently Jung's theory of psychic energy is at times somewhat confusing. Nevertheless, it is relevant to current developments in psychology and psychotherapy and would be even more influential if it were reformulated in modern systems language.

The key difference between the psychologies of Freud and Jung is in their views of the unconscious. For Freud, the unconscious was predominantly personal in nature, containing elements that had never been conscious and others that had been forgotten or repressed. Jung acknowledged those aspects, but he believed that the unconscious was much more. He saw it as the very source of consciousness and held that we begin our lives with our unconscious, not with a *tabula rasa* as Freud believed. The conscious mind, according to Jung, "grows out of an unconscious psyche which is older than it, and which goes on functioning together with it or even in spite of it."[6] Accordingly, Jung distinguished two realms of the unconscious psyche: a personal unconscious belonging to the individual, and a collective unconscious that represents a deeper stratum of the psyche and is common to all humankind.

Jung's concept of the collective unconscious distinguishes his psychology not only from Freud's but from all others. It implies a link between the individual and humanity as a whole—in fact, in some sense, between the individual and the entire cosmos—that cannot be understood within a mechanistic framework but is very consistent with the systems view of mind. In his attempts to describe the collective unconscious, Jung also used concepts that are surprisingly similar to the ones contemporary physicists use in their descriptions of subatomic phenomena. He saw the unconscious as a process involving "collectively present dynamic patterns," which he called archetypes.[7] These patterns, formed by the remote experiences of humanity, are reflected in dreams, as well as in the universal motifs found in myths and fairy tales around the world. Archetypes, according to Jung, are "forms without content, representing merely the possibility of a certain type of perception and action."[8] Although they are relatively distinct, these universal forms are embedded in a web of relationships, in which each archetype, ultimately, involves all the others.

Both Freud and Jung had a deep interest in religion and spirituality, but whereas Freud seemed to be obsessed with the need to find rational and scientific explanations for religious beliefs and behavior, Jung's approach was much more direct. His many personal religious experiences convinced him of the reality of the spiritual dimension in life. Jung considered comparative religion and mythology unique sources of information about the collective unconscious, and saw genuine spirituality as an integral part of the human psyche.

Jung's spiritual orientation gave him a broad perspective on science and rational knowledge. He saw the rational approach as merely one of several approaches, which all resulted in different but equally valid descriptions of reality. In his theory of psychological types, Jung identified four characteristic functions of the psyche—sensation, thinking, feeling, and intuition—that are manifest to different degrees in different individuals. Scientists operate predominantly from the thinking function, but Jung was well aware that his own explorations of the human psyche sometimes made it necessary to go beyond rational understanding. For example, he emphasized repeatedly that the collective unconscious and its patterns, the archetypes, defied precise definition.

In transcending the rational framework of psychoanalysis, Jung also expanded Freud's deterministic approach to mental phenomena by postulating that psychological patterns were connected not only causally but also acausally. In particular, he introduced the term "synchronicity" for acausal connections between symbolic images in the inner, psychic world and events in the external reality.[9] Jung saw these synchronistic connections as specific examples of a more general "acausal orderedness" in mind and matter. Today, thirty years later, this view seems to be supported by several developments in physics. The notion of order—or, more precisely, of ordered connectedness—has recently emerged as a central concept in particle physics, and physicists are now also making a distinction between causal (or "local") and acausal (or "nonlocal") connections.[10] At the same time patterns of matter and patterns of mind are increasingly recognized as reflections of one another, which suggests that the study of order, in causal as well as acausal connectedness, may be an effective way of exploring the relationship between the inner and outer realms.

Jung's ideas about the human psyche led him to a notion of mental illness that has been very influential among psychotherapists in recent years. He saw the mind as a self-regulating or, as we would say today, self-organizing system, and neurosis as a process by which this system tries to overcome various obstructions that prevent it from functioning as an integrated whole. The role of the therapist, in Jung's view, is to support this process, which he saw as being part of a psychological journey along a path of personal development or "individuation." The process of individuation, according to Jung, consists of integrating the conscious and unconscious aspects of our psyche, which will involve encounters with the archetypes of the collective unconscious and will result, ideally, in the experience of a new center of personality, which Jung called the Self.

Jung's views of the therapeutic process reflected his ideas about mental illness. He believed that psychotherapy should flow from a personal encounter between therapist and patient that involves the entire being of both: "By no device can the treatment be anything but the product of mutual influence, in which the whole being of the doctor as well as that of his patient plays its part."[11] This process involves an interaction between the unconscious of the therapist and that of the patient, and Jung advised therapists to communicate with their own unconscious when dealing with patients:

> The therapist must at all times keep watch over himself, over the way he is reacting to his patient. For we do not react only with our consciousness. Also we must always be asking ourselves: how is our unconscious experiencing this situation? We must therefore observe our dreams, pay the closest attention and study ourselves just as carefully as we do the patient.[12]

Because of his seemingly esoteric ideas, his emphasis on spirituality, and his attraction to mysticism, Jung has not been taken very seriously in psychoanalytic circles. With the recognition of an increasing consistency between Jungian psychology and modern science, this attitude is bound to change, and Jung's ideas about the human unconscious, the dynamics of psychological phenomena, the nature of mental illness, and the process of psychotherapy are likely to have a strong influence on psychology and psychotherapy in the future.

Around the middle of the twentieth century, many ideas that are important to current developments in psychology began to emerge in the United States. During the 1930s and 1940s there were two distinct and antagonistic schools of American psychology. Behaviorism was the most popular model in the academic field, and psychoanalysis served as the basis of most psychotherapy. During World War II the discipline of clinical psychology emerged as a major professional field, but it was generally limited to psychological testing, and clinical skills were regarded as ancillary to basic scientific training, as engineering skills and those of other applied sciences were.[13] Then, in the late 1940s and early 1950s, clinical psychologists developed theoretical models of the human psyche and of human behavior that differed markedly from both the Freudian and behaviorist models, and psychotherapies that differed from psychoanalysis.

One of the most vital and enthusiastic movements that arose from dissatisfaction with the mechanistic orientation of psychological thought is the school of humanistic psychology spearheaded by Abraham Maslow. Maslow rejected Freud's view of humanity as being dominated by lower instincts and criticized Freud for deriving his theories of human behavior from the study of neurotic and psychotic individuals. Maslow thought that conclusions based on observing the worst in human beings rather than the best were bound to result in a distorted view of human nature. "Freud supplied to us the sick half of psychology," he wrote, "and we must now fill it out with the healthy half."[14] Maslow's criticism of behaviorism was equally vehement. He refused to see human beings simply as complex animals responding blindly to environmental stimuli, and pointed out the problematic nature and limited value of the behaviorists' heavy reliance on animal experiments. He acknowledged the usefulness of the behaviorist approach in learning about the characteristics we share with animals, but he strongly believed that such an approach was useless when it came to understanding capacities like conscience, guilt, idealism, humor, and so on, which are specifically human.

To counteract the mechanistic tendency of behaviorism and the medical orientation of psychoanalysis, Maslow proposed as a "third force" a humanistic approach to psychology. Rather than studying the behavior of rats, pigeons, or monkeys, humanistic psychologists focused on human experience and asserted that feelings, desires, and hopes were as important in a comprehensive theory of human behavior as external influences. Maslow emphasized that human beings should be studied as integral organisms and concentrated specifically on healthy individuals and positive aspects of human behavior—happiness, satisfaction, fun, peace of mind, joy, ecstasy. Like Jung, Maslow was deeply concerned with personal growth and "self-actualization," as he called it. In particular he undertook a comprehensive study of subjects who had spontaneous transcendental or "peak" experiences, which he considered important stages in the process of self-actualization. A similar approach to human growth was advocated by the Italian psychiatrist Roberto Assagioli, who was one of the pioneers of psychoanalysis in Italy but later went beyond the Freudian model and developed an alternative framework that he called psychosynthesis.[15]

In psychotherapy the humanistic orientation encouraged therapists to move away from the biomedical model, and this was reflected in a subtle but significant shift of terminology. Instead of dealing with "patients" therapists were now dealing with "clients," and the interaction between therapist and client, rather than being dominated and manipulated by the therapist, was seen as a human encounter between equals. The main innovator in this development was Carl Rogers, who emphasized the importance of a positive regard for the client and developed a nondirective, "client-centered" psychotherapy.[16] The essence of the humanistic approach is to see the client as a person capable of growth and self-actualization and to recognize the potentials inherent in all human beings.

Believing that most men and women in our culture had become too intellectual and were alienated from their sensations and feelings, psychotherapists focused on experience rather than intellectual analysis, and developed various nonverbal and physical techniques. A number of these techniques sprang up during the 1960s—sensory awareness, encounter, sensitivity training, and many more. They proliferated particularly in California, where Esalen, on the Big Sur coast, became an extremely influential center of the new psychotherapies and bodywork schools, which were referred to collectively as the human potential movement.[17]

While humanistic psychologists criticized Freud's view of human nature as based too much on the study of sick individuals, another group of psychologists and psychiatrists saw the lack of social considerations as the major shortcoming of psychoanalysis.[18] They pointed out that Freud's theory provided no conceptual framework for experiences shared by human beings and that it could not deal with interpersonal relations or with broader social dynamics. To extend psychoanalysis into these new dimensions Harry Stack Sullivan emphasized interpersonal relations in psychiatric theory and practice. He maintained that the human personality cannot be separated from the network of human relations in which it exists, and he defined psychiatry explicitly as a discipline dedicated to the study of interpersonal relations and interactions. Another social school of psychoanalysis developed under the leadership of Karen Horney, who emphasized the importance of cultural factors in the development of neurosis. She criticized Freud for not taking into account the social and cultural determinants of mental illness and also pointed out the lack of cultural perspective in his ideas about female psychology.

These new social orientations brought new therapeutic approaches that focused on the family and on other social groups and used the dynamics in these groups to initiate and support the therapeutic process. Family therapy is based on the assumption that the mental disorders of the "identified patient" reflect an illness of the entire family system and should therefore be treated within the context of the family. The movement of family therapy began in the 1950s and represents today one of the most innovative and successful therapeutic approaches. It has incorporated explicitly some of the new systems concepts of health and illness.[19]

Group therapy had been practiced in various forms for many decades but had been limited to verbal interactions until humanistic psychologists applied their new techniques of nonverbal communication, emotional release, and physical expression to the group process. Rogers had a strong influence on the development of this new kind of group therapy by applying his client-centered approach to it, establishing the relationship between therapist and client as a basis for relationships within the group.[20] The purpose of these groups, usually referred to as "encounter groups," was not limited to therapy. Many encounter groups met for the explicit purposes of self-exploration and personal growth.

By the mid-1960s it was commonly understood that the central emphasis of humanistic psychology, in theory and practice, was on self-actualization. During the subsequent rapid development of the discipline it became increasingly obvious that a new movement was developing within the humanistic orientation that was specifically concerned with the spiritual, transcendental, or mystical aspects of self-actualization. After several conceptual discussions the leaders of this movement gave it the name transpersonal psychology, a term coined by Abraham Maslow and Stanislav Grof.[21]

Transpersonal psychology is concerned, directly or indirectly, with the recognition, understanding, and realization of nonordinary, mystical, or "transpersonal" states of consciousness, and with the psychological conditions that represent barriers to such transpersonal realizations. Its concerns are thus very close to those of spiritual traditions, and, indeed, a number of transpersonal psychologists are working on conceptual systems intended to bridge and integrate psychology and the spiritual quest.[22] They have placed themselves in a position that differs radically from that of most major schools of Western psychology, which have tended to regard any form of religion or spirituality as based on primitive superstition, pathological aberration, or shared delusions about reality inculcated by the family system and the culture. The notable exception, of course, was Jung, who acknowledged spirituality as an integral aspect of human nature and a vital force in human life.

From these psychological schools and movements developing in both the United States and Europe a new psychology is emerging that is consistent with the systems view of life and in harmony with the views of spiritual traditions. The new psychology is still far from being a complete theory, being developed so far in the form of loosely connected models, ideas, and therapeutic techniques. These developments are taking place largely outside our academic institutions, most of which remain too closely tied to the Cartesian paradigm to appreciate the new ideas.

As in all other disciplines, the systems approach of the new psychology has a holistic and dynamic perspective. The holistic view, which is often associated in psychology with the gestalt principle, maintains that the properties and functions of the psyche cannot be comprehended by reducing them to isolated elements, just as the physical organism cannot be fully understood by analyzing it in terms of its parts. The fragmented view of reality not only is an obstacle for understanding the mind, but also is a characteristic aspect of mental illness. The healthy experience of oneself is an experience of the entire organism, body and mind, and mental illnesses often arise from failure to integrate the various components of this organism. From this point of view the Cartesian split between mind and body and the conceptual separation of individuals from their environment appear to be symptoms of a collective mental illness shared by most of Western culture, as they are indeed often perceived by other cultures.

The new psychology sees the human organism as an integrated whole involving interdependent physical and psychological patterns. Although psychologists and psychotherapists deal predominantly with mental phenomena, they must insist that these can be understood only within the context of the whole mind/body system. Hence the conceptual basis of psychology must also be consistent with that of biology. In classical science the Cartesian framework made it difficult for psychologists and biologists to communicate, and it seemed they could not learn much from each other. There were similar barriers between psychotherapists and physicians. But the systems approach provides a common framework for understanding the biological and psychological manifestations of human organisms in health and illness, one that is likely to lead to mutually stimulating exchanges between biologists and psychologists. It also means that if this is the time for physicians to take a closer look at the psychological aspects of illness, it is also time for psychotherapists to increase their knowledge of human biology.

As in the new systems biology, the focus of psychology is now shifting from psychological structures to the underlying processes. The human psyche is seen as a dynamic system involving a variety of functions that systems theorists associate with the phenomenon of self-organization. Following Jung and Reich, many psychologists and psychotherapists have come to think of mental dynamics in terms of a flow of energy, and they also believe that these dynamics reflect an intrinsic intelligence—the equivalent of the systems concept of mentation—that enables the psyche not only to create mental illness but also to heal itself. Moreover, inner growth and self-actualization are seen as essential to the dynamics of the human psyche, in full agreement with the emphasis on self-transcendence in the systems view of life.

Another important aspect of the new psychology is the growing recognition that the psychological situation of an individual cannot be separated from the emotional, social, and cultural environment. Psychotherapists are becoming aware that mental distress often originates in the breakdown of social relations. Accordingly, there has been a gradual tendency to shift from individual therapies to group and family therapies. A special kind of group therapy, which was not developed by psychotherapists but grew out of the women's movement, is practiced in political consciousness-raising groups.[23] The purpose of these groups is to integrate the personal and the political by clarifying the political context of personal experiences. The therapeutic process in such groups is often initiated simply by making the participants aware that they share the same problems because these problems are generated by the society we live in.

One of the most exciting developments in contemporary psychology is an adaptation of the bootstrap approach to the understanding of the human psyche.[24] In the past, schools of psychology proposed personality theories and systems of therapy that differed radically in their views of how the human mind functions in health and illness. These schools typically limited themselves to a narrow range of psychological phenomena—sexuality, the birth trauma, existential problems, family dynamics, and so on. A number of psychologists are now pointing out that none of these approaches is wrong, but that each of them focuses on some part of a whole spectrum of consciousness and then attempts to generalize its understanding of that part to the entire psyche. According to the bootstrap approach, there may not be any one theory capable of explaining the entire spectrum of psychological phenomena. Like physicists, psychologists may have to be content with a network of interlocking models, using different languages to describe different aspects and levels of reality. As we use different maps when we travel to different parts of the world, we would use different conceptual models on our journeys beyond space and time, through the inner world of the psyche.

One of the most comprehensive systems to integrate different psychological schools is the spectrum psychology proposed by Ken Wilber.[25] It unifies numerous approaches, both Western and Eastern, into a spectrum of psychological models and theories that reflects the spectrum of human consciousness. Each of the levels, or bands, of this spectrum is characterized by a different sense of identity, ranging from the supreme identity of cosmic consciousness to the drastically narrowed identity of the ego. As in any spectrum, the various bands exhibit infinite shades and gradations, merging gradually into one another. Nevertheless, several major levels of consciousness can be perceived. Wilber distinguishes, basically, four levels which are associated with corresponding levels of psychotherapy: the ego level, the biosocial level, the existential level, and the transpersonal level.

At the ego level one does not identify with the total organism but only with some mental representation of the organism, known as the self-image or ego. This disembodied self is thought to exist within the body, and thus people would say, "I *have* a body," rather than "I *am* a body." Under certain circumstances such a fragmented experience of oneself may be further distorted by the alienation of certain facets of the ego, which may be repressed or projected onto other people or the environment. The dynamics of these phenomena are described in great detail in Freudian psychology.

Wilber calls the second major level of consciousness "biosocial" because it represents aspects of a person's social environment—family relationships, cultural traditions and beliefs—that are mapped onto the biological organism and profoundly affect the person's perceptions and behavior. The pervasive influence of social and cultural patterns on the individual's sense of identity has been studied extensively by socially oriented psychologists, anthropologists, and other social scientists.

The existential level is the level of the total organism, characterized by a sense of identity which involves an awareness of the entire mind/body system as an integrated, self-organizing whole. The study of this kind of self-awareness and the exploration of its full potential is the aim of humanistic psychology and of various existential psychologies. At the existential level the dualism between mind and body has been overcome, but two other dualisms remain: the dualism of subject versus object, or self versus other; and that of life versus death. The questions and problems arising from these dualisms are a major concern of existential psychologies but cannot be resolved at the existential level. Their resolution requires a state of mind in which individual existential problems are perceived in their cosmic context. Such an awareness emerges at the transpersonal level of consciousness.

Transpersonal experiences involve an expansion of consciousness beyond the conventional boundaries of the organism and, correspondingly, a larger sense of identity. They may also involve perceptions of the environment transcending the usual limitations of sensory perception.[26] The transpersonal level is the level of the collective unconscious and the phenomena associated with it, as described in Jungian psychology. It is a mode of consciousness in which the individual feels connected to the cosmos as a whole, and so may be identified with the traditional concept of the human spirit. This mode of consciousness often transcends logical reasoning and intellectual analysis, approaching the direct mystical experience of reality. The language of mythology, which is much less restricted by logic and common sense, is often more appropriate to describe transpersonal phenomena than factual language. As the Indian scholar Ananda Coomaraswamy wrote, "Myth embodies the nearest approach to absolute truth that can be stated in words."[27]

At the end of the spectrum of consciousness, the transpersonal bands merge into the level of Mind, as Wilber calls it. This is the level of cosmic consciousness at which one identifies with the entire universe. One may *perceive* the ultimate reality at all transpersonal levels, but one *becomes* this reality only at the level of Mind. Conscious awareness, at this level, corresponds to the true mystical state, in which all boundaries and dualisms have been transcended and all individuality dissolves into universal, undifferentiated oneness. The level of Mind has been the overriding concern of spiritual or mystical traditions in the East as in the West. Although many of these traditions have been well aware of the other levels and have often mapped them out in great detail, they have always emphasized that the identities associated with all levels of consciousness are illusory, except for the ultimate level of Mind, where one finds one's supreme identity.

Another map of consciousness, which is fully consistent with Wilber's spectrum psychology, has been developed through a very different approach by Stanislav Grof. Whereas Wilbur approached the study of consciousness as a psychologist and philosopher, and derived his insights partly from his meditative practice, Grof approached it as a psychiatrist, basing his models on many years of clinical experience. For seventeen years Grof's clinical research was concerned with psychotherapy using LSD and other psychedelic substances. During this time he conducted some three thousand psychedelic sessions and studied the records from almost two thousand sessions run by his colleagues in Europe and the United States.[28] Later on the public controversy surrounding LSD and the resulting legal restrictions led Grof to abandon his practice of psychedelic therapy and to develop therapeutic techniques that induce similar states without the use of drugs.

Grof's extensive observations of psychedelic experiences convinced him that LSD is a nonspecific catalyst or amplifier of mental processes, which brings to the surface various elements from the depth of the unconscious. A person who takes LSD does not experience a toxic psychosis, as many psychiatrists believed in the early days of LSD research, but rather goes on a journey into the normally unconscious realms of the psyche. Psychedelic research, then, according to Grof, is not the study of special effects induced by psychoactive substances, but the study of the human mind with the help of powerful chemical facilitators. "It does not seem inappropriate and exaggerated," he writes, "to compare their potential significance for psychiatry and psychology to that of the microscope for medicine or the telescope for astronomy."[29]

The view that psychedelics merely act as amplifiers of mental processes is supported by the fact that the phenomena observed in LSD therapy are in no way unique and limited to psychedelic experimentation. Many of them have been observed in meditative practice, in hypnosis, and in the new experiential therapies. On the basis of many years of careful observations of this kind, with and without the use of psychedelics, Grof has constructed what he calls a cartography of the unconscious, a map of mental phenomena, which shows great similarities to Wilber's spectrum of consciousness. Grof's cartography encompasses three major domains: the domain of psychodynamic experiences, associated with events in a person's past and present life; the domain of perinatal* experiences, related to the biological phenomena involved in the process of birth; and the domain of transpersonal experiences that go beyond individual boundaries.

The psychodynamic level is clearly autobiographical and individual in origin, involving memories of emotionally relevant events and unresolved conflicts from various periods of the individual's life history. Psychodynamic experiences include the psychosexual dynamics and conflicts described by Freud and can be understood, to a large extent,

* "Perinatal," from the Greek *peri* ("round about") and Latin *natus* ("birth"), is a medical term referring to phenomena surrounding the birth process.

in terms of basic psychoanalytic principles. But, Grof has added an interesting concept to the Freudian framework. According to his observations, experiences in this domain tend to occur in specific memory constellations, which he calls COEX systems (systems of condensed experience.)[30] A COEX system is composed of memories from different periods of the person's life that have a similar basic theme, or contain similar elements, and are accompanied by a strong emotional charge of the same quality. The detailed interrelations among the constituent elements of a COEX system are, in most instances, in basic agreement with Freudian thinking.

The domain of perinatal experiences may be the most fascinating and most original part of Grof's cartography. It exhibits a variety of rich and complex experiential patterns related to the problems of biological birth. Perinatal experiences involve an extremely realistic and authentic reliving of various stages of one's actual birth process—the serene bliss of existence in the womb, in primal union with the mother, as well as disturbances of this peaceful state by toxic chemicals and muscular contractions; the "no exit" situation of the first stage of delivery, when the cervix is still closed while uterine contractions encroach on the fetus, creating a claustrophobic situation accompanied by intense physical discomfort; the propulsion through the birth canal, involving an enormous struggle for survival under crushing pressures, often with a high degree of suffocation; and, finally, the sudden relief and relaxation, the first breath, and the cutting of the umbilical cord, completing the physical separation from the mother.

In perinatal experiences the sensations and feelings associated with the birth process may be relived in a direct, realistic way and may also emerge in the form of symbolic, visionary experiences. For example, the experience of enormous tensions that is characteristic of the struggle in the birth canal is often accompanied by visions of titanic fights, natural disasters, sado-masochistic sequences, and various images of destruction and self-destruction. To facilitate an understanding of the great complexity of physical symptoms, imagery, and experiential patterns, Grof has grouped them into four clusters, called perinatal matrices, which correspond to consecutive stages of the birth process.[31] Detailed studies of the interrelations among the various elements of these matrices have resulted in profound insights into many psychological conditions and patterns of human experience.

One of the most striking aspects of the perinatal domain is the close relationship between the experiences of birth and death. The encounter with suffering and struggle, and the annihilation of all previous reference points in the birth process, are so close to the experience of death that Grof often refers to the entire phenomenon as the death-rebirth experience. Indeed, the visions associated with this experience frequently involve symbols of death, and the corresponding physical symptoms may provoke feelings of an ultimate existential crisis that can be so vivid they are confused with real dying. The perinatal level of the unconscious, then, is the level of both birth and death, a domain of existential experiences that exert a crucial influence on our mental and emotional life. "Birth and death," Grof writes, "appear to be the alpha and omega of human existence, and any psychological system that does not incorporate them has to remain superficial and incomplete."[32]

The experiential encounter with birth and death in the course of psychotherapy often amounts to a true existential crisis, forcing people to examine seriously the meaning of their lives and the values they live by. Worldly ambitions, competitive drives, the longing for status, power, or material possessions all tend to fade away when viewed against the background of potentially imminent death. As Carlos Castaneda has put it, recounting the teachings of the Yaqui sorcerer Don Juan, "An immense amount of pettiness is dropped if your death makes a gesture to you, or if you catch a glimpse of it . . . Death is the only wise adviser that we have."[33]

The only way to overcome the existential dilemma of the human condition, ultimately, is to transcend it by experiencing one's existence within a broader cosmic context. This is achieved in the transpersonal domain, the last major domain of Grof's cartography of the unconscious. Transpersonal experiences seem to offer deep insights into the nature and relevance of the spiritual dimension of consciousness. Like psychodynamic and perinatal experiences, they tend to occur in thematic clusters, but their organization is much more difficult to describe in factual language, as Jung and numerous mystics have emphasized, because the logical basis of our language is seriously challenged by these experiences. In particular, transpersonal experiences may involve so-called paranormal, or psychic, phenomena, which have been notoriously difficult to handle within the framework of rational thinking and scientific analysis. In fact, there seems to be a complementary relationship between psychic phenomena and the scientific method. Psychic phenomena seem to manifest themselves in full strength only outside the framework of analytic thought, and to diminish progressively as their observation and analysis become more and more scientific.[34]

Wilber's and Grof's models both indicate that the ultimate understanding of human consciousness goes beyond words and concepts. This raises the important question whether it is at all possible to make scientific statements about the nature of consciousness; and, furthermore, since consciousness is of central concern to psychology, whether psychology should be regarded as a science. The answers obviously depend on one's definition of science. Traditionally science has been associated with measurement and with quantitative statements, ever since Galileo banned quality from the sphere of scientific knowledge, and most scientists today still take this view. The philosopher and mathematician Alfred North Whitehead expresses the essence of the scientific method in the following rule: "Search for measurable elements among your phenomena, and then search for relations between these measures of physical quantities."[35]

A science concerned only with quantity and based exclusively on measurement is inherently unable to deal with experience, quality, or values. It will therefore be inadequate for understanding the nature of consciousness, since consciousness is a central aspect of our inner world and thus, first of all, an experience. Indeed, both Grof and Wilber describe their maps of consciousness in terms of domains of experience. The more scientists insist on quantitative statements, the less they are able to describe the nature of consciousness. In psychology the extreme case is given by behaviorism, which deals exclusively with measurable functions and behavior patterns and, consequently, cannot make any statement about consciousness at all, denying, in fact, even its existence.

The question, then, will be: Can there be a science that is not based exclusively on measurement; an understanding of reality that includes quality and experience and yet can be called scientific? I believe that such an understanding is, indeed, possible. Science, in my view, need not be restricted to measurements and quantitative analyses. I am prepared to call any approach to knowledge scientific that satisfies two conditions: all knowledge must be based on systematic observation, and it must be expressed in terms of self-consistent but limited and approximate models. These requirements—the empirical basis and the process of model making—represent to me the two essential elements of the scientific method. Other aspects, such as quantification or the use of mathematics, are often desirable but are not crucial.

The process of model making consists of forming a logically consistent network of concepts to interconnect the observed data. In classical science the data were quantities, obtained through measurement, and the conceptual models were expressed, whenever possible, in mathematical language. The purpose of quantification was twofold: to gain precision, and to guarantee scientific objectivity by eliminating any reference to the observer. Quantum theory has changed the classical view of science considerably by revealing the crucial role of the observer's consciousness in the process of observation and thus invalidating the idea of an objective description of nature.[36] Nevertheless, quantum theory is still based on measurement and is, in fact, the most quantitative of all scientific disciplines, reducing all the properties of atoms to sets of integral numbers.[37] Quantum physicists therefore cannot make any statements about the nature of consciousness within the framework of their science, even though human consciousness has been recognized as an inseparable part of that framework.

A true science of consciousness will deal with qualities rather than quantities, and will be based on shared experience rather than verifiable measurements. The patterns of experience constituting the data of such a science cannot be quantified or analyzed into fundamental elements, and they will always be subjective to varying degrees. On the other hand, the conceptual models interconnecting the data must be logically consistent, like all scientific models, and may even include quantitative elements. Grof's and Wilber's maps of consciousness are excellent examples of this new kind of scientific approach. They are characteristic of a new psychology, a science that will quantify its statements whenever this method is appropriate, but will also be able to deal with qualities and values based on human experience.

The new bootstrap, or systems, approach to psychology includes a conception of mental illness that is fully consistent with the general views of health and illness outlined in the previous chapter. Like all illness, mental illness is seen as a multidimensional phenomenon involving interdependent physical, psychological, and social aspects. When Freud developed psychoanalysis, the nervous disorders known as neuroses were central in his thinking, but since then the main attention of psychiatrists has shifted to the more serious disturbances named psychoses, and especially to a broad category of severe mental disorders that have been designated, rather arbitrarily, as schizophrenia.* Unlike the neuroses, these mental illnesses go far beyond the psychodynamic level and cannot be fully understood unless the biosocial, existential, and transpersonal domains of the psyche are taken into account. Such a multilevel approach is certainly needed, since half of all hospital beds available to mentally ill patients in the United States are occupied by people diagnosed as schizophrenics.[38]

Most current psychiatric treatments deal with the biomedical mechanisms associated with a specific mental disorder and, in doing so, have been very successful in suppressing symptoms with psychoactive drugs. This approach has not helped psychiatrists understand mental illness any better, nor has it allowed their patients to solve the underlying problems. In view of these shortcomings of the biomedical approach, over the past twenty-five years a number of psychiatrists and psychologists have developed a systemic view of psychotic disorders that takes into account the multiple facets of mental illness; this view is both social and existential.

Failure to evaluate one's perception and experience of reality and to integrate them into a coherent world view seems to be central to serious mental illness. In current psychiatric practice many people are diagnosed as psychotics, not on the basis of their behavior but rather on the basis of the content of their experiences. These experiences, typically, are of a transpersonal nature and in sharp contradiction to all common sense and to the classical Western world view. However, many of them are well known to mystics, occur frequently in deep meditation, and can also be induced quite easily by various other methods. The new definition of what is normal and what is pathological is not based on the content and nature of one's experiences, but rather on the way in which they are handled and on the degree to which a person is able to integrate these unusual experiences into his life. Research by humanistic and transpersonal psychologists has shown that the spontaneous occurrence of nonordinary experiences of reality is much higher than suspected in conventional psychiatry.[39] Harmonious integration of these experiences is therefore crucial to

* From the Greek *skhizein* ("to split") and *phren* ("mind").

mental health, and sympathetic support and assistance in this process, based on an understanding of the full spectrum of human consciousness, will be of vital importance in dealing with many forms of mental illness.

The inability of some people to integrate transpersonal experiences is often aggravated by a hostile environment. Immersed in a world of symbols and myth, they feel isolated and unable to communicate the nature of their experience. The fear of this isolation can be so overwhelming that it causes a wave of existential panic, and it is this panic more than anything else which produces many of the signs of "insanity."[40] The feeling of isolation and the expectation of hostility are further accentuated by psychiatric treatment, which often involves a degrading examination, stigmatizing diagnosis, and forced hospitalization, which completely invalidate the person as a human being. As a recent investigator of the psychological effects of mental institutions noted, "Neither anecdotal nor 'hard' data can convey the overwhelming sense of powerlessness which invades the individual as he is continually exposed to the depersonalization of the psychiatric hospital."[41]

Among the experiences that psychotic people fail to integrate, those relating to their social environment seem to play a crucial role. Recent major advances in the understanding of schizophrenia have been based on the recognition that the disorder cannot be understood by focusing on individual patients, but has to be perceived in the context of their relations to other people. Numerous studies of families of schizophrenics have shown that the person who is diagnosed as being psychotic is, almost without exception, part of a network of extremely disturbed patterns of communication within the family.[42] The illness manifested in the "identified patient" is really a disorder of the entire family system.

The central characteristic in the communication patterns of families of diagnosed schizophrenics was identified by Gregory Bateson as a "double bind" situation.[43] Bateson found that the behavior labeled schizophrenic represents a special strategy which a person invents in order to live in an unlivable situation. Such a person finds himself facing a situation within his family that seems to put him into an untenable position, a situation in which he "can't win," no matter what he does. For example, the double bind may be set up for a child by contradictory verbal and nonverbal messages, either from one or from both parents, with both kinds of messages implying punishment or threats to the child's emotional security. When these situations occur repeatedly the double-bind structure may become a habitual expectation in the child's mental life, and this is likely to generate schizophrenic experiences and behavior. This does not mean that everybody becomes schizophrenic in such a situation. What exactly makes one person psychotic while another remains normal under the same external circumstances is a complex question, likely to involve biochemical and genetic factors that are not yet well understood. In particular the effects of nutrition on mental health need much further exploration.

R. D. Laing has pointed out that the strategy designed by a so-called schizophrenic can often be recognized as an appropriate response to severe social stress, representing the person's desperate efforts to maintain his integrity in the face of paradoxical and contradictory pressures. Laing extends this observation to an eloquent critique of society as a whole, in which he sees the condition of alienation, of being asleep, unconscious, "out of one's mind," as the condition of the normal person.[44] Such "normally" alienated men and women are taken to be sane, says Laing, simply because they act more or less like anyone else, whereas other forms of alienation, which are out of step with the prevailing one, are labeled psychotic by the "normal" majority. Laing offers the following observation:

> A child born today in the United Kingdom stands a ten times greater chance of being admitted to a mental hospital than to a university . . . This can be taken as an indication that we are driving our children mad more effectively than we are genuinely educating them. Perhaps it is our way of educating them that is driving them mad.[45]

Laing succinctly exposes the dual role of cultural factors in the development of mental illness. On the one hand, the culture generates much of the anxiety that leads to psychotic behavior, and on the other hand it sets up the norms for what is considered sane. In our culture the criteria used to define mental health—sense of identity, image, recognition of time and space, perception of the environment, and so on—require that a person's perceptions and views be compatible with the Cartesian-Newtonian framework. The Cartesian world view is not merely the principal frame of reference but is regarded as the only accurate description of reality. This restrictive attitude is reflected in the tendency of mental health professionals to use rather rigid diagnostic systems. The dangers of such cultural conditioning are well illustrated by a recent experiment in which eight volunteers gained admission to various American mental institutions by saying they had been hearing voices.[46] These pseudopatients found themselves irrevocably labeled as schizophrenics in spite of their subsequent normal behavior. Ironically, many of the other inmates soon recognized that the pseudopatients were normal, but the hospital personnel were unable to acknowledge their normal behavior once they had been diagnosed as psychotic.

It would seem that the concept of mental health should include a harmonious integration of the Cartesian and the transpersonal modes of perception and experience. To perceive reality exclusively in the transpersonal mode is incompatible with adequate functioning and survival in the everyday world. To experience an incoherent mixture of both modes of perception without being able to integrate them is psychotic. But to be limited to the Cartesian mode of perception alone is also madness; it is the madness of our dominant culture.

A person functioning exclusively in the Cartesian mode may be free from manifest symptoms but cannot be considered mentally healthy. Such individuals typically lead ego-centered, competitive, goal-oriented lives. Overpreoccupied with their past and their future, they tend to have a limited awareness of the present and thus a limited ability to derive satisfaction from ordinary activities in everyday life. They concentrate on manipulating the external world and measure their living standard by the quantity of material possessions, while they become ever more alienated from their inner world and unable to appreciate the process of life. For people whose existence is dominated by this mode of experience no level of wealth, power, or fame will bring genuine satisfaction, and thus they become infused with a sense of meaninglessness, futility, and even absurdity that no amount of external success can dispel.

The symptoms of this cultural madness are all-pervasive throughout our academic, corporate, and political institutions, with the nuclear arms race perhaps its most psychotic manifestation. The integration of the Cartesian mode of perception into a broader ecological and transpersonal perspective has now become an urgent task, to be carried out at all individual and social levels. Genuine mental health would involve a balanced interplay of both modes of experience, a way of life in which one's identification with the ego is playful and tentative rather than absolute and mandatory, while the concern with material possessions is pragmatic rather than obsessive. Such a way of being would be characterized by an affirmative attitude toward life, an emphasis on the present moment, and a deep awareness of the spiritual dimension of existence. Indeed, these attitudes and values have been emphasized throughout the ages by saints and sages who experienced reality in the transpersonal mode. It is well known that the experiences of these mystics are often strikingly similar to those of schizophrenics. Yet mystics are not insane, because they know how to integrate their transpersonal experiences with their ordinary modes of consciousness. In Laing's profound metaphor, "Mystics and schizophrenics find themselves in the same ocean, but the mystics swim whereas the schizophrenics drown."[47]

The view of mental illness as a multidimensional phenomenon that may involve the entire spectrum of consciousness implies a corresponding multileveled approach to psychotherapy. By using the languages of different schools—Freudian, Jungian, Reichian, Rogerian, Laingian, and others—to describe different facets of the psyche, psychotherapists should be able to integrate these schools into a coherent framework for interpreting the range of phenomena encountered in the therapeutic process. Therapists know that different clients will exhibit different symptoms which often require different terminologies. Jung, for example, wrote in his autobiography, "To my mind, in dealing with individuals, only individual understanding will do. We need a different language for every patient. In one analysis I can be heard talking the Adlerian dialect, in another the Freudian."[48] Indeed, the same client often goes through different phases in the course of therapy, each characterized by different symptoms and a different sense of identity. When therapeutic work at one level of consciousness has resulted in improved integration, the person may find herself spontaneously at another level. The new framework will make it much easier, in dealing with such cases, to apply a whole spectrum of therapies as the client moves through the spectrum of consciousness.

At the ego or psychodynamic level, pathological symptoms seem to result from a breakdown of communication between various conscious and unconscious facets of the psyche. The main goal of ego-level therapies is to integrate these facets, to heal the split between ego-consciousness and the unconscious, and thus to achieve a fuller sense of identity. To interpret the multitude of experiences at the psychodynamic level Freudian theory seems to be the ideal framework. It allows therapist and client to understand the manifestation of various psychosexual dynamics, regressions to childhood, the reliving of psychosexual traumas, and many other phenomena of a clearly autobiographical nature. However, the Freudian model is limited to the psychodynamic domain and proves inadequate when deeper existential and transpersonal experiences emerge. Nor can it deal with the social origins of individual problems, which are often crucial. The social context is emphasized by a number of approaches that address themselves, in Wilber's terminology, to the biosocial domain of consciousness. In socially oriented therapies the client's problems and symptoms are seen as originating in the pattern of relationships between the individual and other people, and in interactions with social groups and institutions. Transactional analysis, family therapy, and various forms of group therapy, including those with explicit political orientations, use this approach.

While the therapies operating at the ego level aim to expand the person's sense of identity by integrating various unconscious facets of the psyche, those operating at the existential level go one step further. They deal with the integration of mind and body, and their aim is the self-actualization of the total human being. Therapeutic approaches of this kind are not psychotherapies in the strict sense of the term, since they often involve a combination of psychological and physical techniques. Examples include gestalt therapy, Reichian therapy, and the various bodywork therapies. Many of these involve powerful stimulations of the total organism, which often lead to profound experiences related to birth and death, the two outstanding existential phenomena. Grof's perinatal matrices represent a comprehensive conceptual framework for interpreting existential experiences of this kind.

At the transpersonal level, finally, the aim of the therapy is to help clients integrate their transpersonal experiences with their ordinary modes of consciousness in the process of inner growth and spiritual development. Conceptual models dealing with the transpersonal realm include Jung's analytical psychology, Maslow's psychology of being, and Assagioli's psychosynthesis. At the deep end of the transpersonal domain of consciousness, which Wilber calls the level of Mind, the aims of transpersonal therapy merge with those of spiritual practice.

The idea that the human organism has an inherent tendency to heal itself and to evolve is as central to psychotherapy as to any other therapy. In the systems approach the therapist aims, first, to initiate the healing process by helping the client get into a state in which the natural healing forces become active. Contemporary schools of psychotherapy all seem to share this notion of a special healing state. Some call it a resonance phenomenon, others speak of energizing the organism, and most therapists agree that it is virtually impossible to describe exactly what happens in those crucial moments. Thus Laing: "The really decisive moments in psychotherapy, as every patient or therapist who has ever experienced them knows, are unpredictable, unique, unforgettable, always unrepeatable and often indescribable."[49]

Mental illnesses often involve the spontaneous emergence of unusual experiences. In such cases no special techniques are necessary to initiate the healing process, and the best therapeutic approach is to provide a sympathetic and supportive environment in which these experiences are allowed to unfold. This has been practiced very successfully with schizophrenics in therapeutic communities, for example in England by Laing and in California by John Perry.[50] Therapists who use such an approach have often remarked that the experiential drama that is part of the process of healing seems to unfold as an orderly sequence of events that can be interpreted as a journey through the schizophrenic's inner world. As Bateson described the situation:

> It would appear that once precipitated into psychosis the patient has a course to run. He is, as it were, embarked upon a voyage of discovery which is only completed by his return to the normal world, to which he comes back with insights different from those of the inhabitants who never embarked on such a voyage. Once begun, a schizophrenic episode would appear to have as definite a course as an initiation ceremony.[51]

It has often been pointed out that our current mental hospitals are quite inadequate to deal with psychotic journeys of this kind. What we need instead, according to Laing, is "an initiation ceremonial through which the person will be guided with full social encouragement and sanction into inner space and time, by people who have been there and back again."[52]

In many cases of mental illness the resistance to change is so strong that it is necessary to use specific techniques to stimulate the organism—some form of catalyst to induce the healing process. Such catalysts may be pharmacological, or they may be physical or psychological techniques; one of the most important catalysts will always be the personality of the therapist. Once the therapeutic process has been initiated, the therapist's role is to facilitate the emerging experiences and help the client overcome resistances. The full unfolding of experiential patterns can be extremely dramatic and challenging for both client and therapist, but the originators of this experiential approach believe that one should encourage and support the therapeutic process no matter what form and intensity it assumes. Their motivation to do so is based on the idea that the symptoms of mental illness represent frozen elements of an experiential pattern that needs to be completed and fully integrated if the symptoms are to disappear. Rather than suppressing symptoms with psychoactive drugs, the new therapies activate and intensify them to bring about their full experience, conscious integration, and final resolution.

A great many new therapeutic techniques have been developed to mobilize blocked energy and transform symptoms into experiences. In contrast to the traditional approaches, which were mostly limited to verbal interactions between therapist and client, the new therapies encourage nonverbal expression and emphasize direct experience involving the total organism. Hence they are often referred to as experiential therapies. The elemental nature and intensity of the experiential patterns underlying the manifest symptoms have convinced most practitioners of the new therapies that the chances of drastically influencing the psychosomatic system through verbal channels alone are quite remote, and thus great emphasis is placed on therapeutic approaches that combine psychological and physical techniques.

Many therapists believe that one of the most important events in psychotherapy is a certain resonance between the unconscious of the client and that of the therapist. Such a resonance will be most powerful if both therapist and client are willing to drop their roles, masks, defenses, and any other barriers standing between them, so that the therapeutic encounter becomes, as Laing describes it, an "authentic meeting between human beings."[53] Perhaps the first to perceive psychotherapy in such a way was Jung, who strongly emphasized the mutual influence between therapist and client and compared their relationship to an alchemical symbiosis. More recently Carl Rogers affirmed the need to create a special supportive atmosphere to enhance the client's experience and potential for self-actualization. Rogers suggested that the therapist should be with the client in a state of intense awareness, focusing fully on the client's experience and deeply reflecting all verbal and nonverbal expressions from a position of empathy and unconditional positive regard.

One of the most popular approaches among the new experiential therapies is the one developed by Fritz Perls, known as gestalt therapy.[54] It shares with gestalt psychology the basic assumption that human beings do not perceive things as unrelated and isolated elements but organize them during the perceptual process into meaningful wholes. Accordingly the orientation of gestalt therapy is explicitly holistic, emphasizing the tendency, inherent in all individuals, to integrate their experiences and to actualize themselves in harmony with their environment. Psychological symptoms represent blocked elements of experience, and the aim of therapy is to facilitate the process of personal integration by helping the client complete the experiential gestalt.

To unlock the client's blocked experiences the gestalt therapist will direct attention toward various patterns of communication, both interpersonal and internal, with the aim of enhancing the client's awareness of the detailed physical and emotional processes involved. This sharpening of awareness is meant to bring about the special state in which experiential patterns become fluid and the organism begins the process of self-healing and integration. The emphasis is not on interpreting problems, nor on dealing with past events, but on experiencing conflicts and traumas in the present moment. Individual work is often done within the context of a group, and many gestalt therapists are combining psychological approaches with some form of bodywork. This multilevel approach seems to encourage profound existential and, occasionally, even transpersonal experiences.

The most powerful way of activating experiences from all levels of the unconscious, and historically one of the oldest forms of experiential therapy, is the therapeutic use of psychedelics. The basic principles and practical aspects of psychedelic therapy have been set forth in great detail by Stanislav Grof,[55] in view of possible future applications once the legal restrictions caused by widespread misuse of LSD are relaxed. In addition, a number of neo-Reichian approaches can be used to energize the organism in similar ways through physical manipulations.

Grof himself, with his wife Christina, has integrated hyperventilation, evocative music, and bodywork into a therapeutic method that can induce surprisingly intense experiences after a relatively short period of fast, deep breathing.[56] The basic principle is to encourage the client to concentrate on breathing and other physical processes within the body, and to turn off the intellectual analysis as much as possible while surrendering to sensations and emotions. In most instances breathing and music alone lead to successful resolution of the encountered problems. Residual issues, if any, are handled by focused bodywork, during which the therapist tries to facilitate experiences by amplifying the manifest symptoms and sensations and helping to find appropriate modes of expressing them—through sounds, movements, postures, or any other nonverbal ways. After experimenting with this method for a number of years, Grof is convinced that it represents one of the most promising approaches to psychotherapy and self-exploration.

Another form of experiential therapy, which is essentially a neo-Reichian approach, is the primal therapy developed by Arthur Janov.[57] It is based on the idea that neuroses are symbolic behavior patterns that represent the person's defenses against excessive pain associated with childhood traumas. The purpose is to overcome the defenses and to work through the primal pains by experiencing them fully while reliving the memories of the events that caused them. The main method of inducing these experiences is the "primal scream," an involuntary, deep, rattling sound that expresses in a condensed form the person's reaction to past traumas. According to Janov, successive layers of blocked pain can gradually be eliminated in this way by repeated sessions of primal screaming.

Although Janov's initial enthusiastic statements about the efficacy of his method have not withstood the test of time, primal therapy represents an extremely powerful experiential approach. Unfortunately, Janov's conceptual sytem is not broad enough to account for the transpersonal experiences his technique is likely to trigger. For this reason a number of primal therapists have recently dissociated themselves from Janov and formed alternative schools that continue to use Janov's basic techniques while seeking a more open-minded theoretical framework.

Modern psychotherapists have clearly gone far beyond the biomedical model from which psychotherapy originally emerged. The therapeutic process is no longer seen as a treatment of disease but as an adventure in self-exploration. The therapist does not play a dominant role but becomes the facilitator of a process in which the client is the chief protagonist and bears full responsibility. The therapist creates an environment conducive to self-exploration and acts as a guide while this process unfolds. To assume such a role psychotherapists need qualities very different from those required in conventional psychiatry. Medical training may be useful but is by no means sufficient, and even the knowledge of specific therapeutic techniques will not be critical since these can be acquired in a relatively short time. The essential attributes of a good psychotherapist will be such personal qualities as warmth and genuineness, the ability to listen and to show empathy, and willingness to participate in another person's intense experiences. In addition the therapist's own stage of self-actualization and experiential knowledge of the full spectrum of consciousness will be vital.

The basic strategy of the new experiential psychotherapy requires that, to achieve the best therapeutic results, both therapist and client suspend as much as possible their conceptual frameworks, anticipations, and expectations during the experiential process. Both should be open and adventurous, ready to follow the flow of experience with a deep trust that the organism will find its own way to heal itself and evolve. Experience has shown that if the therapist is willing to encourage and support such a healing journey without fully understanding it, and the client is open to venture into unknown territory, they will often be rewarded by extraordinary therapeutic achievements.[58] After the experience is completed, they may try to analyze what happened if they feel inclined to do so, but they should be aware that such an analysis and conceptualization, even though it may be intellectually stimulating, will have little therapeutic relevance. In general, therapists have observed that the more complete an experience, the less analysis and interpretation it will require. A complete experiential pattern, or gestalt, tends to be self-evident and self-validating to the person whose psyche produces it. Ideally, then, the conversation following a therapeutic session will have the form of a happy sharing instead of a painful struggle to understand what happened.

In venturing far into the existential and transpersonal domains of human consciousness, psychotherapists will have to be prepared to deal with experiences so unusual that they defy any attempt at rational explanation.[59] Experiences of such an extraordinary nature are relatively rare, but even milder forms of existential and transpersonal experience will present serious challenges to conventional conceptual frameworks of psychotherapists and their clients, and intellectual resistance to emerging experiences will tend to impede the healing process. Clinging to a mechanistic conception of reality, a linear notion of time, or a narrow concept of cause and effect may turn into a powerful mechanism of defense against the emergence of transpersonal experiences and thus interfere with the therapeutic process. As Grof has pointed out, the ultimate obstacle for experiential therapies is no longer of an emotional or physical nature, but takes the form of a cognitive barrier.[60] The practitioners of experiential psychotherapies will therefore be much more successful if they are familiar with the new paradigm that is now emerging from modern physics, systems biology, and transpersonal psychology, so that they can offer their clients not only powerful stimulations of experience but also the corresponding cognitive expansion.

12 · The Passage to the Solar Age

The systems view of life is an appropriate basis not only for the behavioral and the life sciences but also for the social sciences, and especially for economics. The application of systems concepts to describe economic processes and activities is particularly urgent because virtually all our current economic problems are systemic problems that can no longer be understood via Cartesian science.

Conventional economists, whether neoclassical, Marxist, Keynesian, or post-Keynesian, generally lack an ecological perspective. Economists tend to dissociate the economy from the ecological fabric in which it is embedded, and to describe it in terms of simplistic and highly unrealistic theoretical models. Most of their basic concepts, narrowly defined and used without the pertinent ecological context, are no longer appropriate for mapping economic activities in a fundamentally interdependent world.

The situation is further aggravated because most economists, in a misguided striving for scientific rigor, avoid explicitly acknowledging the value system on which their models are based and tacitly accept the highly imbalanced set of values that dominates our culture and is embodied in our social institutions. These values have led to an overemphasis on hard technology, wasteful consumption, and rapid exploitation of natural resources, all motivated by the persistent obsession with growth. Undifferentiated economic, technological, and institutional growth is still regarded by most economists as the sign of a "healthy" economy, although it is now causing ecological disasters, widespread corporate crime, social disintegration, and an ever increasing likelihood of nuclear war.

Paradoxically, economists are generally unable to adopt a dynamic view in spite of their insistence on growth. They tend to freeze the economy arbitrarily in its current institutional structure instead of seeing it as a continually changing and evolving system, dependent on the changing ecological and social systems in which it is embedded. Today's economic theories perpetuate past configurations of power and unequal distributions of wealth, both within national economies and between developed countries and the Third World. Giant corporate institutions dominate the global and national scenes, their economic and political power permeating virtually every facet of public life, while some economists still seem to believe that Adam Smith's free markets and perfect competition exist. Many corporate giants are now obsolete institutions that generate polluting and socially disruptive technologies and lock up capital, energy, and resources, unable to adapt their uses to the changing needs of our time.

The systems approach to economics will make it possible to bring some order into the present conceptual chaos by giving economists the urgently needed ecological perspective. According to the systems view, the economy is a living system composed of human beings and social organizations in continual interaction with one another and with the surrounding ecosystems on which our lives depend. Like individual organisms, ecosystems are self-organizing and self-regulating systems in which animals, plants, microorganisms, and inanimate substances are linked through a complex web of interdependencies involving the exchange of matter and energy in continual cycles. Linear cause-and-effect relationships exist only very rarely in these ecosystems, nor are linear models very useful to describe the functional interdependencies of the embedded social and economic systems and their technologies. The recognition of the nonlinear nature of all systems dynamics is the very essence of ecological awareness, the essence of "systemic wisdom," as Bateson called it.[1] This kind of wisdom is characteristic of traditional, nonliterate cultures but has been sadly neglected in our overrational and mechanized society.

Systemic wisdom is based on a profound respect for the wisdom of nature, which is totally consistent with the insights of modern ecology. Our natural environment consists of ecosystems inhabited by countless organisms which have coevolved over billions of years, continuously using and recycling the same molecules of soil, water, and air. The organizing principles of these ecosystems must be considered superior to those of human technologies based on recent inventions and, very often, on short-term linear projections. The respect for nature's wisdom is further supported by the insight that the dynamics of self-organization in ecosystems is basically the same as in human organisms, which forces us to realize that our natural environment is not only alive but also mindful. The mindfulness of ecosystems, as opposed to many human institutions, manifests itself in the pervasive tendency to establish cooperative relationships that facilitate the harmonious integration of systems components at all levels of organization.

The nonlinear interconnectedness of living systems immediately suggests two important rules for the management of social and economic systems. First, there is an optimal size for every structure, organization, and institution, and maximizing any single variable—profit, efficiency, or GNP, for example—will inevitably destroy the larger system. Second, the more an economy is based on the continual recycling of its natural resources, the more it is in harmony with the surrounding environment. Our planet is now so densely populated that virtually all economic systems are thoroughly interwoven and interdependent: today's most important problems are global problems. The vital social choices we face are no longer local—choices between more roads, schools, and hospitals—nor do they affect merely a small part of the population. They are choices between principles of self-organization—centralization or decentralization, capital-intensity or labor-intensity, hard technology or soft technology—that affect the survival of humanity as a whole.

In making these choices it will be useful to keep in mind that the dynamic interplay of complementary tendencies is another important characteristic of self-organizing systems. As E. F. Schumacher has noted, "the whole crux of economic life—and indeed of life in general—is that it constantly requires the living reconciliation of opposites which, in strict logic, are irreconcilable."[2] The global interconnectedness of our problems and the virtue of small-scale, decentralized enterprises represent such a pair of complementary opposites. The need to balance the two has found eloquent expression in the slogan "Think globally—act locally!"

A second insight facilitated by the systems approach is the realization that the dynamics of an economy, like that of any other living system, is likely to be dominated by fluctuations. Indeed, several cyclical economic patterns with different periodicities have recently been observed and analyzed, in addition to the short-term oscillations that Keynes studied. Jay Forrester and his Systems Dynamics Group have identified three distinct cycles: a five-to-seven-year cycle that is influenced very little by changes in interest rates and other Keynesian manipulations but instead reflects the interaction between employment and inventories; an eighteen-year cycle related to the investment process; and a fifty-year cycle, which, according to Forrester, has the strongest effect on the economy's behavior but is of an entirely different nature, reflecting the evolution of technologies, such as railroads, automobiles, and computers.[3]

Another example of important economic fluctuations is the well-known cycle of growth and decay, the continual breaking down and building up of structures involving the recycling of all component parts. Hazel Henderson spells out the lesson to be drawn from this basic phenomenon of life: "Just as the decay of last year's leaves provides humus for new growth the following spring, some institutions must decline and decay so that their components of capital, land and human talents can be used to create new organizations."[4]

According to the systems view, an economy, like any living system, will be healthy if it is in a state of dynamic balance, characterized by continual fluctuations of its variables. To achieve and maintain such a healthy economic system it is crucial to preserve the ecological flexibility of our natural environment as well as to create the social flexibility needed to adapt to environmental changes. To Bateson, "Social flexibility is a resource as precious as oil."[5] Furthermore, we will need much greater flexibility of ideas, because economic patterns keep changing and evolving and hence cannot be described adequately except in a conceptual framework that itself is capable of change and evolution.

To describe the economy appropriately within its social and ecological context, the basic concepts and variables of economic theories must be related to those used to describe social and ecological systems. This implies that the task of mapping the economy will require a multidisciplinary approach. It can no longer be left to economists alone, but must be supplemented by insights from ecology, sociology, political science, anthropology, psychology, and other disciplines. Like health care professionals, investigators of economic phenomena need to work in multidisciplinary teams, using different methods and perspectives and focusing on different systems levels to highlight different aspects and implications of economic activities. Such a multidisciplinary approach to economic analyses is already visible in a number of recent books by noneconomists on subjects that formerly belonged exclusively to the domain of economics. Innovative contributions of this kind include those of Richard Barnet (political scientist), Barry Commoner (biologist), Jay Forrester (systems analyst), Hazel Henderson (futurist), Frances Moore Lappé (sociologist), Amory Lovins (physicist), Howard Odum (engineer), and Theodore Roszak (historian), to name just a few.[6]

As Kenneth Boulding, Hazel Henderson, and several others have pointed out, the necessity of multidisciplinary approaches to our current economic problems calls for the end of economics as the predominant basis of national policy. Economics is likely to remain an appropriate discipline for accounting purposes and various analyses of micro-areas, but its methods are no longer adequate to examine macroeconomic processes. An important new role for economics will be that of estimating, as accurately as possible, the social and environmental costs of economic activities—in money, health, or safety—to internalize them within the accounts of private and public enterprises. Economists will be expected to identify the relationships between specific activities in the private sectors of the economy and the social costs generated by these activities in the public sector. For example, the new way of accounting would involve assigning to tobacco companies a reasonable portion of the medical costs caused by cigarette smoking and to the distillers a corresponding portion of the social costs of alcoholism. Work on new economic models of this kind is now in progress and will lead, eventually, to a redefinition of the gross national product and other related concepts. In fact, Japanese economists have already begun to reformulate their GNP in terms of a new indicator in which social costs are deducted.[7]

Macroeconomic patterns will have to be studied within a framework based on the systems approach and using a new set of concepts and variables. One of the main errors in all current schools of economic thought is their insistence on using money as the only variable to measure the efficiency of production and distribution processes. With this sole criterion economists neglect the important fact that most of the world's economic activities consist of informal, use-value production, exchange systems, and reciprocal arrangements of sharing goods and services, all of which occur outside the monetary economies.[8] As more and more of these activities—housework, child care, looking after the sick and the old—become monetized and institutionalized, the values that allow people to provide services to one another free of charge become distorted; social and cultural cohesion dissolves; and the economy, not surprisingly, begins to suffer from "declining productivity." This process is accelerated by the fact that the entire concept of money is becoming ever more abstract and detached from economic realities. While in today's global banking and financing system the units of money can be distorted almost at will by the power of large institutions, the widespread use of credit cards, electronic banking and funds transfer systems, and other tools of modern computer and communications technology have added successive layers of complexity that make it almost impossible to use money as an accurate tracking system of economic transactions in the real world.[9]

In the new conceptual framework, energy, so essential to all industrial processes, will be one of the most important variables for measuring economic activities. As industrial countries with similar standards of living show growing disparities in energy consumption, questions about their relative efficiencies in energy conversion are naturally being raised. Energy modeling, pioneered by the engineer and environmentalist Howard Odum, is now pursued in many countries by imaginative scientists from various disciplines.[10] In spite of many unresolved problems and differences in methods, the mapping of flows of energy is rapidly becoming a more reliable method for macroeconomic analyses than conventional monetary approaches.

Measurement of the efficiency of production processes in terms of net energy, which is now being widely accepted, suggests entropy—a quantity related to the dissipation of energy[11]—as another important variable for the analysis of economic phenomena. The entropy concept was introduced into economic theory by Nicholas Georgescu-Roegen, whose work has been described as the first comprehensive reformulation of economics since Marx and Keynes.[12] According to Georgescu-Roegen, the dissipation of energy, as described by the second law of thermodynamics, is not only relevant to the performance of steam engines but also to the functioning of an economy. As the thermodynamic efficiency of engines is limited by friction and other forms of energy dissipation, so production processes in industrial societies will inevitably generate social frictions and dissipate some of the economy's energy and resources into unproductive activities.

Henderson has pointed out that the dissipation of energy has reached such proportions in many of today's advanced industrial societies that the costs of unproductive activities—maintaining complex technologies, managing large bureaucracies, mediating conflicts, controlling crime, protecting consumers and the environment, and so on—make up an ever increasing portion of the GNP and thus drive inflation to ever increasing heights. Henderson has coined the term "entropy state" for the stage of economic development in which the costs of bureaucratic coordination and maintenance exceed society's productive capabilities, and the whole system winds down of its own weight and complexity.[13] To avoid such a grim future it will be necessary to judge economic activities and technologies not in terms of narrowly defined economic efficiency but in terms of thermodynamic efficiency, which will amount to a radical change of priorities. For example, an economic analysis in terms of energy and entropy makes it clear that our current military expenditures support the most energy-intensive and dissipative activities humans are capable of, as they convert large amounts of stored energy and materials directly into waste and destruction without fulfilling any basic human needs.

Like the concepts of efficiency and the GNP, those of productivity and profit will have to be defined within a broad ecological context and related to the two basic variables of energy and entropy. However, in doing so it will be important to keep in mind that, although entropy is extremely useful as a variable for economic analyses, the framework of classical thermodynamics in which it originated is quite limited. Specifically, it is not adequate to describe living, self-organizing systems—whether individual organisms, social systems, or ecosystems—for which Prigogine's theory provides a much more appropriate description.[14] Recent economic analyses in terms of entropy have sometimes erroneously regarded the second law of thermodynamics as an absolute law of nature,[15] and should be modified to make them consistent with the new theory of self-organization. For example, the concept of technological and organizational complexity will have to be refined and related to the dynamic state of the system under consideration. According to Erich Jantsch, the complexity of a system is limited only if the system is rigid, inflexible, and isolated from its environment.[16] Self-organizing systems in continual interaction with their environment are capable of tremendously increasing their complexity by abandoning structural stability in favor of flexibility and open-ended evolution. Hence the efficiency of our technologies and social institutions will depend not only on their complexity but also on their flexibility and potential for change.

When we adopt an ecological perspective and use the appropriate concepts to analyze economic processes, it becomes evident that our economy, our social institutions, and our natural environment are seriously out of balance. Our obsession with growth and expansion has led us to maximize too many variables for prolonged periods—GNP, profits, the size of cities and social institutions, and others—and the result has been a general loss of flexibility. As in individual organisms, such an imbalance and lack of flexibility can be described in terms of stress, and the various aspects of our crisis can be seen as multiple symptoms of this social and ecological stress. To restore a healthy balance we shall have to return those variables which have been overstrained to manageable levels. This will include, among many other measures, the decentralization of populations and industrial activities, the dismantling of large corporations and other social institutions, the redistribution of wealth, and the creation of flexible, resource-conserving technologies. As in every self-organizing system, the restoration of balance and flexibility may often be achieved through self-transcendence—breaking through a state of instability or crisis to new forms of organization.

Undifferentiated growth tends to go hand in hand with fragmentation, confusion, and widespread breakdown of communication. The same phenomena are characteristic of cancer at the cellular level, and the term "cancerous growth" is very appropriate for the excessive growth of our cities, technologies, and social institutions. Because there is a continual interplay between individuals and their natural and social environment, the consequences of this cancerous growth are unhealthy for individual men and women, as well as for the economy and the ecosystem. The restoration of social and ecological balance will also contribute to improving individual health. Roszak summed up the interdependence between the well-being of the individual person and that of the planetary ecosystem: "The needs of the planet are the needs of the person . . . the rights of the person are the rights of the planet."[17]

The restoration of balance and flexibility in our economies, technologies, and social institutions will be possible only if it goes hand in hand with a profound change of values. Contrary to conventional beliefs, value systems and ethics are not peripheral to science and technology but constitute their very basis and driving force. Hence the shift to a balanced social and economic system will require a corresponding shift of values—from self-assertion and competition to cooperation and social justice, from expansion to conservation, from material acquisition to inner growth. Those who have begun to make this shift have discovered that it is not restrictive but, on the contrary, liberating and enriching. As Walter Weisskopf writes in his book *Alienation and Economics*, the crucial dimensions of scarcity in human life are not economic but existential.[18] They are related to our needs for leisure and contemplation, peace of mind, love, community, and self-realization, which are all satisfied to much greater degrees by the new system of values.

Because our current state of imbalance is largely a consequence of undifferentiated growth, the question of scale will play a central role in the reorganization of our economic and social structures. The criterion of scale has to be the comparison to human dimensions. What is too large, fast, or crowded in comparison to human dimensions is too big. People who have to deal with structures, organizations, or enterprises of such inhuman dimensions will invariably feel threatened, alienated, deprived of their individuality, and this will significantly affect the quality of their lives. The importance of scale is becoming ever more apparent even from the strictly economic point of view, as more and more large enterprises suffer from excessive centralization and the vulnerabilities of complex, interlinked technologies. The heat wasted by big American power plants in the processes of generation and transmission to the point of use would be more than enough to heat every house in the United States.[19] Similarly, the rising costs of transporting goods across the country will soon make it possible for regional and local enterprises to compete again with national companies. At the same time the creation of small-scale, decentralized technologies will be the only solution to the problem of excessive federal regulation, which has become one of the most troublesome consequences of undifferentiated growth.

In the process of decentralization many of our obsolete, resource-intensive corporations will have to be allowed to undergo profound transformations and, in some cases, go out of business. And we shall need a new legal framework to clarify and redefine the nature of private enterprise and of corporate responsibility. In all these considerations the most important thing will be to achieve balance. Not everything needs to be decentralized. Some big systems, such as the telephone and other communication systems, must be maintained; others, like mass transit, need to grow. But all growth must be qualified, and a dynamic balance must be maintained between growth and decline, so that the system as a whole remains flexible and open to change.

Among the many examples of excessive growth, the growth of cities is one of the greatest threats to social and ecological balance, and deurbanization will therefore be a crucial aspect of the return to a more human scale. As Roszak has argued convincingly, the process of deurbanization is not something that needs to be enforced; it need only be allowed to happen.[20] Several opinion polls have shown that only a small minority of city dwellers live in the city because they like it. The overwhelming majority would rather live in small towns, suburbs, or on farms, but cannot afford to do so. What we need, then, to curb the growth of cities, is to create the appropriate economic incentives, technologies, and programs of assistance that will allow people who wish to to make the transition from urban to rural life.

Similar considerations apply to the decentralization of political power. During the second half of our century it has become increasingly apparent that the nation-state is no longer workable as an effective unit of governance. It is too big for the problems of its local populations and, at the same time, confined by concepts too narrow for the problems of global interdependence. Today's highly centralized national governments are able neither to act locally nor to think globally. Thus political decentralization and regional development have become urgent needs of all large countries. This decentralization of economic and political power will have to include a redistribution of production and wealth, to balance foods and populations within countries and between the industrial nations and the Third World. Finally, at the planetary level, the recognition that we cannot "manage" the planet but have to integrate ourselves harmoniously into its multiple self-organizing systems calls for a new planetary ethic and new forms of political organization.

To return to a more human scale will not mean a return to the past but, on the contrary, will require the development of ingenious new forms of technology and social organization. Much of our conventional, resource-intensive, and highly centralized technology is now obsolete. Nuclear power, gas-guzzling cars, petroleum-subsidized agriculture, computerized diagnostic tools, and many other high-technology enterprises are antiecological, inflationary, and unhealthy. Although these technologies often involve the latest discoveries in electronics, chemistry, and other fields of modern science, the context in which they are developed and applied is that of the Cartesian conception of reality. They must be replaced by new forms of technology that incorporate ecological principles and are consistent with the new system of values.

Many of these alternative technologies are already being developed. They tend to be small-scale and decentralized, responsive to local conditions and designed to increase self-sufficiency, thus providing a maximum degree of flexibility. They are often called "soft" technologies because their impact on the environment is greatly reduced by the use of renewable resources and constant recycling of materials. Solar energy collectors, wind generators, organic farming, regional and local food production and processing, and recycling of waste products are examples of such soft technologies. Rather than being based on the principles and values of Cartesian science, they incorporate the principles observed in natural ecosystems and thus reflect systemic wisdom. As Schumacher has observed, "Wisdom demands a new orientation of science and technology towards the organic, the gentle, the non-violent, the elegant and beautiful."[21] Such a redirection of technology offers tremendous opportunities for human creativity, entrepreneurship, and initiative. The new technologies are by no means less sophisticated than the old ones, but their sophistication is of a different kind. To increase complexity simply by letting everything grow is not difficult, but to recapture elegance and flexibility requires wisdom and creative insight.

As our physical resources become scarcer, it is becoming evident that we should invest more in people—the only resource we have plenty of. Indeed, ecological awareness makes it obvious that we have to conserve our physical resources and develop our human resources. In other words, ecological balance requires full employment. This is precisely what the new technologies facilitate. Being small-scale and decentralized, they tend to be labor-intensive and thus help to establish an economic system that is noninflationary and environmentally benign.

The shift from hard to soft technologies is most urgently needed in the areas related to energy production. As emphasized in a previous chapter,[22] the deepest roots of our current energy crisis lie in the patterns of wasteful production and consumption that have become characteristic of our society. To solve the crisis we do not need more energy, which would only aggravate our problems, but profound changes in our values, attitudes, and life styles. However, while we pursue this long-term goal we also need to shift our energy production from nonrenewable to renewable resources, and from hard to soft technologies, to achieve ecological balance. The energy policies of most industrialized countries reflect what Amory Lovins, physicist and energy consultant to numerous organizations, has called the "hard energy path,"[23] in which energy is produced from nonrenewable resources—oil, natural gas, coal, and uranium—by means of highly centralized technologies that are rigidly programmed, and are uneconomical and unhealthy. Nuclear power is by far the most dangerous component of the hard energy path.[24] At the same time it is rapidly becoming the most inefficient and uneconomical source of energy. A prominent utilities investment adviser recently concluded a thorough investigation of the nuclear industry with the following devastating statement: "The conclusion that must be reached is that, from an economic standpoint alone, to rely upon nuclear fission as the primary source of our stationary energy supplies will constitute economic lunacy on a scale unparalleled in recorded history."[25]

As the nuclear option is becoming ever more unrealistic and the heavy dependence of the industrialized countries on petroleum increases the risk of military confrontations, governments and representatives of the energy industry are eagerly pursuing a number of alternatives. In doing so, however, they still blindly stick to the obsolete principles of the hard energy path. Production of synthetic fuels from coal and oil shale, which has recently been vigorously promoted, involves yet another resource-intensive technology that is extremely wasteful and causes large-scale environmental disturbances. Nuclear fusion is often talked about, but is far too uncertain to be an acceptable solution. Besides, it seems to be pursued by the nuclear industry mainly to produce plutonium, which would then be used in fission reactors.[26] All these forms of energy production require massive capital investments and centralized power stations with complex technologies. They are inefficient and highly inflationary without creating significant numbers of jobs. Conservation measures and solar energy could generate several times as many jobs as produced by the nuclear industry, while every new power station destroys about 4,000 net jobs.[27]

The only way out of the energy crisis is to follow a "soft energy path," which, in Lovins' thinking, has three main components: conservation of energy by more efficient use, intelligent use of present nonrenewable energy sources as "bridging fuels" during the transition period, and rapid development of soft technologies for energy production from renewable sources. Such a threefold approach would not only be environmentally benign and ecologically balanced; it would also be the most efficient and cheapest energy policy. A recent Harvard Business School study has stated authoritatively that efficiency improvements and soft technologies are the most economical of all available energy sources, besides providing more and better jobs than any of the other options.[28] The soft energy path should be embarked upon without further delay. Since the role of fossil fuels as a bridge to the new, renewable energy sources is a vital element of the necessary transition, it will be crucial to start the transition process while we still have enough fossil fuels to guarantee a smooth passage.

In the long run the greatest conservation of energy will be achieved by abandoning our present unhealthy and wasteful patterns of production and consumption in favor of an ecologically harmonious way of life. But while this profound change takes place, enormous energy savings can be accomplished by improving energy efficiency throughout the economy. This can be done right now by means of available technologies, while maintaining our present levels of economic activity. In fact, conservation turns out to be our best short-term energy source, outstripping all conventional fuels combined. This is dramatically confirmed by the observation that, during 1973–78, 95 percent of all new energy supplies in Europe came from more efficient use. Thus millions of individual conservation measures added up to supply almost twenty times as much energy as all other new sources combined, including the entire European nuclear program. During the same time the United States, without trying very hard, obtained 72 percent of its new energy supplies from conservation measures—two and a half times as much as the energy from all other new sources.[29]

An important part of using energy more efficiently is to use the appropriate kind for each task, which means applying the type of energy that will allow that particular task to be carried out in the cheapest and most effective way. In the United States 58 percent of all energy needs are for heating and cooling, 34 percent for liquid fuels to run vehicles, and only 8 percent for the special uses that require electricity. This electric energy is by far the most expensive, with electricity from a new power station costing about three times as much as the 1980 crude oil price set by the Organization of Petroleum Exporting Countries. Thus electricity is grossly wasteful for most of our energy needs, and since we already produce more of it than we can use appropriately, building more centralized power stations would vastly increase the inefficiency of the entire system. As Lovins says, "Arguing about what kind of new power station to build is like shopping for . . . antique furniture to burn in the furnace."[30] What we need is not more electricity but a greater variety of energy sources that can be matched more appropriately with our needs.

Because we use more than half of our energy supply for heat, the greatest savings can be achieved by insulating our buildings more efficiently. It is now technically possible and highly cost effective to make buildings so heat-tight that they need virtually no space heating, even in cold climates, and many existing buildings can be brought to nearly that standard. A further important means of increasing energy efficiency is the so-called cogeneration of useful heat and electricity. A cogenerator is a device which makes use of the heat that is inevitably produced in the generation of electricity instead of wastefully ejecting it into the environment. Any engine that produces motion by burning a fuel can also be used as a cogenerator. Installed in a building, it can efficiently operate the building's heating and cooling systems while, at the same time, driving its electric appliances. In this way the energy contained in the fuel can be converted into useful forms with up to 90 percent efficiency, whereas conventional generation of electricity alone would use at most 30 to 40 percent of the fuel's energy.[31] Several recent studies have found that the combined effect of cogeneration and improved insulation, along with improved efficiencies in cars and machines, would result in energy savings of 30 to 40 percent without any changes in our standards of living and economic activities.[32]

In the long run, we need an energy source that is renewable, economically efficient, and environmentally benign. Solar energy is the only kind of energy that satisfies all these criteria. The sun has been the planet's main energy source for billions of years, and life in its myriad forms has become exquisitely adapted to solar energy during the long course of planetary evolution. All the energy we use, except for nuclear power, represents some form of stored solar energy. Whether we burn wood, coal, oil, or gas, we use energy originally radiated to the earth from the sun and converted into chemical form by photosynthesis. The wind pushing our sailboats and driving our windmills is a flow of air caused by the upward movement of other air masses heated by the sun. The falling water that drives our turbines is part of the ongoing water cycle sustained by solar radiation. Thus virtually all our energy sources supply us with solar energy of one form or another. Not all these forms of energy are renewable, however. In the current energy debate the term "solar energy" is used more specifically to refer to the forms of energy that come from inexhaustible or renewable sources. Solar energy in this sense is available in forms as varied as the planet itself.[33] In forested areas it is present as a solid fuel (wood); in agricultural areas it can be produced as a liquid or gaseous fuel (alcohol or methane produced from plant products), in mountainous regions as hydroelectric power, and in windy places as wind-generated electricity; in sunny areas it can be transformed into electricity by photovoltaic cells, and almost everywhere it can be collected as direct heat.

Most of these forms of solar energy have been exploited by human societies throughout the ages by means of time-honored technologies. The U.S. Department of Energy likes to call solar energy an "exotic" new energy source, but in fact the solar transition does not require any major technological innovations. It simply involves the judicious integration of long-known agricultural and technological processes into the activities of our modern society. Contrary to a widespread misconception, the problem of storing energy from these renewable sources has already been solved, and several studies have shown that the existing soft technologies are sufficient to meet all our long-term energy needs.[34] In fact, many of them are already being used successfully by solar-conscious communities. The most distinctive feature of these technologies is their decentralized nature. Since the energy radiated from the sun is diffused across the entire planet, centralized solar power stations do not make sense. In fact, they are inherently uneconomical.[35] The most efficient solar technologies involve small-scale devices, to be used by local communities, which generate a wide variety of jobs and are benign in their effects. As Barry Commoner reminds us, "When a pump fails in a solar device there is no need to call upon the President to visit the scene in order to calm the fear of catastrophe."[36]

One of the main arguments against solar energy is the claim that it is not economically competitive with conventional energy sources. This is not true. Certain forms of solar energy are competitive already; others can be within a few years. This can be shown even without questioning the narrow notion of economic competitiveness, which disregards most of the social costs generated by conventional energy production. One form of solar energy that can already be used with great advantage is solar heating. It can either be "passive," with the building itself capturing and storing the heat, or "active," as when special solar collectors are used. Energy from the sun can also be used to cool buildings during the summer. Solar heating and cooling systems have been developed intensively over the past few years and now represent a vibrant and rapidly expanding industry, as documented in the Harvard Business School report: "Many people still assume that solar energy is something for the future, awaiting a technological breakthrough. That assumption represents a great misunderstanding, *for active and passive solar heating is a here-and-now alternative to conventional energy sources.*"[37]

Another solar technology with tremendous potential is the local production of electricity by means of photovoltaic cells.[38] A photovoltaic* cell is a silent and motionless device that converts sunlight into electricity. The principal raw material used for its manufacture is silicon, which is present abundantly in common sand, and the manufacturing processes are similar to those used by the semiconductor industry to build transistors and integrated circuits ("chips"). At present photovoltaic cells are still too expensive for residential use, but so were transistors at the beginning of their development. In fact the photovoltaic industry is now going through the same stages as the semiconductor industry did two decades earlier. When the American space and military

* The term "photovoltaic" refers to the fact that an electric voltage is generated when light falls on the cell.

programs needed lightweight electronic equipment, massive federal investments led to a great reduction of production costs. This was the beginning of the industry that is now producing millions of low-cost transistor radios, pocket calculators, and digital watches.

Similarly, photovoltaic cells were first used to provide electricity for orbiting space satellites and were very expensive at that time. Since then their costs have dropped sharply, although their market is still quite restricted. For them to become competitive with conventional electricity, a further reduction of costs to $500 per kilowatt—about one-tenth of their present price—will be necessary, and this could easily be achieved with a substantial federal investment in photovoltaic technology. A recent study by the Federal Energy Administration estimated that the required price reduction to $500 per kilowatt would be achieved with a government order of 152,000 kilowatts of photovoltaic cells, to be delivered over a five-year period, for a total price of less than half a billion dollars.[39] This compares more than favorably with the two billion dollars of federal funds slated for the Clinch River nuclear breeder reactor, which has been estimated to produce electricity at a cost of $5,000 per kilowatt.[40] Obviously a major investment of public funds in photovoltaic technology would launch a huge industry capable of producing electricity in efficient and benign ways, to the great benefit of all consumers. Similar estimates have shown that the generation of electricity from wind could be started almost at once, at economically competitive costs, if sufficient funds were invested in windmill technology.[41]

These developments would bring about fundamental structural changes in the utility industry, since photovoltaics and wind generators, like solar heating, are used most efficiently on site with no need for centralized power stations. The political power of the utility companies, reluctant to give up their monopoly in the production of electricity, is today the major obstacle to the rapid development of the new solar technologies.

Any realistic solar energy program will have to come up with enough liquid fuel to operate airplanes and at least some of our ground transport, and with liquid or gaseous fuel to be used for cogenerators where the local supply of solar energy is inadequate. The solar technology most readily available to obtain these fuels is also the oldest—the production of energy from biomass. The term "biomass" refers to the organic matter produced by green plants, which represents stored solar energy. This energy can not only be retrieved as heat by burning the material; it can also be converted into liquid or gaseous fuels by distilling alcohol from fermented grain or fruit or by capturing the methane that bacteria generate from manure, sewage, or garbage. Both these fuels can be used to run internal combustion engines without any pollution, and both can be produced by well-known and relatively simple methods. Alcohol production from biomass is most developed in Brazil, where all gasoline contains up to 20 percent alcohol; and simple methane generators, producing fuel from manure and sewage, have been built by the millions in India and China.[42]

Of all the solar technologies, the production of methane—a major component of natural gas—with the help of bacterial activity seems to come closest to the principles observed in natural ecosystems. It involves the cooperation of other organisms—a characteristic aspect of all life—and can be used very effectively to recycle garbage, sewage, and underwater sludge, which are some of our major pollutants. The organic residue from methane production is an excellent fertilizer, ideally suited to replace at least part of our resource-consuming and polluting synthetic fertilizers. Like other forms of solar energy, biomass is widely dispersed and thus very appropriate for small-scale, local fuel production.

Here we need to keep in mind that the production of liquid fuels from agricultural products will not sustain our transportation system at its present level. To do so would require massive alcohol production from farms, which would be an irresponsible use of our soils because it would cause their rapid erosion, as Wes Jackson has argued emphatically.[43] Although biomass is a renewable resource, the soil in which it grows is not. We can certainly expect significant alcohol production from biomass, including crops, but a massive alcohol fuel program designed to sustain present needs would deplete our soil at the same rate as we are now depleting our coal, oil, and other natural resources. The way out of this dilemma will be to thoroughly redesign our transportation system, especially in the United States, together with many other aspects of our wasteful and resource-consuming life styles. This will not mean lowering our living standards. On the contrary, it will enhance the quality of our lives.

The authoritative studies of our energy options cited above show that the road to a solar future is open. Although significant technological advances are to be expected in several areas, we do not have to wait for any technological breakthroughs to embark on this historic transition. What we need most is accurate public information about the potential of solar energy, along with corresponding social and economic policies to facilitate the passage to the solar age. Barry Commoner has outlined a detailed scenario for replacing most of the nonrenewable energy sources in the United States with solar energy within fifty years.[44] His proposal does not assume any major technological innovations nor depend on any drastic measures of energy conservation. Either of these developments, both of which will almost certainly take place, would significantly shorten and ease the transition period.

The key to Commoner's sketch of the solar transition is the role of natural gas as the main bridging fuel. The basic idea is to expand the present production and distribution network of natural gas, and then to gradually replace the natural gas with solar methane. To do so, methane-generating plants would be built wherever sufficient biomass is readily available—in the form of garbage and sewage in cities, of crops, manure, and agricultural residues in farming areas, of wood in forest areas and seaweed along the coasts, and so on. Like natural gas, solar methane could easily be stored as a fuel reserve to balance the natural variations of other solar energy sources, and it would also be used for cogeneration of heat and electricity to conserve energy and reduce environmental pollution. Cogenerators could easily be produced on a large scale by the auto industry, as Fiat has already begun to do in Italy. The transition from natural gas to solar methane could be so smooth that it would hardly be noticed. In fact, it is already under way in some parts of the United States, for example in Chicago.

According to Commoner's outline, which is of course only one of many possible plans, the initial phase of the transition would consist of installing natural gas-fired generators wherever possible and building more extended gas distribution systems to supply them. At the same time active and passive solar heating would expand, alcohol produced from waste and crops would begin to replace gasoline, and increasing amounts of solar methane, produced from biomass, would be added to the natural gas in the expanding pipeline system. Within several years the use of photovoltaic cells and wind generators would expand significantly, while the total production of solar energy would gradually increase, until it made up about 20 percent of the total energy budget after the first twenty-five years. At that stage, halfway through the transition period, solar energy and natural gas together would account for slightly more than half the total United States energy budget, which would make it possible to completely eliminate dependence on nuclear power. During the second half of the transition, the production of oil and coal would gradually be reduced to zero and natural gas production would fall to about half its present rate. At that point the system would be about 90 percent solar. In subsequent years the 10 percent contribution of natural gas could be eliminated, but it would be important to keep this energy source as a back-up fuel to compensate for irregularities due to unexpected fluctuations in climate. To carry out the entire transition, according to Commoner's estimates, the United States would need a supply of natural gas equivalent to about 250 billion barrels of oil over the fifty-year period, which represents a quantity somewhere between 10 and 30 percent of the estimated reservoir of natural gas in the United States.[45]

The main obstacles to the solar transition are not technical but political. The shift from nonrenewable to renewable resources will force the oil companies to give up their dominant roles in the world economy and to change their functions in fundamental ways. One solution, suggested by Commoner, would be to convert those companies that wished to remain in the oil and natural gas business into public utilities, while the major oil companies would be likely to invest their cash in more attractive enterprises, as many of them have already begun to do. Similar problems will arise in other industrial sectors as the solar transition generates clashes between social and private interests. The soft energy path would clearly be in the interest of the overwhelming majority of energy users, but a reasonably smooth passage to the solar age will be possible only if we are able, as a society, to put long-term social returns before short-term private gains.

The transition to the solar age is really under way now, not merely in terms of new technologies but, in a broader sense, as a profound transformation of our entire society and culture. The shift from the mechanistic to the ecological paradigm is not something that will happen sometime in the future. It is happening right now in our sciences, in our individual and collective attitudes and values, and in our patterns of social organization. The new paradigm is better understood by individuals and small communities than by large academic and social institutions, which often tend to be locked into Cartesian thinking. To facilitate the cultural transformation, it will therefore be necessary to restructure our system of information and education, so that the new knowledge can be presented and discussed appropriately.

Much of this restructuring of information is already being done successfully by citizen movements and public-interest groups, and by numerous alternative networks. However, if the new ecological awareness is to become part of our collective consciousness, it will have to be transmitted, eventually, through the mass media. These are presently dominated by business, especially in the United States, and their contents are censored accordingly.[46] The public's right of access to the mass media will thus be an important aspect of the current social change. Once we succeed in reclaiming our mass media, we can then decide what needs to be communicated and how to use the media effectively to build our future. This means that journalists, too, will change their thinking from fragmentary to holistic modes and develop a new professional ethic based on social and ecological awareness. Instead of concentrating on sensational presentations of aberrant, violent, and destructive happenings, reporters and editors will have to analyze the complex social and cultural patterns that form the context of such events, as well as reporting the quiet, constructive, and integrative activities going on in our culture. That such a mature kind of journalism is not only socially beneficial but can also be good business is proven by the recent growth of alternative media that promote new values and life styles.[47]

An important part of the necessary restructuring of information will be the curtailing and reorganization of advertising. Since product advertisements tend to obscure the social costs generated by the patterns of consumption they stimulate, it is vital that information provided by environmental and consumer groups be given "equal time." Moreover, legal restrictions on advertising resource-intensive, wasteful, and unhealthy products would be our most effective way of reducing inflation and moving toward ecologically harmonious ways of living.

Finally, the restructuring of information and knowledge will involve a profound transformation of our system of education. Indeed, this transformation too is well on its way. It is not occurring in our academic institutions so much as among the general population, in thousands of spontaneous adult-education efforts undertaken by the social movements that emerged during the 1960s and 1970s. In the United States many of these movements have shown their durability in spite of repeated predictions of their early demise, and the values and life styles they promote are being adopted by ever increasing numbers of people. Although the movements sometimes fail to communicate and cooperate with one another, they all go in the same direction. In their concerns for social justice, ecological balance, self-realization, and spirituality, they emphasize different aspects of the gradually emerging new vision of reality.[48]

The last decade saw a proliferation of citizen movements formed around social and environmental issues, following the pioneering efforts of Ralph Nader. In recent years there has been a broad convergence of these movements and a tendency to go beyond single issues to address fundamental systemic concerns. Many organizations have been particularly concerned with corporate accountability and with the influence of large corporations on government policies. The political strength of these citizen movements is considerable, and opinion polls have shown that the overwhelming majority of the population considers them a positive social force.[49] Closely related to their efforts are the activities of a number of organizations referred to collectively as the ecology movement. These groups maintain information centers and publish newsletters on environmental protection, organic farming, recycling of waste, and other ecological concerns. Some also provide practical assistance in developing and applying soft technologies, and many of them belong to antinuclear alliances and coalitions.

Citizen and consumer movements are also the sources of emerging countereconomies based on decentralized, cooperative, and ecologically harmonious life styles, and involving the bartering of skills and home-produced goods and services. These alternative economies—also known as "informal," "dual," or "convivial" economies—cannot be centrally planned and installed but have to grow and develop organically, which usually involves a great deal of pragmatic experimentation and requires considerable social and cultural flexibility. Interesting and significant patterns of countereconomies have grown in this way in the United States, Canada, the United Kingdom, the Scandinavian countries, the Netherlands, Japan, Australia, and New Zealand.[50]

The new emphasis on dual economies is based on the realization that these informal, cooperative and nonmonetized sectors are predominant in the world's economies, and that the institutionalized and monetized sectors grew out of them and rest upon them, rather than the reverse. This fact can be documented even in the industrialized countries, although the bias of economic statistics makes it almost impossible to carry out such an analysis.[51] It is clearly necessary for any modern society to have both formal and informal sectors in its economy, but our overemphasis on money—dollars, yen, or rubles—to measure economic efficiency has created huge imbalances and is now threatening to destroy the informal sectors. To counteract this trend more and more people are now trying to drop out of the monetized economy, working only a few hours a week to earn a minimum of cash and adopting more communal, reciprocal, and cooperative ways of living to satisfy their other, nonmonetary needs. There has been a growing interest in household economies based on use value rather than market value, and a significant rise in the numbers of self-employed people. Household economies are ideally suited to develop small-scale soft technologies and to practice the various crafts now being revived in many countries. All these activities enhance the autonomy and security of families, households, and neighborhoods, and improve social cohesion and stability.

Another important contribution to the reorganization of economic patterns comes from worker-participation and self-management movements, which are active in Canada and several European countries. The first successful model of worker self-management was achieved in Yugoslavia and has since inspired similar developments in Sweden, Germany, and other Western European countries. In the United States and Japan, the idea that workers should participate in their own management is taking hold more slowly, owing to the different political traditions of these countries, but even there it is now beginning to be accepted.[52] Following the principle of thinking globally and acting locally, we now have the unique possibility of synthesizing and adapting to our needs the strategies of creative communities around the world—from the Chinese model of self-reliant communal development and the traditional values and life styles of numerous communities in the Third World to the Yugoslavian model of worker self-management and the informal economies that are now being developed in the United States and many other countries.

The new vision of reality is an ecological vision in a sense which goes far beyond the immediate concerns with environmental protection. To emphasize this deeper meaning of ecology, philosophers and scientists have begun to make a distinction between "deep ecology" and "shallow environmentalism."[53] Whereas shallow environmentalism is concerned with more efficient control and management of the natural environment for the benefit of "man," the deep ecology movement recognizes that ecological balance will require profound changes in our perception of the role of human beings in the planetary ecosystem. In short, it will require a new philosophical and religious basis.

Deep ecology is supported by modern science, and in particular by the new systems approach, but it is rooted in a perception of reality that goes beyond the scientific framework to an intuitive awareness of the oneness of all life, the interdependence of its multiple manifestations and its cycles of change and transformation. When the concept of the human spirit is understood in this sense,[54] as the mode of consciousness in which the individual feels connected to the cosmos as a whole, it becomes clear that ecological awareness is truly spiritual. Indeed, the idea of the individual being linked to the cosmos is expressed in the Latin root of the word religion, *religare* ("to bind strongly"), as well as in the Sanskrit *yoga,* which means union.

The philosophical and spiritual framework of deep ecology is not something entirely new but has been set forth many times throughout human history. Among the great spiritual traditions Taoism offers one of the most profound and most beautiful expressions of ecological wisdom,[55] emphasizing both the fundamental oneness and the dynamic nature of all natural and social phenomena. Thus Huai Nan Tzu: "Those who follow the natural order flow in the current of the Tao."[56]

While such ecological principles were expounded by even earlier Taoist sages, a very similar philosophy of flow and change was being taught by Heraclitus in ancient Greece.[57] Later, the Christian mystic Saint Francis had views and ethics that were profoundly ecological and presented a revolutionary challenge to the traditional Judeo-Christian view of "man" and nature. The wisdom of deep ecology is also apparent in many works of Western philosophy, including those of Baruch Spinoza and Martin Heidegger. It is found throughout Native American culture, and has been expressed by poets ranging from Walt Whitman to Gary Snyder. It has even been argued that the world's greatest pieces of literature, such as Dante's *Divine Comedy,* are structured according to the ecological principles observed in nature.[58]

The deep ecology movement, then, is not proposing an entirely new philosophy but is reviving an awareness which is part of our cultural heritage. What is new, perhaps, is the extension of the ecological vision to the planetary level, supported by the powerful experience of the astronauts and expressed in images like "spaceship Earth" and the "Whole Earth," as well as the new maxim "Think globally and act locally." This new awareness is being generated specifically by numerous individuals, groups, and networks, but a significant shift of values has also been observed in large sections of the general population, a shift from material consumption to voluntary simplicity, from economic and technological growth to inner growth and development. In 1976 a study by the Stanford Research Institute estimated that four to five million adult Americans had drastically reduced their incomes and withdrawn from their former positions in the consumer economy in favor of a life style embracing the principle of voluntary simplicity.[59] SRI further estimated that another eight to ten million American adults lived according to some, but not all, of the tenets of the voluntary simplicity approach—frugal consumption, ecological awareness, and concern with personal, inner growth. This value shift has since been confirmed by several opinion polls and has been widely discussed in the media. In other countries, such as Canada, the voluntary simplicity theme has emerged officially,[60] as it has in California in the speeches of Governor Jerry Brown.

The shift from material growth to inner growth is being promoted by the human potential movement, the holistic health movement, the feminist movement, and various spiritual movements. While economists have seen human needs in terms of material acquisitions and have postulated that these needs are in principle insatiable, humanistic psychologists have concentrated on the nonmaterial needs of self-actualization, altruism, and loving interpersonal relationships. In doing so they have outlined a radically different image of human nature, which transpersonal psychologists have extended by emphasizing the value of a direct, experiential understanding of oneness with the entire human family and the cosmos at large. At the same time the holistic health movement is pointing out the impact of the materialistic value system on our well-being and promoting healthy attitudes and living habits, together with a new conceptual basis and new practical approaches to health care.

The forces promoting the new ideas about health and healing work both inside and outside the medical system. Physicians in the United States, Canada, and Europe are forming associations and holding conferences to discuss the merits of holistic medicine. As a result of these discussions doctors are trying to eliminate unnecessary surgery, diagnostic tests, and prescriptions, recognizing that this will be the most effective way to bring down health costs. Others are advocating restoring the integrity of the medical profession by getting their information about drugs from sources independent of the drug industry, for example by subscribing to independent medical newsletters and establishing closer links with pharmacists.

As to the organization of health care, there is now a strong trend toward decentralization and general practice, with a veritable renaissance of primary care in Europe and North America in the last few years. The emphasis on family practice has become much stronger in medical schools, where a new generation of medical students realizes that primary health care, motivated by the prevention of ill health and an awareness of its environmental and social origins, not only brings greater human satisfaction but is also intellectually more challenging and more rewarding than the biomedical approach. At the same time there has been a revival of psychosomatic medicine, generated by recognition of the crucial role of stress in the onset and development of disease, and numerous research projects are focusing on the interplay between mind and body in health and illness.

With this growing interest in the broader context of health, nonmedical health professionals and institutions have been able to improve their status and increase their influence. Nurses, who have long perceived the shortcomings of the biomedical approach, are expanding their role in health care and fighting for full recognition of their qualifications as healers and health educators. They are also investigating various unorthodox therapeutic techniques in an attempt to develop a truly holistic approach to primary care. Public health organizations committed to prevention and health education are growing and gaining recognition in medical circles. In addition, some governments are taking a new interest in disease prevention and health maintenance, and various government agencies are being set up to study the development of holistic health care.

The most important force of all in this health care revolution is a strong grass-roots movement of individuals and newly formed organizations dissatisfied with the existing system of medical care. They have embarked on an extensive exploration of alternative approaches, including promotion of healthy living habits, combined with recognition of personal responsibility for health and of the individual's potential for self-healing; a strong interest in traditional healing arts from various cultures that integrate physical and psychological approaches to health; and formation of holistic health care centers, many of them experimenting with unorthodox and esoteric therapies.

The shift to the value system that the holistic health movement, the human potential movement, and the ecology movement advocate is further supported by a number of spiritual movements that reemphasize the quest for meaning and the spiritual dimension of life. Some individuals and organizations among these "New Age" movements have shown clear signs of exploitation, fraud, sexism, and excessive economic expansion, quite similar to those observed in the corporate world, but these aberrations are transitory manifestations of our cultural transformation and should not prevent us from appreciating the genuine nature of the current shift of values. As Roszak has pointed out, one must distinguish between the authenticity of people's needs and the inadequacy of the approaches that may be offered to meet those needs.[61]

The spiritual essence of the ecological vision seems to find its ideal expression in the feminist spirituality advocated by the women's movement, as would be expected from the natural kinship between feminism and ecology, rooted in the age-old identification of woman and nature.[62] Feminist spirituality is based on awareness of the oneness of all living forms and of their cyclical rhythms of birth and death, thus reflecting an attitude toward life that is profoundly ecological. As numerous feminist authors have recently pointed out, the image of a female deity seems to embody this kind of spirituality more accurately than that of a male god. Indeed, worship of the Goddess predates that of male deities in many cultures, including our own, and may also have been closely connected with the nature mysticism of the ancient Taoist tradition.[63]

According to Beatrice Bruteau, different images of the Divine can be seen as reflecting different solutions to the fundamental metaphysical problem of "the One and the Many."[64] The male god typically represents the One that can exist alone, independent and absolute, while the Many exist only by the will of God, dependent and relative. In human society such a situation is exemplified by the conventional father-child relationship. Fatherhood, as Bruteau points out, is characterized by separation. The father is at no time physically united with the child, and the relationship tends to be one of confrontation and conditional love. When this image of the father is applied to God, it naturally evokes the notions of obedience, loyalty, and faith, and often includes some image of challenge, with subsequent reward or punishment.

The image of the Goddess, on the other hand, according to Bruteau, represents a solution to the One/Many problem in terms of union and mutual embodiment, with the One manifest in the Many and the Many dwelling within the One. In such a relationship of union, which is not imposed or attained but is organically given, there is no sense of opposition between God and the world. Their relationship is characterized by harmony, warmth, and affection, rather than challenge and drama. Such an image is clearly maternal, reflecting the mother's unconditional love, mother and child being physically united and participating in life together.

With renaissance of the Goddess image, the feminist movement is also creating a new self-image for women, along with new modes of thinking and a new system of values. Thus feminist spirituality will have a profound influence not only on religion and philosophy but also on our social and political life.[65] One of the most radical contributions men can make to developing our collective feminist awareness will be to get fully involved in raising our children from the moment of birth, so that they can grow up with the experience of the full human potential inherent in women and men. John Lennon, always a step ahead of his time, did just that during the last five years of his life.

While men will become more active as fathers, the full participation of women in all areas of public life, which will undoubtedly be achieved in the future, is bound to bring about far-reaching changes in our attitudes and behavior. Thus the feminist movement will continue to assert itself as one of the strongest cultural currents of our time. Its ultimate aim is nothing less than a thorough redefinition of human nature, which will have the most profound effect on the further evolution of our culture.

Conventional stereotypic images of human nature are challenged today not only by the women's movement but also by a great number of ethnic liberation movements in revolt against the oppression of minorities through ethnic prejudice and racism. Their protest is amplified by the struggles of several other kinds of minorities—homosexuals, old people, single parents, the physically handicapped, and many more—who have been stigmatized by rigidly assigned social roles and identities. The roots of these protests lie in the 1960s, the decade that saw the simultaneous emergence of several powerful social movements all of which began to question authority. While civil rights leaders demanded that black citizens be included in the political process, the free speech movement demanded the same for students. At the same time the women's movement questioned patriarchal authority, and humanistic psychologists undermined the authority of doctors and therapists.

Today a similar questioning of authority is being initiated at the global level, as Third World countries challenge the conventional notion that they are "less developed" than the industrialized countries. An increasing number of their leaders now perceive the multifaceted crisis of the Northern Hemisphere with great clarity, and are resisting the industrialized world's attempts to export its problems to the Southern Hemisphere. Some Third World leaders are discussing how the countries of the Southern Hemisphere might decouple themselves and develop their own indigenous technologies and economic patterns. Others have proposed shifting the definition of "development" from the development of industrial production and the distribution of material goods to the development of human beings.[66]

Because feminism is a major force in our cultural transformation, especially in North America and Europe, it is likely that the women's movement will play a pivotal role in the coalescence of various social movements. Indeed, it may well become the catalyst that will allow the various movements to flow together during the 1980s. Today many of these movements still operate separately, without perceiving how their purposes interrelate, but several significant coalitions have recently begun to form. Not surprisingly, women are playing important roles in contacts among environmental groups, consumer groups, ethnic liberation movements, and feminist organizations. Helen Caldicott, who has helped to provide the antinuclear movement with a sound scientific basis, as well as a sense of urgency and compassion, and Hazel Henderson, who has lucidly analyzed the shortcomings of the Cartesian framework in current economic thinking, are examples of women in leading positions who are forging valuable coalitions.

The new alliances and coalitions, which already interlink hundreds of groups and networks, aim to be nonhierarchical, nonbureaucratic, and nonviolent. Some of them function very effectively around the world. An example of such a world-wide coalition is Amnesty International's great campaign for human rights. These new, effective organizations demonstrate how the world-wide implementation of vital functions, such as environmental protection or the fight for economic justice, can be achieved through the coordination of local and regional actions, based on agreed-upon global principles. The multiple networks and coalitions have not yet asserted themselves decisively in the political arena, but as they continue to give substance to the new vision of reality, a critical mass of awareness will be reached that will allow them to coalesce into new political parties. The members of these parties, some of them already being formed in various countries, will include environmentalists, consumer groups, feminists, ethnic minorities, and all those for whom the corporate economy is no longer working. Together they clearly represent a winning majority at a time when most voters are so disenchanted they do not even bother to participate in elections. By bringing this nonvoting population back into the electoral process the new coalitions should be able to turn the paradigm shift into political reality.

Such predictions may seem rather idealistic, especially in view of the current political swing to the right in the United States and the crusades of Christian fundamentalists promoting medieval notions of reality. But when we look at the situation from a broad evolutionary perspective, these phenomena become understandable as inevitable aspects of cultural transformation. In the regular pattern of rise, culmination, decline, and disintegration, which seems to be characteristic of cultural evolution, the decline occurs when a culture has become too rigid—in its technologies, ideas, or social organization—to meet the challenge of changing conditions.[67] This loss of flexibility is accompanied by a general loss of harmony, leading to the outbreak of social discord and disruption. During the process of decline and disintegration the dominant social institutions are still imposing their outdated views but are gradually disintegrating, while new creative minorities face the new challenges with ingenuity and rising confidence.

This process of cultural transformation, shown schematically in the diagram below, is what we are now observing in our society. The Democratic and Republican parties, as well as the traditional Right and Left in most European countries, the Chrysler Corporation, the Moral Majority, and most of our academic institutions are all part of the declining culture. They are in the process of disintegration. The social

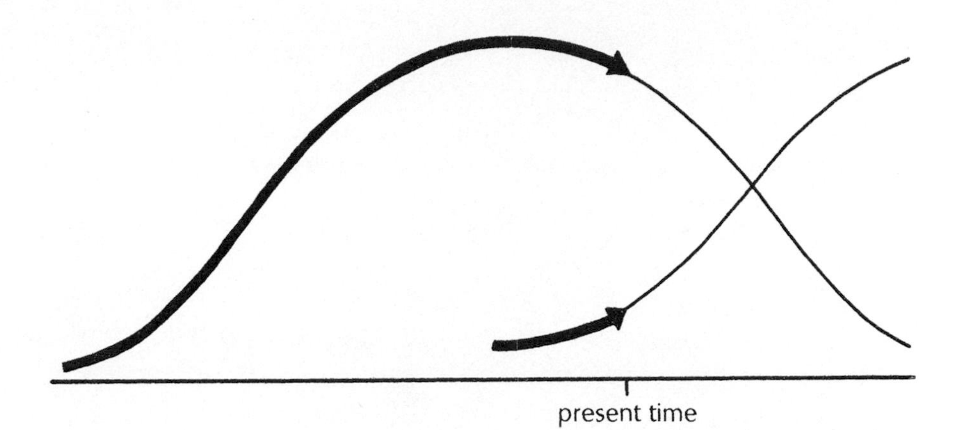

Schematic depiction of the rising and declining cultures in the current process of cultural transformation.

movements of the 1960s and 1970s represent the rising culture, which is now ready for the passage to the solar age. While the transformation is taking place, the declining culture refuses to change, clinging ever more rigidly to its outdated ideas; nor will the dominant social institutions hand over their leading roles to the new cultural forces. But they will inevitably go on to decline and disintegrate while the rising culture will continue to rise, and eventually will assume its leading role. As the turning point approaches, the realization that evolutionary changes of this magnitude cannot be prevented by short-term political activities provides our strongest hope for the future.

NOTES

[*Full publication information for these citations will be found in the Bibliography.*]

1. *The Turning of the Tide*

1. See Rothschild (1980).
2. See *Mother Jones*, July 1979.
3. See Sivard (1979).
4. See Chapter 8.
5. See Chapter 8.
6. Quoted in Ehrlich and Ehrlich (1972), p. 147.
7. Ibid., Chapter 7.
8. Fuchs (1974), p. 42.
9. *Washington Post*, May 20, 1979.
10. See Harman (1977).
11. This graph is not meant to give an exact representation of the civilizations shown, but has been drawn merely to illustrate their general pattern of development. Approximate dates for the beginning, culmination, and end of each civilization have been used, but the individual curves have been given equal and arbitrary height. They have been displaced vertically for the sake of clarity.
12. Toynbee (1972).
13. For references, see ibid., p. 89.
14. See Henderson (1981).
15. For a comprehensive discussion of the multiple facets of patriarchy, see Rich (1977).
16. Ibid., p. 40.
17. For an extensive discussion of paradigms and paradigm shifts, see Kuhn (1970).
18. Sorokin (1937–41).
19. Ibid., vol. 4, pp. 775 ff.
20. Mumford (1956).
21. *I Ching*, comments on the hexagram "The Turning Point," Wilhelm (1968), p. 97.
22. For an extremely lucid analysis of materialist dialectics that shows striking similarities with ancient Chinese thought, without ever acknowledging them, see Mao Zedong's famous essay "On Contradiction"; Mao (1968).
23. See Barzun (1958), p. 186.
24. Wang Ch'ung, quoted in Capra (1975), p. 106.
25. Porkert (1974), pp. 9 ff. For a good introduction, see Porkert (1979).
26. See Goleman (1978) for a review of recent research on sex differences.
27. See Merchant (1980), p. 13.
28. Quoted in Capra (1975), p. 114.
29. Wilhelm (1960), p. 18.
30. Quoted in Capra (1975), p. 117.
31. Quoted ibid.
32. Merchant (1980), p. xvii.
33. See Dubos (1968), p. 34.
34. See Chapter 9.
35. Koestler (1978), p. 57.
36. See Mumford (1970).
37. Roszak (1969).
38. Toynbee (1972), p. 228.
39. Quoted in Capra (1975), p. 28.

2. *The Newtonian World-Machine*

1. Quoted in Randall (1976), p. 237.
2. See, for example, Crosland (1971), p. 99.
3. Laing (1982).
4. Huai Nan Tzu, quoted in Capra (1975), p. 117.
5. For references to these Baconian metaphors, see Merchant (1980), p. 169.
6. This point has been argued convincingly by Carolyn Merchant, ibid.

7. Russell (1961), p. 542.
8. See Vrooman (1970), pp. 54–60.
9. Quoted ibid., p. 51.
10. Quoted in Garber (1978).
11. Quoted ibid.
12. Quoted in Vrooman (1970), p. 120.
13. Quoted in Garber (1978).
14. Ibid.
15. Quoted in Sommers (1978).
16. Heisenberg (1962), p. 81.
17. Merchant (1980), p. 3.
18. Quoted in Randall (1976), p. 224.
19. Quoted in Rodis-Lewis (1978).
20. Quoted ibid.
21. Quoted in Vrooman (1970), p. 258.
22. Quoted in Capra (1975), p. 56.
23. Quoted in Randall (1976), p. 263.
24. Keynes (1951).
25. Quoted in Capra (1975), p. 55.
26. Ibid.
27. Ibid., p. 56.
28. Quoted in Vrooman (1970), p. 189.
29. See Capra (1975), p. 59.
30. Quoted in Randall (1976), p. 486.
31. Bateson (1972), p. 427.

3. *The New Physics*

1. W. Heisenberg, quoted in Capra (1975), p. 50.
2. W. Heisenberg, quoted ibid., p. 67.
3. W. Heisenberg, quoted ibid., p. 53.
4. A. Einstein, quoted ibid., p. 42.
5. See Chapter 9.
6. For a definition and concise description of mysticism, see Stace (1960), Chapter 1.
7. At present some properties of subatomic particles, like electric charge or magnetic moment, seem to be independent of the experimental situation. However, recent developments in particle physics, to be discussed below, indicate that these properties too may well depend on our framework of observation and measurement.
8. See Capra (1975), p. 160.
9. N. Bohr, quoted ibid., p. 137.
10. W. Heisenberg, quoted ibid., p. 139.
11. Stapp (1971).
12. Bateson (1979), p. 17.
13. I am indebted to Henry Stapp for a discussion of this point; see also Stapp (1972).
14. See Schilpp (1951); see also Stapp (1972).
15. See Bohm (1951), pp. 614 ff.
16. See Stapp (1971); for a discussion of the implications of Bell's theorem in relation to the philosophy of A. N. Whitehead, see Stapp (1979).
17. The following presentation is based on the comprehensive discussion of the EPR experiment given by David Bohm in Bohm (1951), pp. 614 ff.
18. Stapp (1971).
19. See Bohm (1951), p. 167.
20. Bohm (1951), pp. 169 ff.
21. Jeans (1930).
22. For a more detailed discussion of this phenomenon and its relation to the uncertainty principle, see Capra (1975), p. 192.
23. The interactions between subatomic particles fall into four basic categories with markedly different interaction strengths: the strong, electromagnetic, weak, and gravitational interactions; see Capra (1975), pp. 228 ff.
24. See Capra (1975) for a more detailed discussion of both quantum field theory and S-matrix theory.
25. Ibid., pp. 286 ff.
26. G. F. Chew, quoted ibid., p. 295.
27. See Capra (1979a).
28. Bohm (1980).
29. Holography is a technique of lensless photography based on the interference property of light waves. The resulting "picture" is called a hologram; see Collier (1968). For a comprehensive nontechnical introduction to the subject, see Outwater and van Hamersveld (1974).

4. *The Mechanistic View of Life*

1. Quoted in Dubos (1968), p. 76.
2. Handler (1970), p. 55.

3. Weiss (1971), p. 267.
4. Dubos (1968), p. 117.
5. A small number of scientists, most of them in the older generation, have attempted to address biological problems within a broader, holistic or systemic framework. The writings I have found most inspiring are those by Gregory Bateson (1972, 1979), George Coghill, as discussed by Herrick (1949), René Dubos (1959, 1965, 1968, 1976, 1979), Lewis Thomas (1975, 1978, 1979), and Paul Weiss (1971, 1973).
6. For an introduction to the history of biology, including an extensive bibliography, see Magner (1979), on which much of the following discussion is based.
7. La Mettrie (1960); the passage quoted is my own translation from the French original.
8. Needham (1928).
9. Ibid., p. 90.
10. Ibid., p. 66.
11. Ibid., p. 86.
12. Quoted in Magner (1979), p. 330.
13. Quoted in Dubos (1968).
14. Cannon (1939).
15. See Chapter 9 for further details.
16. We may note, however, that the recently discovered phenomenon of "jumping genes," known technically as transposable genetic elements (see Cohen and Shapiro, 1980), may represent a Lamarckian aspect of evolution.
17. Quoted in Magner (1979) p. 357.
18. See Chapter 9. Darwin himself emphasized that, although he saw natural selection as the most important evolutionary mechanism, it was by no means the only one; see Gould and Lewontin (1979).
19. Monod (1971), p. 122.
20. Wilson (1975).
21. See Caplan (1978).
22. Quoted in Randall (1976), p. 479.
23. Quoted ibid., p. 480.
24. See Ruesch (1978).
25. For a nontechnical review of the historical development of molecular biology, see Stent (1969), Chapters 1–4.
26. See Judson (1979).
27. For example, Bohr suggested that our knowledge of a cell's being alive may be complementary to the complete knowledge of its molecular structure.
28. Quoted in Judson (1979), p. 218.
29. Weiss (1971), p. 270.
30. See Stent (1969), p. 10.
31. Quoted in Judson (1979), p. 209.
32. Quoted in ibid., p. 220.

5. *The Biomedical Model*

1. Engel (1977).
2. See Chapter 9 for the systems view of living organisms, and Chapter 11 for the corresponding systems view of health.
3. See Dubos (1979).
4. See Dunn (1976).
5. See Corea (1977); Ehrenreich and English (1978); see also Rich (1977), pp. 117 ff.
6. See Vrooman (1970), pp. 173 ff.
7. See Chapter 11 for a more detailed discussion of homeopathy.
8. Dubos (1976), pp. xxvii–xxxix. The following quotations of Pasteur's statements are taken from this source. Some of them are my own translations from the French originals.
9. See Chapter 6.
10. See, for example, Knowles (1977a).
11. See Dubos (1965), pp. 369 ff.
12. See "Development of Medical Technology," Report of the United States Congress Office of Technology Assessment, August 1976.
13. See Chapter 11.
14. See Knowles (1977b).
15. See Richmond (1977).
16. See Fuchs (1974), pp. 31 ff.
17. See Knowles (1977a); the statements

quoted are on pp. 7 (Knowles), 87 (Rogers), 29 (Callahan), 37 (Thomas), and 105 (Wildavsky).
18. See Fuchs (1974), pp. 104 ff.
19. McKeown (1976).
20. See Dubos (1968), p. 78.
21. See Chapter 7 for a discussion of the relation between birth rates and living standards.
22. See Haggerty (1979).
23. For an example of a concise and thoughtful critique from within the medical profession, see Holman (1976).
24. This discussion focuses on health care in the United States, but similar trends can be observed in Canada and in most European countries.
25. See Illich (1977).
26. Fredrickson (1977).
27. See, for example, Seldin (1977).
28. Knowles (1977b).
29. See Simonton, Simonton, and Creighton (1978), p. 56; for a detailed discussion of the mind-body approach to cancer that the Simontons have developed, see Chapter 11.
30. See Melzack (1973).
31. L. Shlain, private communication, 1979.
32. See Chapter 11.
33. Szasz (1961).
34. Dubos (1959).
35. See Feifel (1967).
36. See Kübler-Ross (1969, 1975); Cohen (1979).
37. See Powles (1979).
38. See Shortt (1979).
39. Thomas (1977).
40. See Ref. 12.
41. See Holman (1976).
42. See Culliton (1978).
43. Ibid.; see also Bunker, Hinkley, and McDermott (1978).
44. See Illich (1977), p. 23.
45. See Tancredi and Barondess (1978).
46. Thomas (1979), pp. 168 ff.
47. McKeown (1976), p. 128.
48. See Dubos (1968), pp. 74 ff.
49. See Cassell (1976); Kleinman, Eisenberg, and Good (1978).
50. See Kleinman, Eisenberg, and Good (1978).
51. See Chapter 10.
52. Thomas (1975), p. 88.
53. See Dubos (1965), p. 134.
54. Thomas (1975), p. 90.
55. See Dubos (1965), pp. 171 ff.
56. See Thomas (1978).
57. See Fuchs (1974), p. 120.
58. See Holman (1976).
59. See Lock (1980), p. 136.
60. See Corea (1977); Ehrenreich and English (1978).
61. See Fuchs (1974), p. 56.
62. See Ehrenreich and Englisch (1978), pp. 74 ff.
63. See Seldin (1977).
64. See David E. Rogers (1977).
65. See Eisenberg (1977).
66. David E. Rogers (1977).
67. See Fuchs (1974), pp. 70 ff.
68. May (1978).
69. See Knowles (1977b).
70. See Chapter 8.

6. *Newtonian Psychology*

1. See, for example, Murphy and Kovach (1972).
2. For a brief introduction to Eastern mystical traditions, see Capra (1975), Chapters 5–9.
3. See Wilber (1977), pp. 164 ff.
4. See Fromm, Suzuki, and De Martino (1960); Watts (1961); Rama, Ballentine and Weinstock (1976).
5. See Chapter 2.
6. For a discussion of the relation between the Leibnizian theory of monads and the bootstrap theory of subatomic particles, see Capra (1975), pp. 298 ff.
7. James (1961), p. 305.
8. See Murphy-Kovach (1972), p. 238.
9. Watson (1970), p. ix.
10. Watson (1914), p. 27.
11. Quoted in Capra (1975), p. 300.
12. See Chapter 2.
13. See Murphy-Kovach (1972), p. 320.
14. Skinner (1953), pp. 30–31.
15. Weiss (1971), p. 264.

16. Skinner (1975), p. 3.
17. See Murphy-Kovach (1972), p. 278.
18. Freud (1914), p. 78.
19. See Murphy-Kovach (1972), p. 282.
20. The relation between psychoanalysis and physics has been explored in great detail by D. C. Levin in a comprehensive paper upon which much of the following discussion is based; see Levin (1977).
21. Freud (1921), pp. 178 ff.
22. See Chapter 2.
23. See, for example, Fenichel (1945).
24. See Levin (1977) for a more detailed discussion of this intriguing parallel between Newton's and Freud's theories.
25. Freud (1933), p. 80.
26. Freud (1938), p. 181.
27. Freud (1926), pp. 224 ff.
28. See Murphy-Kovach (1972), pp. 296–297.
29. See Strouse (1974).
30. Freud (1926), p. 212.
31. See Chapter 10.
32. See Chapter 11.
33. See Deikman (1978).

7. *The Impasse of Economics*

1. Henderson (1978).
2. See Weiss (1973), p. 71.
3. Navarro (1977), p. x.
4. Schumacher (1975), p. 46.
5. Ibid., pp. 53 ff.
6. Quoted by Myrdal (1973), p. 149.
7. See Henderson (1978), p. 78.
8. See Myrdal (1973), p. 150.
9. *Washington Post,* May 20, 1979.
10. For references to these opinion polls, see Henderson (1978), pp. 13, 155.
11. *Harvard Business Review,* December 1975.
12. Quoted by Henderson (1978), p. 63.
13. Quoted ibid.
14. Quoted in *Fortune,* September 11, 1978.
15. Interview in the *Washington Post,* November 4, 1979.
16. See Madden (1972).
17. See Chapter 1.
18. See Polanyi (1968).
19. See Polanyi (1944), p. 50.
20. Weber (1958).
21. References to the works of these authors are listed in the Bibliography.
22. See Henderson (1981).
23. See Rich (1977), p. 100.
24. Quoted in Routh (1975), p. 45.
25. See Chapter 2.
26. See Soule (1952), p. 51.
27. See Dickinson (1974), pp. 79–81.
28. Lucia F. Dunn, private communication, 1980.
29. See Henderson (1978), p. 94.
30. Ibid., p. 76.
31. See Kapp (1971).
32. Heilbroner (1978).
33. Marx (1888), p. 109.
34. Heilbroner (1980), p. 134.
35. Marx (1891), pp. 317 ff.
36. See Sombart (1976).
37. See Harrington (1976), p. 85.
38. Ibid., p. 106.
39. Quoted ibid., p. 126.
40. Marx (1844), p. 58.
41. Harrington (1976), p. 77.
42. Marx (1844), p. 61.
43. Marx (1970), p. 254.
44. Quoted by Heilbroner (1980), p. 148.
45. See Marx (1844), pp. 93 ff.
46. Keynes (1934), p. 249.
47. See Henderson (1978), p. 36.
48. Quoted ibid., p. 3.
49. See Horney (1937); Galbraith (1958).
50. Hubbert (1974).
51. See Commoner (1980).
52. See Chapter 8.
53. See Goldsen (1977); Mander (1978).
54. See Rothschild (1980).
55. See Aldridge (1978), pp. 14 ff.
56. Henderson (1978), p. 158.
57. Schumacher (1975), p. 146.
58. Theodore Roszak, in his book *Person/Planet,* has provided a comprehensive and eloquent discussion of the nature and consequences of institutional growth, focusing in par-

ticular on the growth of cities; see Roszak (1978), pp. 241 ff.
59. See Navarro (1977), p. 153; see also Schwartz (1980).
60. Walter B. Wriston, interview in *The New Yorker*, January 5, 1981.
61. The investigation of criminal corporate activities has been one of the main purposes of the San Francisco–based magazine *Mother Jones*. For reports on corporate practices in the Third World, see, for example, the issues of August 1977 (agribusiness and world hunger), December 1977 (baby bottle scandal), and November 1979 ("dumping" of dangerous products).
62. See, for example, Grossman and Daneker (1979).
63. Roszak (1978), p. 33.
64. See Navarro (1977), p. 83.
65. See Henderson (1978), p. 73.
66. Quoted by Navarro (1977), pp. 137 ff.
67. *Wall Street Journal*, August 5, 1975.
68. See Galbraith (1979).
69. For a concise account of the history of the debate between ecologists and economists, see Henderson (1978), pp. 63 ff.
70. Henderson (1978), p. 319.
71. Quoted by Commoner (1979), p. 72.
72. See Chapter 12.
73. See Robertson (1979), pp. 88 ff.; see also Roszak (1978), pp. 205 ff.
74. See Burns (1975), p. 23.
75. Roszak (1978), p. 220.
76. See Henderson (1981).
77. See Chapter 12.

8. The Dark Side of Growth

1. Brown (1980).
2. Ibid., pp. 294–298.
3. See Dumanoski (1980).
4. See Chapter 12 for a discussion of the necessity and feasibility of the transition to solar energy.
5. Ellsberg (1980).
6. Quoted in Sivard (1979), p. 14.
7. Aldridge (1978).
8. Ibid., pp. 71 ff.
9. For a brief but comprehensive review of the whole issue of nuclear power, see Caldicott (1978); for a more detailed presentation of the case against nuclear power, see Nader and Abbotts (1977).
10. See Woollard and Young (1979).
11. See Ellsberg (1980).
12. See Nader and Abbotts (1977), p. 80.
13. For a detailed discussion of these issues, see Nader and Abbotts (1977).
14. Ibid., p. 365.
15. See, for example, Airola (1971).
16. See Winikoff (1978).
17. See Illich (1977), p. 63.
18. See Silverman and Lee (1974), p. 293.
19. See Fuchs (1974), p. 109.
20. See Woodman (1977).
21. See Bekkanen (1976).
22. See Woodman (1977).
23. See Hughes and Brewin (1980); see also Mosher (1976).
24. See Brooke (1976).
25. See Woodman (1977).
26. See Commoner (1977), p. 152.
27. Quoted by Berry (1977), p. 66.
28. See Zwerdling (1977).
29. Commoner (1977), p. 161.
30. Ibid.
31. Ibid., p. 163.
32. See Zwerdling (1977).
33. Jackson (1980), p. 69.
34. Quoted by Berry (1977), p. 61.
35. See Zwerdling (1977).
36. See Weir and Shapiro (1981).
37. Moore Lappé and Collins (1977a); for summaries of their arguments, see Moore Lappé and Collins (1977b, c). My discussion of agribusiness and world hunger closely follows these two articles.
38. See Culliton (1978).
39. Quoted by Navarro (1977), p. 161.

9. The Systems View of Life

1. For a brief introduction to systems thinking, see Laszlo (1972b); for

more extensive treatments, see von Bertalanffy (1968) and Laszlo (1972a).

2. The study of transactions actually predates systems theory; see Dewey and Bentley (1949), pp. 103 ff.
3. Weiss (1971), p. 284.
4. Ibid., pp. 225 ff.
5. See Jantsch (1980).
6. Weiss (1973), p. 25.
7. Prigogine (1980).
8. See Laszlo (1972), p. 42.
9. See Bateson (1972), pp. 351 ff.
10. Thomas (1975), p. 86.
11. See, for example, Locke (1974).
12. See Chapter 4.
13. See Goreau, Goreau, and Goreau (1979).
14. See Thomas (1975), pp. 26 ff., 102 ff.
15. See Dubos (1968), pp. 7 ff.
16. See Thomas (1975), p. 83.
17. Ibid., p. 6.
18. Ibid., p. 9.
19. See Chapter 1.
20. See Laszlo (1972), p. 67.
21. For a discussion of hierarchical thinking as a culture-bound phenomenon, see Maruyama (1967, 1979); for a feminist critique of hierarchies, see Dodson Gray (1979).
22. Weiss (1971), p. 276.
23. Thomas (1975), p. 113.
24. L. Shlain, lecture at College of Marin, Kenfield, California, January 23, 1979.
25. See Lovelock (1979); for a discussion of the original myth of Gaia, see Spretnak (1981a).
26. Jantsch (1980).
27. See Chapter 4.
28. See Jantsch (1980), p. 48.
29. The relation of this indeterminacy to the unpredictability of individual events in atomic physics and to the so-called non local connections between such events (see Chapter 3) remains to be explored.
30. Laszlo (1972), p. 51.
31. See Bateson (1972), p. 451.
32. Livingston (1978), p. 4.
33. Jantsch (1980), p. 75.
34. See ibid, pp. 121 ff.
35. Bateson (1979), pp. 92 ff.
36. G. Bateson, private communication, 1979.
37. See Herrick (1949), pp. 195 ff.
38. See Chapter 11.
39. Jantsch (1980), p. 308.
40. For a recent review, see the special issue of *Scientific American*, September 1979.
41. See Jantsch (1980), p. 61.
42. See Kinsbourne (1978).
43. See Russell (1979).
44. The fact that I have kept the conventional description of the psychological realm as an "inner" world should not be taken to mean that it is located somewhere inside the body. It refers to a form of mentation that transcends space and time and hence cannot be associated with any location.
45. See Dubos (1968), p. 47; see also Herrick (1949).
46. See Livingston (1963).
47. See Chapter 11.
48. See, for example, Edelman and Mountcastle (1978), p. 74.
49. See Capra (1975), p. 29.
50. For testimonies of transpersonal experiences, see, for example, Bucke (1969); for a further discussion of the limitations of the current scientific framework in regard to consciousness, see Chapter 11.
51. Onslow-Ford (1964), p. 36.
52. See Jantsch (1980), pp. 165 ff.
53. Quoted in Koestler (1978), p. 9.
54. See Leonard (1981), pp. 48 ff.
55. Pribram (1977, 1979).
56. See Chapter 3.
57. See Chapter 3.
58. See Capra (1975), p. 292.
59. See *Re-Vision*, special issue on the holographic theories of Karl Pribram and David Bohm, Summer/Fall 1978; see also the special issue of *Dromenon*, Spring/Summer 1980.

60. See Leonard (1981), pp. 14 ff.
61. See Towers (1968, 1977).

10. *Wholeness and Health*

1. See, for example, Eliade (1964).
2. See Glick (1977).
3. See Janzen (1978).
4. Lévi-Strauss (1967), pp. 181 ff.
5. See Graves (1975), vol. I, p. 176.
6. See Spretnak (1981a).
7. See Dubos (1968), p. 55.
8. See, for example, Meier (1949); for a detailed description of the Asclepian ritual, see Edelstein and Edelstein (1945).
9. See Dubos (1968), p. 56 ff.
10. Dubos (1979b).
11. Dubos (1968), p. 58.
12. See Capra (1975), p. 102.
13. See Veith (1972).
14. Needham (1962), p. 279.
15. For an introduction to the philosophy of classical Chinese medicine, see Porkert (1979).
16. Ibid.
17. For an extensive list of these correspondences, see Lock (1980), p. 32.
18. See Veith (1972), p. 105.
19. For a detailed explanation of some of the many pulse qualities recognized by Chinese doctors, see Manaka (1972), Appendix C.
20. See Lock (1980), p. 217.
21. Lock (1980).
22. See Kleinman, Eisenberg, and Good (1978).
23. See Selye (1974).
24. For an extensive discussion of the nature of stress and its role in various illnesses, see Pelletier (1977).
25. For an overview of the history and current state of psychosomatic medicine, see Lipowski (1977).
26. See Dubos (1968), p. 64.
27. See Chapter 11.
28. See Pelletier (1977), p. 42.
29. See below for further details.
30. See Cousins (1977).
31. Ibid.
32. See Knowles (1977b).
33. See White (1978).
34. For more details, see Knowles (1977b), White (1978).
35. Eisenberg (1977).
36. White (1978).
37. See White (1978).
38. Fuchs (1974), p. 104.
39. Rasmussen (1975).
40. Ibid.
41. For a brief outline of such a national health insurance plan, see White (1978).
42. See Fuchs (1974), p. 76.
43. For a review of various traditions of psychic healing and their relation to modern psychosomatic medicine and psychotherapy, see Krippner (1979); for recent experimental approaches to healing by the laying-on of hands, see Krieger (1975) and Grad (1979).
44. See Chapter 3; in particular, the transfer of energy is always associated with a transfer of matter (particles or collections of particles). In phenomena involving so-called nonlocal connections, no energy is transferred.
45. Vithoulkas (1980).
46. Ibid., p. 140.
47. See Chapter 11.
48. Reich (1979); see especially the chapter entitled "The Expressive Language of the Living," pp. 136–182.
49. Ibid., p. 177.
50. See Mann (1973), pp. 24–25.
51. Reich (1979), pp. 279 ff.
52. See Mann (1973), pp. 270 ff.
53. See Thie (1973).
54. For an annotated bibliography of bodywork literature, see Popenoe (1977), pp. 17–53.
55. See Bartenieff (1980).
56. See Chapter 8.
57. See Randolph and Moss (1980).
58. See Chapter 5.
59. For a more detailed discussion of these techniques, see Pelletier (1977).

60. See ibid., pp. 197 ff.
61. See Green and Green (1977).
62. For a detailed description of the Simonton approach, see Simonton, Matthews-Simonton, and Creighton (1978).
63. C. Simonton, private communication, 1978.
64. See Simonton, Matthews-Simonton, and Creighton (1978), pp. 57 ff.
65. LeShan (1977), pp. 49 ff.

11. *Journeys Beyond Space and Time*

1. Jung (1951a), p. 261.
2. For a brief introduction to Jung's psychology, see Fordham (1972).
3. See Chapter 6.
4. Jung (1928), p. 17.
5. In his paper "On Psychic Energy," ibid., Jung draws numerous analogies to classical physics. In particular, he introduces the concept of entropy in the context of Boltzmann's thermodynamics, which is quite inadequate to describe living organisms.
6. Jung (1939), p. 71.
7. Jung (1965), p. 352.
8. Jung (1936), p. 48; for an interesting extension of the concept of archetypal forms to numbers and other mathematical structures, see von Franz (1974), pp. 15 ff.
9. Jung (1951b).
10. See Chapter 3.
11. Jung (1929), p. 71.
12. Jung (1965), p. 133.
13. See Murphy and Kovach (1972), p. 432.
14. Maslow (1962), p. 5.
15. Assagioli (1965).
16. Carl Rogers (1951).
17. For a lively account of the colorful history of the Esalen Institute, see Tomkins (1976).
18. See Murphy and Kovach (1972), pp. 298 ff.
19. See, for example, Goldenberg and Goldenberg (1980).
20. Carl Rogers (1970).
21. See Sutich (1976).
22. See Walsh and Vaughn (1980); see also Pelletier and Garfield (1976).
23. See Mander and Rush (1974); see also Roszak (1978), pp. 16 ff.
24. S. Grof., *Journeys Beyond the Brain*, unpublished manuscript.
25. Wilber (1977); for a brief introduction, see Wilber (1975).
26. See Grof (1976), pp. 154 ff.
27. Quoted in Capra (1975), p. 43.
28. Grof (1976).
29. Ibid., pp. 32 ff.
30. Ibid., pp. 46 ff.
31. Ibid., pp. 101 ff.
32. S. Grof, *Journeys Beyond the Brain*, unpublished manuscript.
33. Castaneda (1972), p. 55.
34. See Capra (1979b).
35. Whitehead (1926), p. 66.
36. See Chapter 3.
37. See Capra (1975), p. 71.
38. See Berger, Hamburg, and Hamburg (1977).
39. See, for example, Maslow (1964) and McCready (1976), pp. 129 ff.
40. See Perry (1974), pp. 8 ff.
41. Rosenhan (1973).
42. See Laing (1978), p. 114.
43. Bateson (1972), pp. 201 ff.
44. Laing (1978), p. 28.
45. Ibid., p. 104.
46. See Rosenhan (1973).
47. R. D. Laing, private communication, 1978.
48. Jung (1965), p. 131.
49. Laing (1978), p. 56.
50. See Laing (1972); Perry (1974), pp. 149 ff.
51. Quoted by Laing (1978), p. 118.
52. Ibid., p. 128.
53. Ibid., p. 46.
54. Perls (1969).
55. Grof (1980).
56. Ibid.
57. Janov (1970).

58. Grof, *Journeys Beyond the Brain,* unpublished manuscript.
59. For a striking example of a most extraordinary and, at the same time, highly therapeutic experience of this kind, see Laing (1982).
60. Grof, *Journeys Beyond the Brain,* unpublished manuscript.

12. *The Passage to the Solar Age*

1. Bateson (1972), p. 434.
2. Schumacher (1975), p. 258.
3. Forrester (1980).
4. Henderson (1978), p. 226.
5. Bateson (1972), p. 497.
6. See Bibliography for references to books by these authors.
7. See Henderson (1978), p. 52.
8. See Henderson (1981).
9. Ibid.
10. Odum (1971).
11. See Chapter 2.
12. Georgescu-Roegen (1971).
13. Henderson (1978), p. 83.
14. See Chapter 9.
15. See, for example, Rifkin (1980).
16. Jantsch (1980), p. 255.
17. Roszak (1978), p. xxx.
18. Weisskopf (1971), p. 24.
19. See Cook (1971).
20. Roszak (1978), pp. 254 ff.
21. Schumacher (1975), p. 34.
22. See Chapter 8.
23. Lovins (1977); for a more recent, updated summary, see Lovins (1980).
24. See Chapter 8.
25. Quoted by Commoner (1979), p.46.
26. See *Mother Jones,* September/October 1979.
27. See Lovins (1977), p. 9; Grossman and Daneker (1979).
28. Stobaugh and Yergin (1979).
29. See Lovins (1980).
30. Ibid.
31. See Commoner (1979), p. 56.
32. See, for example, Stobaugh and Yergin (1979), p. 167.
33. See Commoner (1979), p. 54.
34. See Lovins (1978).
35. See Commoner (1979), p. 44.
36. Ibid., p. 64.
37. Stobaugh and Yergin (1979), p. 238.
38. Ibid., pp. 258 ff.
39. See Commoner (1979), p. 36.
40. See Stobaugh and Yergin (1979), p. 262.
41. See Commoner (1979), p. 38.
42. Ibid., pp. 41 ff.
43. Jackson (1980), pp. 62 ff.
44. Commoner (1979), pp. 58 ff.
45. Ibid., p. 62.
46. See Chapter 7.
47. See Henderson (1978), p. 387.
48. For a list of people and organizations actively promoting the ideas, values and activities discussed in the following paragraphs, see Robertson (1979), pp. 135 ff.; for an extensive discussion of various informal educational networks, see Ferguson (1980).
49. See Henderson (1978), p. 359.
50. Ibid., pp. 387 ff.
51. See Huber (1979).
52. See Henderson (1978), p. 391.
53. See Sessions (1981).
54. See Chapter 11.
55. See Chapter 9; for a more detailed discussion of Taoist principles, see Capra (1975), pp. 113 ff.
56. Quoted in Capra (1975), p. 117.
57. See ibid., p. 116.
58. See Meeker (1980).
59. See *Co-Evolutionary Quarterly,* Summer 1977; see also Elgin (1981).
60. See Henderson (1978), p. 395.
61. Roszak (1978), p. xxiv.
62. See Chapter 1.
63. See Stone (1976) for a history of Goddess worship and its suppression, Spretnak (1981a) for a discussion of prepatriarchal Greek Goddess mythology, and Chen (1974) for a discussion of a possible connection between Taoism and Goddess spirituality.
64. Bruteau (1974).
65. See Spretnak (1981b).
66. See Henderson (1980).
67. See Chapter 1.

BIBLIOGRAPHY

Airola, Paavo. 1971. *Are You Confused?* Phoenix, Arizona: Health Plus.

Aldridge, Robert C. 1978. *The Counterforce Syndrome.* Washington, D.C.: Institute for Policy Studies.

Assagioli, Roberto. 1965. *Psychosynthesis.* New York: Viking.

Barnet, Richard J., and Muller, Ronald E. 1974. *Global Reach: The Power of the Multinational Corporations.* New York: Simon and Schuster.

Bartenieff, Irmgard. 1980. *Body Movement: Coping with the Environment.* New York: Gordon and Breach.

Barzun, Jacques. 1958. *Darwin, Marx, Wagner.* New York: Doubleday/Anchor.

Bateson, Gregory. 1972. *Steps to an Ecology of Mind.* New York: Ballantine.

———. 1979. *Mind and Nature.* New York: Dutton.

Bekkanen, John. 1976. "The Impact of Promotion on Physicians' Prescribing Patterns." *Journal of Drug Issues,* Winter.

Berger, Philip, Hamburg, Beatrix, and Hamburg, David. 1977. "Mental Health: Progress and Problems." In Knowles, John H., ed. *Doing Better and Feeling Worse.* New York: Norton.

Berry, Wendell. 1977. *The Unsettling of America.* San Francisco: Sierra Club.

von Bertalanffy, Ludwig. 1968. *General Systems Theory.* New York: Braziller.

Bohm, David. 1951. *Quantum Theory,* New York: Prentice-Hall.

———. 1980. *Wholeness and the Implicate Order.* London: Routledge & Kegan Paul.

Boulding, Kenneth E. 1968. *Beyond Economics.* Ann Arbor: University of Michigan Press.

Brooke, Paul. 1976. "Promotional Parameters: A Preliminary Examination of Promotional Expenditures." *Journal of Drug Issues,* Winter.

Brown, Michael. 1980. *Laying Waste.* New York: Pantheon.

Bruteau, Beatrice. 1974. "The Image of the Virgin-Mother." In Plaskow, J., and Romero, J. A., eds. *Women and Religion.* Missoula, Mont.: Scholars Press.

Bucke, Richard. 1969. *Cosmic Consciousness.* New York: Dutton.

Bunker, J., Hinkley, D., and McDermott, W. 1978. "Surgical Innovation and Its Evaluation." *Science,* May 26.

Burns, Scott. 1975. *Home Inc.* New York: Doubleday.

Caldicott, Helen. 1978. *Nuclear Madness.* Brookline, Mass.: Autumn Press.

Cannon, Walter. 1939. *The Wisdom of the Body.* New York: Norton.

Caplan, Arthur L., ed. 1978. *The Sociobiology Debate.* New York: Harper & Row.

Capra, Fritjof. 1975. *The Tao of Physics.* Berkeley: Shambhala.

———. 1979a. "Quark Physics Without Quarks: A Review of Recent Developments in S-Matrix Theory." *American Journal of Physics*, January.
———. 1979b. "Can Science Explain Psychic Phenomena?" *Re-Vision*, Winter/Spring.
Cassell, Eric J. 1976. "Illness and Disease." *Hastings Center Report*, April.
Castaneda, Carlos. 1972. *Journey to Ixtlan*. New York: Simon and Schuster.
Chen, Ellen Marie. 1974. "Tao as the Great Mother and the Influence of Motherly Love in the Shaping of Chinese Philosophy." *History of Religions*, August.
Cohen, Kenneth P. 1979. *Hospice: Prescription for Terminal Care*. Germantown, Md.: Aspen.
Cohen, Stanley N., and Shapiro, James A. 1980. "Transposable Genetic Elements." *Scientific American*, February.
Collier, Robert J. 1968. "Holography and Integral Photography." *Physics Today*, July.
Commoner, Barry. 1977. *The Poverty of Power*. New York: Bantam.
———. 1979. *The Politics of Energy*. New York: Knopf.
———. 1980. "How Poverty Breeds Overpopulation." In Arditti, Rita, Brennan, Pat, and Cavrak, Steve, eds. *Science and Liberation*. Boston: South End Press.
Cook, Earl. 1971. "The Flow of Energy in an Industrial Society." *Scientific American*, September.
Corea, Gena. 1977. *The Hidden Malpractice*. New York: Morrow.
Cousins, Norman. 1977. "The Mysterious Placebo." *Saturday Review*, October 1.
Crosland, M. P., ed. 1971. *The Science of Matter*. Baltimore: Penguin.
Culliton, B. J. 1978. "Health Care Economics: The High Cost of Getting Well." *Science*, May 26.

Deikman, Arthur. 1978. "Comments on the GAP Report on Mysticism." AHP *Newsletter*, San Francisco, January.
Dewey, John, and Bentley, Arthur F. 1949. *Knowing and the Known*. Boston: Beacon Press.
Dickson, David. 1974. *Alternative Technology*. London: Fontana.
Dodson Gray, Elizabeth. 1979. *Why the Green Nigger?* Wellesley, Mass.: Roundtable Press.
Dubos, René. 1959. *Mirage of Health*. New York: Harper.
———. 1965. *Man Adapting*. New Haven: Yale University Press.
———. 1968. *Man, Medicine and Environment*. New York: Praeger.
———. 1976. *Louis Pasteur*. New York: Scribner. Introduction to the 1976 edition.
———. 1979a. Preface to Sobel, David S., ed. *Ways of Health*, New York: Harcourt Brace Jovanovich.
———. 1979b. "Hippocrates in Modern Dress." In Sobel, David S., ed. *Ways of Health*. New York: Harcourt Brace Jovanovich.
Dumanoski, Dianne. 1980. "Acid Rain." *Sierra*, The Sierra Club Bulletin, May/June.
Dunn, Fred L. 1976. "Traditional Asian Medicine and Cosmopolitan Medicine as Adaptive Systems." In Leslie, Charles, ed. *Asian Medical Systems*. Berkeley: University of California Press.

BIBLIOGRAPHY

Edelman, Gerald, and Mountcastle, Vernon. 1978. *The Mindful Brain.* Cambridge, Mass.: MIT Press.

Edelstein, Emma J., and Edelstein, Ludwig. 1945. *Asclepius.* Baltimore: Johns Hopkins University Press.

Ehrenreich, Barbara, and English, Deidre. 1978. *For Her Own Good.* New York: Doubleday.

Ehrlich, Paul R., and Ehrlich, Anna H. 1972. *Population Resources Environment.* San Francisco: Freeman.

Eisenberg, Leon. 1977. "The Search for Care." In Knowles, John H., ed. *Doing Better and Feeling Worse.* New York: Norton.

Elgin, Duane. 1981. *Voluntary Simplicity.* New York: Morrow.

Eliade, Mircea. 1964. *Shamanism.* Princeton: Princeton University Press.

Ellsberg, Daniel. 1980. Interview in *Not Man Apart,* Friends of the Earth, San Francisco, February.

Engel, George L. 1977. "The Need for a New Medical Model: A Challenge for Biomedicine." *Science,* April 8.

Feifel, Herman. 1967. "Physicians Consider Death." *Proceedings of the American Psychological Association.*

Fenichel, Otto. 1945. *The Psychoanalytic Theory of Neurosis.* New York: Norton.

Ferguson, Marilyn, 1980. *The Aquarian Conspiracy.* Los Angeles: Tarcher.

Fordham, Frieda. 1972. *An Introduction to Jung's Psychology.* Penguin.

Forrester, Jay W. 1971. *World Dynamics.* Cambridge, Mass.: Wright Allen.

———. 1980. "Innovations and the Economic Long Wave." *Planning Review,* November.

von Franz, Marie-Louise. 1974. *Number and Time.* London: Rider.

Frederickson, Donald S. 1977. "Health and the Search for New Knowledge." In Knowles, John H., ed. *Doing Better and Feeling Worse.* New York: Norton.

Freud, Sigmund. 1914. "On Narcissism." In Strachey, James, ed. *Standard Edition of the Complete Works of Sigmund Freud,* vol. 14. New York: Hogarth Press.

———. 1921. "Psychoanalysis and Telepathy." *SE,* vol. 18.

———. 1926. "The Question of Lay Analysis." *SE,* vol. 20.

———. 1933. "Dissection of the Psychical Personality." *SE,* vol. 22.

———. 1938. "An Outline of Psychoanalysis." *SE,* vol. 23.

Fromm, Erich. 1976. *To Have or To Be?* New York: Harper & Row.

Fromm, Erich, Suzuki, D. T., and De Martino, Richard. 1960. *Zen Buddhism and Psychoanalysis.* New York: Harper & Row.

Fuchs, Victor R. 1974. *Who Shall Live?* New York: Basic Books.

Galbraith, John Kenneth. 1958. *The Affluent Society.* Boston: Houghton Mifflin.

———. 1979. *The Nature of Mass Poverty.* Cambridge, Mass.: Harvard University Press.

Garber, Daniel. 1978. "Science and Certainty in Descartes." In Hooker, Michael, ed. *Descartes.* Baltimore: Johns Hopkins University Press.

Georgescu-Roegen, Nicholas. 1971. *The Entropy Law and the Economic Process.* Cambridge, Mass.: Harvard University Press.

Glick, Leonard B. 1977. "Medicine as an Ethnographic Category: The Gimi of the New Guinea Highlands." In Landy, David, ed. *Culture, Disease, and Healing: Studies in Medical Anthropology.* New York: Macmillan.

Goldenberg, Irene, and Goldenberg, Herbert. 1980. *Family Therapy: An Overview.* Belmont, Calif.: Brooks/Cole.

Goldsen, Rose. 1977. *The Show and Tell Machine.* New York: Dial.

Goleman, Daniel, 1978. "Special Abilities of the Sexes: Do They Begin in the Brain?" *Psychology Today,* November.

Goreau, Thomas F., Goreau, Nora I., and Goreau, Thomas J. 1979. "Corals and Coral Reefs." *Scientific American,* August.

Gould, S. J., and Lewontin, R. C. 1979. "The spandrels of San Marco and the Panglossian paradigm: a critique of the adaptionist programme." *Proceedings of the Royal Society,* London, September 21.

Grad, Bernard. 1979. "Healing by the Laying On of Hands: A Review of Experiments." In Sobel, David, ed. *Ways of Health.* New York: Harcourt Brace Jovanovich.

Graves, Robert. 1975. *The Greek Myths,* 2 vols. Penguin.

Green, Elmer, and Green, Alyce. 1977. *Beyond Biofeedback.* San Francisco: Delacorte Press.

Grof, Stanislav. 1976. *Realms of the Human Unconscious.* New York: Dutton.

———. 1980. *LSD Psychotherapy.* Pomona, Calif.: Hunter House.

———. *Journeys Beyond the Brain,* unpublished manuscript.

Grossman, Richard, and Daneker, Gail. 1979. *Energy, Jobs and the Economy.* Boston: Alyson Publications.

Haggerty, Robert J. 1979. "The Boundaries of Health Care." In Sobel, David, ed. *Ways of Health.* New York: Harcourt Brace Jovanovich.

Handler, Philip, ed. 1970. *Biology and the Future of Man.* New York: Oxford University Press.

Harman, Willis W. 1977. "The Coming Transformation." *The Futurist,* April.

Harrington, Michael. 1976. *The Twilight of Capitalism.* New York: Simon and Schuster.

Heilbroner, Robert. 1978. "Inescapable Marx." *The New York Review of Books,* June 29.

———. 1980. *The Worldly Philosophers.* New York: Simon and Schuster.

Heisenberg, Werner. 1962. *Physics and Philosophy.* New York: Harper & Row.

Henderson, Hazel. 1978. *Creating Alternative Futures.* New York: Putnam.

———. 1980. "The Last Shall Be First, 1980s Style." *Christian Science Monitor,* May 16.

———. 1981. *The Politics of the Solar Age.* New York: Doubleday/Anchor.

Herrick, C. Judson, 1949. *George Ellett Coghill: Naturalist and Philosopher.* Chicago: University of Chicago Press.

Holman, Halsted R. 1976. "The 'Excellence' Deception in Medicine." *Hospital Practice,* April.

Horney, Karen. 1937. *The Neurotic Personality of Our Time.* New York: Norton.

Hubbert, M. King. 1974. "World Energy Resources." *Proceedings of the Tenth Commonwealth Mining and Metallurgical Congress.* Ottawa, Canada.

Huber, Joseph, ed. 1979. *Anders arbeiten—anders wirtschaften.* Frankfurt, Germany: Fischer.

Hughes, Richard, and Brewin, Robert. 1980. *The Tranquilizing of America.* New York: Harcourt Brace Jovanovich.

Illich, Ivan. *Medical Nemesis.* New York: Bantam.

Jackson, Wes. 1980. *New Roots for Agriculture.* San Francisco: Friends of the Earth.

James, William. 1961. *The Varieties of Religious Experience.* New York: Collier Macmillan.

Janov, Arthur. 1970. *The Primal Scream.* New York: Dell.

Jantsch, Erich. 1980. *The Self-Organizing Universe.* New York: Pergamon.

Janzen, John M. 1978. *The Quest for Therapy in Lower Zaire.* Berkeley: University of California Press.

Jeans, James. 1930. *The Mysterious Universe.* New York: Macmillan.

Jerison, Harry J. 1973. *Evolution of the Brain and Intelligence.* New York: Academic Press.

Judson, Horace Freeland. 1979. *The Eighth Day of Creation.* New York: Simon and Schuster.

Jung, Carl Gustav. 1928. "On Psychic Energy." In Read, Herbert, Fordham, Michael, and Adler, Gerhard, eds. *The Collected Works of Carl G. Jung,* vol. 8. Princeton: Princeton University Press.

———. 1929. "Problems of Modern Psychotherapy." *CW,* vol. 16.

———. 1936. "The Concept of the Collective Unconscious." *CW,* vol. 9, i.

———. 1939. "Conscious, Unconscious and Individuation." *CW,* vol. 9, i.

———. 1951a. "Aion." *CW,* vol. 9, ii.

———. 1951b. "On Synchronicity." *CW,* vol. 8.

———. 1965. *Memories, Dreams, Reflections.* New York: Random House/Vintage.

Kapp, Karl William. 1971. *Social Costs of Private Enterprise.* New York: Schocken.

Keynes, John Maynard. 1934. *General Theory of Employment, Interest and Money.* New York: Harcourt Brace.

———. 1951. "Newton the Man." In *Essays in Biography.* London: Hart-Davis.

Kinsbourne, Marcel, ed. 1978. *Asymmetrical Function of the Brain.* New York: Cambridge University Press.

Kleinman, Arthur, Eisenberg, Leon, and Good, Byron. 1978. "Culture, Illness, and Care." *Annals of Internal Medicine,* February.

Knowles, John H., ed. 1977a. *Doing Better and Feeling Worse.* New York: Norton.

———. 1977b. "The Responsibility of the Individual." In Knowles, John H., ed. *Doing Better and Feeling Worse,* New York: Norton.

Koestler, Arthur. 1978. *Janus.* London: Hutchinson.

Krieger, Dolores. 1975. "Therapeutic Touch: The Imprimatur of Nursing." *American Journal of Nursing,* May.

Krippner, Stanley. 1979. "Psychic Healing and Psychotherapy." *Journal of Indian Psychology*, vol. 1.
Kübler-Ross, Elisabeth. 1969. *On Death and Dying*. New York: Macmillan.
———, ed. 1975. *Death: The Final Stage of Growth*. Englewood Cliffs, N.J.: Prentice-Hall.
Kuhn, Thomas S. 1970. *The Structure of Scientific Revolutions*. Chicago: University of Chicago Press.

Laing, R. D. 1972. "Metanoia: Some Experiences at Kingsley Hall." In Ruitenbeek, H. M., ed. *Going Crazy: The Radical Therapy of R. D. Laing and Others*. New York: Bantam.
———. 1978. *The Politics of Experience*. New York: Ballantine.
———. 1982. The Voice of Experience. New York: Pantheon.
La Mettrie. 1960. *L'Homme Machine—A Study in the Origins of an Idea*. Edited by Vartanian, A. Princeton: Princeton University Press.
Laszlo, Ervin. 1972a. *Introduction to Systems Philosophy*. New York: Harper Torchbooks.
———. 1972b. *The Systems View of the World*. New York: Braziller.
Leonard, George. 1981. *The Silent Pulse*. New York: Bantam.
LeShan, Lawrence L. 1977. *You Can Fight for Your Life*. New York: Evans.
Levin, D. C. 1977. "Physics and Psycho-Analysis: An Epistemological Study," unpublished paper.
Lévi-Strauss, Claude. 1967. *Structural Anthropology*. New York: Doubleday.
Lipowski, Z. J. 1977. "Psychosomatic Medicine in the Seventies: An Overview." *The American Journal of Psychiatry*, March.
Livingston, Robert B. 1963. "Perception and Commitment." *Bulletin of the Atomic Scientists*, February.
———. 1978. *Sensory Processing, Perception, and Behavior*. New York: Raven Press.
Lock, Margaret M. 1980. *East Asian Medicine in Urban Japan*. Berkeley: University of California Press.
Locke, David Millard. 1974. *Viruses*. New York: Crown.
Lovelock, J. E. 1979. *Gaia*. New York: Oxford University Press.
Lovins, Amory B. 1977. *Soft Energy Paths*. New York: Harper & Row.
———. 1978. "Soft Energy Technologies." *Annual Review of Energy*.
———. 1980. "Soft Energy Paths." AHP *Newsletter*, San Francisco, June.

McCready, William C. 1976. *The Ultimate Values of the American Population*. Beverly Hills, Calif.: Sage Publications.
McKeown, Thomas. 1976. *The Role of Medicine: Mirage or Nemesis*. London: Nuffield Provincial Hospital Trust.

Madden, Carl H. 1972. *Clash of Culture. Management in an Age of Changing Values*. Washington, D.C.: National Planning Association.
Magner, Lois N. 1979. *History of the Life Sciences*. New York: Dekker.

Manaka, Yoshio. 1972. *The Layman's Guide to Acupuncture.* New York: John Weatherhill.
Mander, Anica, and Rush, Anne Kent. 1974. *Feminism as Therapy.* New York: Random House.
Mander, Jerry. 1978. *Four Arguments for the Elimination of Television.* New York: Morrow.
Mann, W. Edward. 1973. *Orgone, Reich and Eros.* New York: Simon and Schuster.
Mao Zedong. 1968. *Four Essays on Philosophy.* Beijing: Foreign Languages Press.
Maruyama, Magoroh. 1967. "The Navaho philosophy: an esthetic ethic of mutuality." *Mental Hygiene,* April.
———. 1979. "Mindscapes: The Limits to Thought." *World Future Society Bulletin,* September–October.
Marx, Karl. 1844. *Economic and Philosophic Manuscripts.* In Tucker, Robert C., ed. *The Marx-Engels Reader.* New York: Norton, 1972.
———. 1888. *Theses on Feuerbach.* Ibid.
———. 1891. *Capital.* Ibid.
———. 1970. *Das Kapital, Abridged Edition.* Chicago: Henry Regnery.
Maslow, Abraham. 1962. *Toward a Psychology of Being.* Princeton: Van Nostrand Reinhold.
———. 1964. *Religions, Values, Peak Experiences.* New York: Viking.
May, Scott. 1978. "On My Medical Education: Seeking a Balance in Medicine." *Medical Self-Care.* Fall.
Meeker, Joseph W. 1980. *The Comedy of Survival.* Los Angeles: Guild of Tutors Press.
Meier, Carl Alfred. 1949. *Antike Inkubation und Moderne Psychotherapie.* Zurich: Rascher.
Melzack, Ronald. 1973. *The Puzzle of Pain.* Penguin.
Merchant, Carolyn. 1980. *The Death of Nature.* New York: Harper & Row.
Monod, Jacques. 1971. *Chance and Necessity.* New York: Knopf.
Moore Lappé, Frances, and Collins, Joseph. 1977a. *Food First: Beyond the Myth of Scarcity.* New York: Houghton Mifflin.
———. 1977b. "Six Myths of World Hunger." *New West,* June.
———. 1977c. "Still Hungry After All These Years." *Mother Jones,* August.
Mosher, Elissa Henderson. 1976. "Portrayal of Women in Drug Advertising: A Medical Betrayal." *Journal of Drug Issues,* Winter.
Mumford, Lewis. 1956. *The Transformations of Man.* New York: Harper.
———. 1970. "Closing Statement." In Disch, Robert, ed. *The Ecological Conscience.* New York: Prentice-Hall.
Murphy, Gardner, and Kovach, Joseph K. 1972. *Historical Introduction to Modern Psychology.* New York: Harcourt Brace Jovanovich.
Myrdal, Gunnar. 1973. *Against the Stream.* New York: Pantheon.

Nader, Ralph, and Abbots, John. 1977. *The Menace of Atomic Energy.* New York: Norton.
Navarro, Vicente. 1977. *Medicine Under Capitalism.* New York: Prodist.

Needham, Joseph. 1928. *Man a Machine.* New York: Norton.
———. 1962. *Science and Civilisation in China,* vol. 2. Cambridge, England: Cambridge University Press.

Odum, Howard. 1971. *Environment, Power and Society.* New York: Wiley Interscience.
Onslow-Ford, Gordon. 1964. *Painting in the Instant.* London: Thames & Hudson.
Outwater, Christopher, and van Hamersveld, Eric. 1974. *Practical Holography.* Beverly Hills, Calif.: Pentangle Press.

Pelletier, Kenneth R. 1977. *Mind as Healer, Mind as Slayer.* New York: Delta.
Pelletier, Kenneth R., and Garfield, Charles. 1976. *Consciousness: East and West.* New York: Harper & Row.
Perls, Fritz. 1969. *Gestalt Therapy Verbatim.* New York: Bantam.
Perry, John Weir. 1974. *The Far Side of Madness.* Englewood Cliffs, N.J.: Prentice-Hall.
Polanyi, Karl. 1944. *The Great Transformation.* New York: Rinehart.
———. 1968. *Primitive, Archaic and Modern Economics.* New York: Doubleday/Anchor.
Popenoe, Cris. 1977. *Wellness.* Washington, D.C.: Yes!
Porkert, Manfred. 1974. *The Theoretical Foundations of Chinese Medicine.* Cambridge, Mass.: MIT Press.
———. 1979. "Chinese Medicine, a Traditional Healing Science." In Sobel, David, ed. *Ways of Health.* New York: Harcourt Brace Jovanovich.
Powles, John. 1979. "On the Limitations of Modern Medicine." In Sobel, David, ed. *Ways of Health.* New York: Harcourt Brace Jovanovich.
Pribram, Karl H. 1977. "Holonomy and Structure in the Organization of Perception." In Nicholas, John M., ed. *Images, Perception and Knowledge.* Dordrecht-Holland: Reidel.
———. 1979. "Holographic Memory." Interview by Daniel Goleman, *Psychology Today,* February.
Prigogine, Ilya. 1980. *From Being to Becoming.* San Francisco: Freeman.

Rama, Swami, Ballentine, Rudolf, and Weinstock, Allan. 1976. *Yoga and Psychotherapy.* Glenview, Ill.: Himalaya Institute.
Randall, John Herman. 1976. *The Making of the Modern Mind.* New York: Columbia University Press.
Randolph, T. G., and Moss, R. W. 1980. *An Alternative Approach to Allergies.* New York: Lippincott & Crowell.
Rasmussen, Howard. 1975. "Medical Education—Revolution or Reaction." *Pharos,* April.
Reich, Wilhelm. 1979. *Selected Writings.* New York: Farrar, Straus & Giroux.
Rich, Adrienne. 1977. *Of Woman Born.* New York: Bantam.
Richmond, Julius B. 1977. "The Needs of Children." In Knowles, John H., ed. *Doing Better and Feeling Worse.* New York: Norton.
Rifkin, Jeremy. 1980. *Entropy.* New York: Viking.

Robertson, James. 1979. *The Sane Alternative.* St. Paul, Minn.: River Basin Publishing Company.
Rodis-Lewis, Geneviève. 1978. "Limitations of the Mechanical Model in the Cartesian Conception of the Organism." In Hooker, Michael, ed. *Descartes.* Baltimore: Johns Hopkins University Press.
Rogers, Carl R. 1951. *Client-Centered Therapy.* Boston: Houghton Mifflin.
———. 1970. *On Encounter Groups.* New York: Harper & Row.
Rogers, David E. 1977. "The Challenge of Primary Care." In Knowles, John H., ed. *Doing Better and Feeling Worse.* New York: Norton.
Rosenhan, D. L. 1973. "On Being Sane in Insane Places." *Science,* January 19.
Roszak, Theodore. 1969. *The Making of a Counter Culture.* New York: Doubleday/Anchor.
———. 1978. *Person/Planet.* New York: Doubleday/Anchor.
Rothschild, Emma. 1980. "Boom and Bust." *New York Review of Books,* April 3.
Routh, Guy. 1975. *The Origin of Economic Ideas.* New York: Macmillan.
Ruesch, Hans. 1978. *Slaughter of the Innocent.* New York: Bantam.
Russell, Bertrand. 1961. *History of Western Philosophy.* London: Allen & Unwin.
Russell, Peter, 1979. *The Brain Book.* New York: Dutton.

Schilpp, Paul Arthur, ed. 1951. *Albert Einstein: Philosopher-Scientist.* New York: Tudor.
Schumacher, E. F. 1975. *Small Is Beautiful.* New York: Harper & Row.
Schwartz, Charles. 1980. "Scholars for Dollars." In Arditti, Rita, Brennan, Pat, and Cavrak, Steve, eds. *Science and Liberation.* Boston: South End Press.
Seldin, Donald W. 1977. "The Medical Model: Biomedical Science as the Basis of Medicine." In *Beyond Tomorrow.* New York: Rockefeller University Press.
Selye, Hans. 1974. *Stress Without Distress.* New York: Lippincott.
Sessions, George. 1981. "Shallow and Deep Ecology: A Review of the Philosophical Literature." In Schultz, B., and Hughes, D., eds. *Ecological Consciousness.* Lanham, Md.: University Press of America.
Shortt, S.E.D. 1979. "Psychiatric Illness in Physicians." *CMA Journal,* August 4.
Silverman, Milton, and Lee, Philip R. 1974. *Pills, Profits and Politics.* Berkeley: University of California Press.
Simonton, O. Carl, Matthews-Simonton, Stephanie, and Creighton, James. 1978. *Getting Well Again.* Los Angeles: Tarcher.
Sivard, Ruth Leger. 1979. *World Military and Social Expenditures.* Leesburg, Virginia, Box 1003: World Priorities.
Skinner, B. F. 1953. *Science and Human Behavior.* New York: Macmillan.
———. 1975. *Beyond Freedom and Dignity.* New York: Bantam.
Sombart, Werner. 1976. *Why Is There No Socialism in the United States?* White Plains, N.Y.: International Arts and Sciences Press.
Sommers, Fred. 1978. "Dualism in Descartes: The Logical Ground." In Hooker, Michael, ed. *Descartes.* Baltimore: Johns Hopkins University Press.
Sorokin, Pitirim A. 1937–41. *Social and Cultural Dynamics,* 4 vols. New York: American Book Company.
Soule, George Henry. 1952. *Ideas of the Great Economists.* New York: Viking.

Spretnak, Charlene. 1981a. *Lost Goddesses of Early Greece.* Boston: Beacon Press.
———, ed. 1981b. *The Politics of Women's Spirituality.* New York: Doubleday/Anchor.
Stace, Walter T. 1960. *The Teachings of the Mystics.* New York: New American Library.
Stapp, Henry Pierce. 1971. "S-Matrix Interpretation of Quantum Theory." *Physical Review D,* March 15.
———. 1972. "The Copenhagen Interpretation." *American Journal of Physics,* August.
———. 1979. "Whiteheadian Approach to Quantum Theory and the Generalized Bell's Theorem." *Foundations of Physics,* February.
Stent, Gunther S. 1969. *The Coming of the Golden Age.* New York: Natural History Press.
Stobaugh, Robert, and Yergin, Daniel, eds. 1979. *Energy Future: Report of the Energy Project at the Harvard Business School.* New York: Ballantine.
Stone, Merlin. 1976. *When God Was a Woman.* New York: Harcourt Brace Jovanovich.
Strouse, Jean, ed. 1974. *Women & Analysis.* New York: Grossman.
Sutich, Anthony J. 1976. "The Emergence of the Transpersonal Orientation: A Personal Account." *Journal of Transpersonal Psychology,* 1.
Szasz, Thomas. 1961. *The Myth of Mental Illness.* New York: Hoeber-Harper.

Tancredi, Laurence R., and Barondess, Jeremiah A. 1978. "The Problem of Defensive Medicine." *Science,* May 26.
Thie, John F. 1973. *Touch for Health.* Marina del Rey, Calif.: DeVorss.
Thomas, Lewis. 1975. *The Lives of a Cell.* New York: Bantam.
———. 1977. "On the Science and Technology of Medicine." In Knowles, John H., ed. *Doing Better and Feeling Worse.* New York: Norton.
———. 1978. Interview in *New Yorker,* January 2.
———. 1979. *The Medusa and the Snail.* New York: Viking.
Tomkins, Calvin. 1976. "New Paradigms." *New Yorker,* January 5.
Towers, Bernard. 1968. "Man in Evolution: The Teilhardian Synthesis." *Technology and Society,* September.
———. 1977. "Toward an Evolutionary Ethic." *Teilhard Review,* October.
Toynbee, Arnold. 1972. *A Study of History.* New York: Oxford University Press.

Veith, Ilza. 1972. *The Yellow Emperor's Classic of Internal Medicine.* Berkeley: University of California Press.
Vithoulkas, George. 1980. *The Science of Homeopathy.* New York: Grove.
Vrooman, Jack Rochford. 1970. *René Descartes.* New York: Putnam.

Walsh, Roger N., and Vaughn, Frances, eds. 1980. *Beyond Ego.* Los Angeles: Tarcher.
Ward, Barbara. 1979. *Progress for a Small Planet.* New York: Norton.
Watson, John B. 1914. *Behavior.* New York: Holt.
———. 1970. *Behaviorism.* New York: Norton.
Watts, Alan W. 1961. *Psychotherapy East and West.* New York: Pantheon.

Weber, Max. 1958. *The Protestant Ethic and the Spirit of Capitalism.* New York: Scribner.
Weir, David, and Schapiro Mark. 1981. *Circle of Poison.* San Francisco: Institute for Food and Development Policy.
Weiss, Paul A. 1971. *Within the Gates of Science and Beyond.* New York: Hafner.
———. 1973. *The Science of Life.* Mount Kisco, N.Y.: Futura.
Weisskopf, Walter A. 1971. *Alienation and Economics.* New York: Dutton.
White, Kerr L. 1978. "Ill Health and Its Amelioration: Individual and Collective Choices." In Carlson, Rick J., ed. *Future Directions in Health Care: A New Public Policy.* Cambridge, Mass.: Ballinger.
Whitehead, Alfred North. 1926. *Science and the Modern World.* New York: Macmillan.
Wilber, Ken. 1975. "Psychologia Perennis: The Spectrum of Consciousness." *Journal of Transpersonal Psychology,* Number 2.
———. 1977. *The Spectrum of Consciousness.* Wheaton, Ill.: Theosophical Publishing House.
Wilhelm, Hellmut. 1960. *Change.* New York: Harper Torchbooks.
Wilhelm, Richard. 1968. *The I Ching.* London: Routledge & Kegan Paul.
Wilson, E. O. 1975. *Sociobiology.* Cambridge, Mass.: Harvard University Press.
Winikoff, Beverly. 1978. "Diet Change and Public Policy." In Carlson, Rick J., ed. *Future Directions in Health Care: A New Public Policy.* Cambridge, Mass.: Ballinger.
Woodman, Joseph. 1977. "The Unhealthiest Alliance." *New Age,* October.
Woollard, Robert F., and Young, Eric R., eds. 1979. *Health Dangers of the Nuclear Fuel Chain and Low-Level Ionizing Radiation: A Bibliography/Literature Review.* Watertown, Mass. 02172, Box 144: Physicians for Social Responsibility.

Zwerdling, Daniel. 1977. "The Day of the Locust." *Mother Jones,* August.

INDEX

䷗ 復

After a time of decay comes the turning point. The powerful light that has been banished returns. There is movement, but it is not brought about by force. . . . The movement is natural, arising spontaneously. For this reason the transformation of the old becomes easy. The old is discarded and the new is introduced. Both measures accord with the time; therefore no harm results.

—*I Ching*